Women in Space – Following Valentina

David J. Shayler and Ian A. Moule

Women in Space – Following Valentina

Published in association with
Praxis Publishing
Chichester, UK

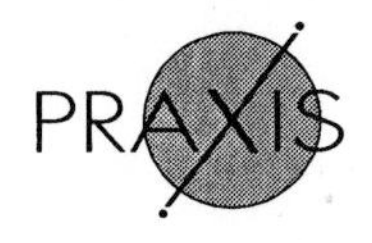

David J. Shayler
Astronautical Historian
Astro Info Service
Halesowen
West Midlands
UK

Ian A. Moule
Aerospace Historian
Raunds
Northamptonshire
UK

SPRINGER–PRAXIS BOOKS IN SPACE EXPLORATION
SUBJECT *ADVISORY EDITOR*: John Mason B.Sc., M.Sc., Ph.D.

ISBN 1-85233-744-3 Springer Berlin Heidelberg New York

Springer is a part of Springer Science + Business Media (*springeronline.com*)

Library of Congress Control Number: 2005922816

Cover design: Jim Wilkie
Copy editing and graphics processing: R. A. Marriott
Typesetting: BookEns Ltd, Royston, Herts., UK

Printed in Germany on acid-free paper

Achieving the dream: Earth orbit.

In memory of

Kalpana Chawla
Lauren Blair Salton Clark
Sharon Christa McAuliffe
Diane Prinz
Judith Arlene Resnik
Patricia Hilliard Robinson

Other books by David J. Shayler in this series

Disasters and Accidents in Manned Spaceflight (2000), ISBN 1-85233-225-5

Skylab: America's Space Station (2001), ISBN 1-85233-407-X

Gemini: Steps to the Moon (2001), ISBN 1-85233-405-3

Apollo: The Lost and Forgotten Missions (2002), ISBN 1-85233-575-0

Walking in Space (2004), ISBN 1-85233-710-9

With Rex D. Hall

The Rocket Men (2001), ISBN 1-85233-391-X

Soyuz: A Universal Spacecraft (2003), ISBN 1-85233-657-9

Table of contents

Foreword

I have often been asked if it was my childhood dream to become an astronaut. The answer is 'no'. The thought never occurred to me until I was thirty-two years old. When Sputnik orbited the Earth in 1957, I was only five years old, and as aware of the significance of the event as most other Japanese were: that is to say, not at all. The 'space firsts' that marked the next decades inspired me to read the biographies of the history-makers; but that was the extent of my interest in space exploration, which seemed to be another world entirely.

My childhood dream was much more immediate and personal. I wanted to be a doctor, and to help those, like my younger brother, suffering from diseases. He had aseptic necroses – a rare disease which made his leg bones brittle. Our family watched him struggle to walk, and the teasing by other children made our hearts heavy with sadness. My parents eventually took him to a big university hospital in Tokyo, and as his condition improved, so did my determination to become a doctor. When I was ten, a composition I wrote in school, entitled 'What will I be in the future?', promised as much.

I left my parents' home at fourteen, and moved to Tokyo to prepare for medical school. After years of education and formal training, I became a doctor, specialising in cardiovascular surgery. Then, one morning in December 1983, as I was relaxing in my office after a night on duty in the intensive care unit, a newspaper article caught me by surprise. The Japanese space agency was looking for candidates to fly onboard the Space Shuttle in 1988. I literally shouted, 'Gee! Can someone from Japan actually fly in space?' I thought (stereotypically) that space travellers had to be either American or Russian. I did not know that a German had flown on Spacelab 1 just days earlier, much less that a Czech, a Pole, another German, a Bulgarian, a Hungarian, a Vietnamese, a Cuban, a Mongolian, a Romanian and a Frenchman had already flown to Soviet space stations. I did not even know that we had had a Japanese space agency – the National Space Development Agency – since 1969.

The article held another surprise for me: the candidates were to have scientific backgrounds and conduct experiments in space. But were not astronauts always pilots and aviators? With a shock, I realised that science and technology had

Chiaki Mukai floats into the International Microgravity Laboratory Spacelab module during mission STS-65 in 1994.

progressed to the point where ordinary people living and working on Earth were actually able to do the same kind of work in space. We were now entering the era of space utilisation.

I became more intrigued with the possibility of seeing our beautiful blue planet from outer space with my own eyes. Would such a magnificent sight deepen my way of thinking and expand my concept of life itself? At the same time, I was fascinated by the possibility of using the spaceflight environment – especially weightlessness – for research purposes. Here was an opportunity to contribute my medical expertise to the space programme.

One thought did not cross my mind, however, and for a long time it surprised me when people asked, 'Did you think that, as a woman, you had any realistic chance of being selected?' That I was a woman had never struck me as either a limitation or an advantage. I saw myself only as one human candidate among hundreds of applicants. Perhaps I would have been reassured to hear that a second Russian woman had flown in space just a year earlier; that the US had already selected eight women as career astronauts; that Sally Ride had flown onboard the Shuttle just six months earlier; that she and five other women were in training for impending Shuttle missions; that a woman was about to be one of the first six Canadian astronauts; and that two other women scientists were already candidates to fly on future Spacelab science missions. But that information never penetrated my intensely focused and sleep-deprived medical bubble.

Not that it would have mattered. In any professional or personal endeavour, my approach has always been, 'If I want to do something – and I believe I can do it well

– then I will overcome any obstacles and challenges, and go for it!' So I applied for my second dream: to travel into and work in outer space.

My path into space had highs and lows. The first high was being one of three candidates, selected from 533 applicants, for a Spacelab-J mission in 1988. But the first low, four months later, was the loss of the Space Shuttle *Challenger*. In the wake of the tragedy and the ensuing uncertainty about the future of the space programme, I spent days in consideration and soul-searching over whether to abandon my second dream and return to my first: the medical field.

Other lows followed, as our mission was delayed repeatedly into 1992, and then when I was selected as the back-up and not as a member of the prime crew. Of course I was disappointed for a while, but the lows became highs as the training put me in an advantageous position to understand the mission as a whole, and to witness how many people it took, all working together, to make the mission successful. This preparation served me well when I finally achieved my second dream and flew into space in 1994.

Now I see how my own experience reflects the progress of women in the field of space exploration: slowly dawning awareness of the possibilities, intensive preparation, repeated disappointments, and finally success.

Still, when Sputnik opened the Space Age, no-one would have believed that thirty-four years later a Japanese woman would fly into space, that she would be only the second Japanese citizen to do so, or that she would do so twice (so far). And even I would not have thought that it would be me. But I would not have doubted that I could do it, or that any woman ought to do it.

Dr Chiaki Mukai
NASDA/JAXA astronaut
Japan

Payload Specialist, STS-65
Payload Specialist, STS-95

Authors' preface

This cooperative venture is the result of the two authors' independent research into the history and activities of women in the space programme. Our research tracks followed separate paths until 2001, when the current volume was proposed through Praxis.

Ian Moule

As the 'rookie' working on this (my first) book, the journey from its conception to fruition has been both long and convoluted. My initial exposure to the contributions made by women in the conquest of space was through articles published by the British Interplanetary Society in *Journal of the British Interplanetary Society* and *Spaceflight*, and in my continuing interest in the development of 'winged', rocket-propelled aerospace vehicles. However, the subject of *Women in Space* really came to the fore when, in the early 1990s, I was invited by Alan Fennell – then Editor of a number of publications based upon the popular 1960s science fiction TV series created by Gerry Anderson – to 'research and originate suitable features relating to space development and exploration, undersea exploration, and other similar activities' for inclusion in the aforementioned publications. Gerry Anderson (who I assert is the UK's equivalent to America's creator of *Star Trek*, Gene Roddenberry) gave women a prominent role in *Fireball XL5* (*c.*1965), which revolved around a reusable 'winged', rocket-propelled, missile-shaped spaceship of the same name, and featured, amongst its three-person crew, a female Doctor of Space Medicine called Venus; *Captain Scarlet* (*c.*1967), which featured five young female pilots (American, English, French, African-American and Japanese) called the Angels, who flew supersonic jet-fighter combat aircraft; and *UFO* (*c.*1969), which was being filmed at the same time as the first Moon-landing (Apollo 11) was taking place, and featured an operational Moon-base commanded and operated by women. I therefore felt that I should write a feature on the realities of female pilots and astronauts. Consequently, I wrote to Frank Winter at the Smithsonian Institution National Air and Space Museum (NASM), who kindly sent me a copy of a dossier containing material on 'Women in Space' that had been complied by Lillian Kozloski, a Research Assistant at the Museum who had also written several articles on this subject. At around the same time I also attended an evening lecture on 'Women in Aviation', which had been organised by the Royal Aeronautical Society's Aviation Medicine Group. This consisted of two presentations: the first of them on 'Women in the Fast Jet Cockpit: Aeromedical Considerations', by Col Terence Lyons USAF, Chief Aerospace Medicine Branch; and the second on 'Concorde: The Differences' (in the design and flying characteristics of this supersonic passenger aircraft to those of its subsonic counterparts), delivered by Barbara Harmer, Senior First Officer with British Airways and Concorde's first female pilot.

Shortly afterwards I received a letter from Col Lyons which contained not only a copy of his presentation but also a commentary on 'Women's Health Issues and Space-Based Research', which set me on another line of enquiry. I also began thinking about compiling a book on 'Women in Space' and the potential benefits from space-related research to health issues of women on Earth. Unfortunately, with the demise of the Anderson-based publications, coupled with an increasing workload from the 'day-job', the impetus for such a project waned.

Then, as the old century drew to a close and the twenty-first century beckoned, a letter arrived from Nicky Humphries – a final-year degree student (BA Hons Photography) who had been researching and creating a documentary-based exhibition on the Mercury 13 'ladies'. During a discussion with a member of staff at the National Space Centre in Leicester, whom I had met earlier, Nicky had been informed of my research, and therefore wrote a speculative letter to arrange a meeting. I duly met her at her interactive exhibition at the London College of Printing, in Clerkenwell, where she informed me that after much Internet surfing, letter-writing, transatlantic telephone calls and some sponsorship from United Airlines, she and a friend (Lucy) had spent a month travelling from Albuquerque, New Mexico, to Orlando, Florida, meeting with people who had been involved in the astronaut testing programmes during the early 1960s. In Albuquerque she had stayed with Dr Don Kilgore – the last surviving doctor involved in the astronaut testing at the Lovelace Foundation – before flying to Dallas, Texas, where she met and interviewed Jerri Truhill. She then travelled to Orlando for the Women in Aviation Tenth Anniversary Conference, where she met and interviewed Wally Funk and Jerrie Cobb (the first woman to be put through the astronaut testing programme). Nicky also informed me that she had been invited to view the launch of Eileen Collins' historic mission – as the first female Shuttle Commander – with the 'ladies'!

As this was a unique event we decided to pool our resources with the intention of co-authoring a book on 'Eve's Journey into Space.' However, upon her return from witnessing Eileen's launch at Cape Canaveral, Nicky began full-time employment, and I began an unrelated MSc course. Work on the book was put on hold!

A few years later, during a discussion with David Shayler about another book project he was planning, he mentioned that he had been given approval by Praxis to write a book on 'Women in Space'. I naturally mentioned the research and stalled book project with Nicky, and Dave, without any hesitation(!), kindly agreed to my joining forces with him. Sadly, Nicky had to withdraw, but without her speculative letter and our subsequent collaboration I believe that this book would have been the poorer. Therefore, I will always think of this book as David's, mine and Nicky's!

David J. Shayler

My interest in female space activities began in the late 1960s, on reading the biography of NASA astronaut Bill Thornton (whose wife was English), and in the accounts of former astronauts of the Apollo era and the support of their wives and families when training for or flying a mission. Subsequent discussions with these astronauts often revealed how much support and encouragement was given by their families. Although several astronauts have penned autobiographies, relatively little information on the role of their wives during their time at NASA has been revealed, and an accurate in-depth account of the families left on the ground is yet to be written. An interest in the astronauts' families led to research into support roles in the space programme – notably during the Apollo programme and in supporting the development of the Shuttle and the activities of Mission Control at Houston. At this time, interest in both the Shuttle and the Soviet space station programme was supplemented with news of the selection of female astronauts for Shuttle crew assignments, and a few years later, the first female cosmonaut to fly to a space station.

Interest in the Russian cosmonaut team prompted further research (with the help of Rex Hall) into the first female cosmonaut selected for space training in 1962, and the members of the subsequent selections – most of whom were relatively unknown at that time. And associated research into Soviet stratospheric balloons led to an interest in early exploits in balloons, biplanes and combat (First and Second World Wars) involving women around the world. This book also discusses some of the women involved in the history of astronomy, to show how, over many centuries, women have been interested in science and technology, and how this interest has extended into the space programme.

This book is not intended to provide biographical accounts of female paths to space, whether on the ground in support or research roles or in participation in spaceflights (although several biographical accounts of female space explorers are listed in the Bibliography). Rather, the aim is to record the long history of women's quest for space, and their competitive nature in matching and at times surpassing male achievements in astronomy, aeronautics, aerial combat, and space programme support roles.

In detailing the activities of women in the role of astronaut or cosmonaut, the purpose is to demonstrate not only the efforts required to achieve flight assignment, but also to record which missions were flown by female crew-members and, more importantly, just how much work they carried out on each mission and the numerous responsibilities of each crew-member. A flight into space by a female astronaut or cosmonaut is as equally difficult, dangerous and challenging as for any male colleague; but without exception they prefer to be measured by their achievements as space explorers and not female space explorers.

Over the years my educational work has increased, and I have certainly seen a growing awareness and interest by young girls in science and technology, and a greater increase of interest in the astronaut or cosmonaut as a role model. Many young girls realise that they too may be able to fly in space, given the right opportunities, the ability to study hard, the discipline to keep fit and healthy, and to be fortunately in the right place at the right time. We await with interest the words and exploits of the first woman to step on the lunar dust or on the surface of Mars,

The medical aspects of long-duration or long-distance spaceflight is also an important issue for both men and women, and as the programme develops the idea of families in space alongside the mother or father space-explorer brings the story full circle as the frontiers of family and human exploration stretches out towards Mars and beyond. This book is an account of the first forty years of small steps in female exploration of the Cosmos. The first giant stride was taken on 16 June 1963, by 26-year-old cosmonaut Valentina Tereshkova – the first woman in space. For all those who have followed her, or plan to follow her, this is their book too.

David J. Shayler
Halesowen
West midlands
England
www.astroinfoservice.co.uk

Ian A. Moule
Raunds
Northamptonshire
England

November 2004

Acknowledgements

This book is the result of an immense project by both authors, spanning several years. Consequently, we extend our thanks and gratitude to many organisations, researchers and individuals, not all of whom can be listed here.

Foremost are the staff of the public affairs offices of NASA, ESA, JAXA and the Canadian Space Agency; Elena Esina, Curator of the museum in the House of Cosmonauts, Star City, Russia; fellow researchers Colin Burgess, Mike Cassutt, Rex Hall and Bert Vis; John Charles, at NASA JSC; and Japanese astronaut Chaiki Mukai, for her excellent Foreword.

Personal interviews were held with Russian cosmonaut Valentina Tereshkova, American astronauts Kathryn Sullivan, Rhea Seddon, Ellen Ochoa, Yvonne Cagle, NASA Flight Director Glenda Laws, and family members (wives and daughters) of several former astronauts. Thanks are also due to James Skipper, for the information on Shuttle EVA vacuum-chamber simulations, and Iva 'Scotty' Scott, formerly of NASA JSC. In addition, the staff of the University of Clear Lake, Houston, provided access to the JSC history collection, and the NARA offices and resources in Fort Worth were invaluable. Janet Kovacevich, Joey Pellerin, formerly of NASA JSC History Office, Joan Ferry, of Rice University, Meg Hacker, of NARA, Fort Worth, Kay Grinter, at NASA KSC, and Barbara Schwartz and Eileen Hawley, of PAO JSC, Houston, provided valuable assistance during many visits over several years.

We also appreciate the immense research efforts of Teresa Kingsnorth, Nicola Humphries and Lillian Kozloski, who have generously shared the results of their own research. Additional thanks go to Dr Jeanne Becker – formerly Associate Professor in the Department of Obstetrics and Gynaecology, and Medical Microbiology and Immunology at the University of South Florida College of Medicine, Tampa, Florida – for her support and her willingness to answer numerous 'medical' questions; Margaret Weitekamp, for granting approval to use her seminal work 'The Right Stuff, The Wrong Sex: The Science, Culture and Politics of the Lovelace Women in Space Program, 1959–1963'; Hartmut Sänger, for supplying photographs of and information on his mother, Dr Irene Sänger-Bredt; Dr Kerry Joels, for his help and assistance in compiling the section on Nichelle Nichols; Joseph

Ruseckas and Mary Mahassel, for their insight and experience in manufacturing pressure suits for the David Clark Company; Thomas Dreschel and Julie Nycum, Fundamental Space Biology Outreach, NASA KSC, for the student/teacher copies of the 'Human Physiology in Space' manual; and Rudy Opitz (via his son Michael), for his invaluable contribution to the section on Hanna Reitsch and her crash in the Me163B Komet.

The following also merit our thanks, as they not only answered the questions posed to them, but also checked draft sections of the book: Dr Ann Whitaker, NASA MSFC; Dr Bonnie Dalton, NASA Ames; Dr Emily Holton, NASA Ames; Dolores Beasley, NASA HQ; Debra Burnham, Hologic, Richard Fry, MD, Cellon SA; Richard Hular, Chairman, BioLuminate; Ted Bateman, Bioengineering Department, Clemson University; and Jessica Wiederhorn, Columbia University Oral History Research Office.

Other valuable help and assistance was provided by Darlene Feikema and Diane Leigh, University of Washington; Bruce Hess, Staff Historian at Wright–Patterson Air Force Base; Al Hartmann, research associate working with the University of California on the Florida Space Coast History; and Linda Plush, of the Space Nursing Society.

Unless otherwise credited, all photographs are either from the authors' collections or courtesy NASA.

Work on the Mercury 13 women would have been much the poorer without the help of one of America's premier space enthusiasts and collectors – and the most genial of characters – Al Hallonquist, to whom thanks are also extended (www.mercury13.com). We are also grateful for the assistance of Garry King, of Autographica (www.autographica.co.uk) – for providing the opportunity to meet Valentina Tereshkova and to ask her about her meeting with Jerrie Cobb – and David Price, who recorded her response.

We also thank Ruth Shayler for transcribing taped interviews, and Mike Shayler and Gary Robinson for their support, skills and talent in transforming the original draft into the submitted text. Once again, thanks are extended to Project Editor Bob Marriott – whose mother Kathleen built Spitfires (with some help) from 1942 to 1944 – for his continued long hours spent editing and preparing the text, and scanning and processing the illustrations. Bob's advice was invaluable in the writing of the section on women astronomers in the first chapter, and he wrote the section on British amateur astronomers.

We must also thank Jim Wilkie, for his cover design; Arthur and Tina Foulser, at BookEns; and, of course, Clive Horwood, of Praxis, who again demonstrated understanding and patience throughout an unavoidably extended period of production.

Thanks also to Teresa Moule and Beryl Edge for their encouragement and support in this project. We are also indebted to all those who are not mentioned by name, but who gave of their time and effort.

List of illustrations

Front cover Kathy Thornton performs EVA during the first Hubble Space Telescope service mission in December 1993; (inset) Valentina Tereshkova reports her readiness for her historic flight in June 1963.

Back cover (Top) Kalpana Chawla inside the Spacehab module onboard *Columbia* during mission STS-107 in January 2003; (centre) Eileen Collins – the Space Shuttle's first female Pilot (1995) and first female Commander (1999) – is scheduled to command the Return-to-Flight mission STS-114 in 2005; (bottom) Japanese astronaut Chiaki Mukai (who provided the Foreword for this book) floats above the aft flight deck controls of the Shuttle during mission STS-95 in October 1998.

Prologue

> 'On 16 June 1963, at 12.30 pm Moscow Time, a spaceship, Vostok 6, was launched into orbit piloted, for the first time in history, by a woman, citizen of the Soviet Union, Communist Comrade Valentina Vladimirovna Tereshkova.'
>
> *TASS news statement, 16 June 1963*

Twenty-six months and four days after Yuri Gagarin became the first man to fly in space, the Soviet Union succeeded in orbiting the first woman on a three-day mission. In a joint flight with the previously launched Valeri Bykovsky onboard Vostok 5, these two missions were the final flights under the Vostok programme, recording not only the first female spaceflight, but also a solo world endurance record of 119 hours by Bykovsky that he still holds more than forty years later. Tereshkova became the first, and for the next nineteen years, the only woman in space, logging more time in space than all the six American Mercury astronauts together. She never flew in space again, but will forever remain a pioneer in space exploration alongside the names of Yuri Gagarin, Alexei Leonov and Neil Armstrong. From 1982, other women began to enter space to continue the journey begun two decades earlier. Following in the trail of Valentina, their missions were varied and challenging, but as with Gagarin, Leonov and Armstrong, only one could claim to be *first*. For Tereshkova, the title 'First Woman in Space' is an honour she has carried with pride and dignity for more than forty years. During that time, many have tried to follow her, several have achieved their dream and orbited the Earth, others have been unable to make the step from Earth to orbit, and a few have made the ultimate sacrifice in the pursuit of the peaceful exploration of space.

Vostok 6 was launched by an R-7 launch vehicle from the Baikonur cosmodrome in Kazakhstan. The objectives of the flight were officially announced as continued studies on the effects of spaceflight on the human organism and, specifically, to provide a comparative analysis on the effects of spaceflight on a woman. Tereshkova would also conduct a number of visual observations, experiments and communication sessions with the Earth and with her colleague in Vostok 5.

Using the call sign Chiaka (Seagull), she soon gave her first impressions upon viewing the Earth from space: 'It is I, Seagull! Everything is fine. I see the horizon; it's a sky blue with a dark strip. How beautiful the Earth is ... everything is going well.'

Flying in an orbital plane 30° from Vostok 5, it was only possible for the two spacecraft to be in close proximity to each other for a few minutes during each orbit, drifting further apart each revolution. The closest approach was only 5 km on the first orbit after the launch of Vostok 6, and neither cosmonaut could clearly confirm that they had spotted the other.

The two cosmonauts established communications with each other, however, and received greetings from Premier Khruschev. Live pictures of Tereshkova in space were beamed to Soviet TV as the propaganda machine went into overdrive, claiming the success of the flight as yet another demonstration of the superior socialist system. There were sceptics who suggested that this was nothing more than a publicity stunt – which in part it was – but others marvelled at the technological skills of a Soviet programme that could send a woman into space for three days when America could only manage a one-day flight.

Onboard her spacecraft, Tereshkova took photographs and film of the terrain and cloud patterns passing below her, as well as observing the Moon and Earth's horizon over the poles. From her position, the beauty of Earth was overwhelming, the blackness of space during the night-side passes scattered with countless stars was most impressive, and even in daylight, with the Earth illuminated by the Sun, the stars could still be seen in the blackness of space. In 2004, when asked of her impressions and memory of viewing Earth, she recalled thinking initially that it was extremely large, but after a few orbits taking less than ninety minutes her view changed, seeing our home planet as a small, fragile, but beautiful place.

Official status reports mentioned her conducting extensive tests of the spacecraft, monitoring the controls and onboard equipment, and supervising a programme of small experiments including the habitation of the capsule – all part of 'her contribution to the space programme'. She also had to log the parameters of the life support system and her condition during the flight. After a period of sleep her condition was reviewed, and she reportedly asked permission to continue the flight, as all was well.

After three days in space, her 'official programme' was completed, and on 19 June she was instructed to begin the return to Earth, a few hours before Bykovsky. Landing at 11.20 Moscow Time, 385 miles north-east of the city of Karaganda in Kazakhstan, Tereshkova logged 2 days 22 hrs 50 min in flight, and completed forty-nine orbits of the Earth. Almost immediately upon entering orbit, the adulation and excitement of her achievement spread around the world, and though the mission away from her home planet lasted just three short days, her 'mission on Earth' as a goodwill ambassador has continued ever since.

The flight of Tereshkova and Vostok 6 – no matter the political intentions of the mission – remains a milestone in the history of space exploration and in the achievements of women in science and engineering. That milestone had its origins centuries earlier, as new technology moved the world into new ages, and in the four decades since Tereshkova's flight, the struggle for orbit continues – as does the sacrifice. We have seen female space explorers from several countries spend months on space stations, deploy and retrieve payloads from the Space Shuttle, walk in space, take leadership roles on space crews, and pilot the Space Shuttle. Yet no matter what their achievement, all of them follow in the trail blazed by Valentina – as will the first woman to walk on the Moon and, eventually, on the surface of Mars.

Into the wide blue yonder

Women have had a shared involvement in aviation and space exploration, alongside their male colleagues, for well over two centuries. They have also contributed to advances and achievements in science, astronomy, medicine, exploration, aviation, engineering and astronautics; and more recently, a fortunate few have left Earth and explored space.

On 16 June 1963 Valentina Tereshkova was launched into space onboard Vostok 6, and became the first woman to leave the Earth and enter orbit. Around the world her achievement was hailed as a milestone; but, pioneering and daring as it was, it was also a politically motivated mission using limited hardware, and there would be no other female-crewed space missions for another twenty years. The journey to space had been long and difficult, and for women following Tereshkova's achievement it would be an equally difficult journey.

WOMEN IN ASTRONOMY

With the development of astronomy during the eighteenth and nineteenth centuries, some private and academic observatories required employees. Some of these were women, but although they carried out the same work as the men they received little recognition for their efforts. However, the situation was the same for men. Many observatory 'assistants' carried out observational work which was afterwards credited to the director or owner of the observatory; and, moreover, the benefactors of private observatories would sometimes take the credit for the observations even if they knew nothing about astronomy.

Astronomy in the ancient world

The history books tell us that the female exploration of space began with Valentina Tereshkova's flight onboard Vostok 6. This, however, was the physical presence of a woman in space. Women had long before been involved in studying the heavens.

Around 2354 BC, Sargon of Akkad – the founder of the Sargonian Dynasty in Babylon – appointed his daughter En Hedu'anna as his chief priestess of the Moon goddess – a title with enormous prestige and power. Part of her role was to create 'observatories' to monitor and record the movement of stars, which were interpreted to foretell the fortunes and events of the coming year.

This, of course, was astrology, but in the real sense of the word – observations of the cycle of the heavens to attain knowledge of forthcoming natural events essential for agriculture, trade and religion. In Egypt, for example, the annual first rising of Sirius – the Dog Star – signified the imminent flooding of the Nile, which was essential in maintaining the fertile land on each side of the river on which the Egyptians depended for survival.

In ancient Greece, Aglaonike studied the Sun and the Moon, and could predict eclipses. She was a natural philosopher, but she was believed, by some, to be a sorceror and mystic – a person to be feared and respected. (It should be noted that 'natural philosophy' is now called 'science' – a comparatively new word.)

The most notable female philosopher of the classical world, however, was Hypatia of Alexandria, the daughter of Theon – a famous scholar, and one of the most educated men of his time. Hypatia developed an interest in mathematics and what would now be called 'physical sciences', including astronomy, and became one of the most notable scholars and teachers of her time. However, she eventually fell victim to the decline in respect for academics and intellectuals. She was publicly flogged and murdered, and soon afterwards the great library at Alexandria was burned to the ground by the mob. Almost two millennia later she was commemorated with the naming of a feature on the Moon: the Hypatia rille.

The beginnings of modern astronomy

The invention of the telescope in 1608, and its subsequent development, presented far greater opportunities for observational astronomy.

In the seventeenth century, Elisabetha Hevelius (1647–1693) – the second wife of Polish astronomer Johannes Hevelius (1611–1687) – often worked with her husband, carrying out observations and preparing the results for publication. They could afford the best instruments, and their observations were the most accurate and reliable of the time. Tragically, many of these records were lost in a house and observatory fire on 26 September 1679; but Hevelius and his wife persevered. The primary results of their labours were *Selenographia*, *Cometographia* and *Atlas Coelestis*.

One of Hevelius's contemporaries, Marie Cunitz (1610–1664), translated (from Latin) the works of Johannes Kepler. (Kepler had spent many years in calculating the orbit of Mars, based on Tycho Brahe's observations, and had eventually formulated his three fundamental laws of planetary motion.) Due to Cunitz's efforts, Kepler's work was made more accessible, and she became known as *Urania Propitia* – 'She who is closest to the muse of astronomy'. Her commitment to her work was such that she often neglected her household, spending most of the daylight hours asleep so that she could spend the night hours observing the heavens.

Maria Winkelmann (1670–1720) was the wife of German astronomer Gottfried Kirch (1639–1710), and worked with her husband in producing calendars and compiling planetary ephemerides. On one occasion in 1702, after her husband had spent the evening observing, Maria decided that she would take advantage of a clear sky and observe the heavens herself. To her surprise, she thought she saw a comet. She therefore woke her husband, who confirmed the observation. Inspired by this, she continued her interest in observing, and even, some years later, had some of her observations published. After the death of her husband in 1710, Maria applied for membership of the Berlin Academy of Sciences, to continue her work on the preparation of calendars in an official capacity. However, the doyens of the establishment were not pleased. In a letter to the Academy's President, Gottfried Leibnitz, Johann Thedor Jablonski wrote: 'I do not believe that Maria Winkelmann should continue to work on our official calendar of observations. It simply will not do. Even before her husband's death the Academy was ridiculed because its calendar was prepared by a woman. If she were to be kept on in such a capacity, mouths would gape even wider.'[1]

'It simply will not do'

In 1667 Samuel Pepys attended a meeting of the Royal Society at which the Duchess of Newcastle was, after much debate, invited to address the audience. To Pepys she seemed to be a 'pleasant looking woman, her dress looked old and her deportment ordinary ... Nor did I hear her say anything that was worth hearing.' However, individual prejudice should not be interpreted as a general concensus, and not all men looked unfavourably on women as scientists. The Duchess had, after all, written and had published (in 1653) a book entitled *A World Made by Atomes*. In 1697 the German mathematician and philosopher Gottfried Leibnitz opined that women who possess an elevated mind 'probably advance in knowledge more effectively than men' – especially since their situation placed them 'above troublesome and laborious cares' and allowed them to become more detached and, consequently, 'more capable of contemplating the good and the beautiful.'

In 1742 Dorothea Erxleben became the first woman to be granted a Doctorate in Medicine by the University of Halle. However, she felt that men would see her advancement as declaring 'war' on the male world or as a devious attempt to deprive them of their privilege, and that women might consider that she was trying to place herself above them. Her study was entitled 'Inquiry into the Causes Preventing the Female Sex from Studying'. Three years later, the head of the university, Johann Junker, was more scathing: 'Learned women attract little attention as long as they limit their study to music and arts. When a woman dares to attend a university, however, or qualifies for and receives a doctorate, she attracts a great deal of attention. The legality of such an undertaking must be investigated.'

Even women cautioned about their own involvement in science. In 1775, Marie Thiroux d'Arconville – a French anatomical illustrator – wrote: 'Women should not study medicine and astronomy. These subjects fall beyond their sphere of competence. Women should be satisfied with the power that their grace and beauty gives them and not extend their empire to include medicine and astronomy.' But in 1787 a Göttingen newspaper reported the award of the first PhD from a German University to a woman, Dorothea Schlozer: 'Usually one thinks of a learned woman as neurotic, and should she ever go beyond the study of literature into higher sciences, one knows in advance that her clothing will be neglected and her hair will be done in antiquarian fashion ... For Madam Scholzer, this is not at all the case. She sews, knits and understands household economy perfectly well. One must gain her confidence before one comes to know the scholar in her.'

Around the same time, a young German woman with musical talents, living in England, was about to make an even greater contribution to science: Caroline Herschel.

Caroline Herschel

Caroline Herschel (1750–1848) came from a musical family in Hanover, Germany, and in 1772 moved to England with her brother Alexander with the intention of working with a second brother in their own musical careers. This second brother was Friedrich Wilhelm Herschel (1738–1822), who had moved to England after a brief career as a musician attached to the Hanoverian Guards (although he was not, as is often believed, in the army), and was earning a living by copying music and giving recitals. It became a highly successful partnership, but William was increasingly being drawn to a new interest. His desire to study the theory of music took him into the world of mathematics and the works of Isaac Newton. In turn, this led to a growing passion for astronomy and optics, and he began to make mirrors and telescopes.

At first, Caroline was not amused by William's change in direction. She had come to England to sing with William, and as the female, with no formal education of her own, was also expected to look after her family. This new venture could have been very risky, but William was naturally gifted and became very successful in his astronomical work, helped by a pension from his years in music, and supportive and influential friends. Caroline continued to support him, and cooked and cleaned for him until he married. Gradually, however, she not only assisted her talented brother in his observations, but also became interested in astronomy herself – and subsequently worked with William for the rest of his life.

Together they made thousands of astronomical observations, but it was while moving in March 1781 that the name Herschel would become linked with history. Their temporary property at 5 Rivers Street, Bath, did not include a garden for the telescopes, so they were in the process of moving back to 19 New King Street. On the night of 13 March 1781, Caroline was dealing with some last-minute issues at Rivers Street, while William was at his telescope in New King Street. On that night he observed what he thought was a new comet, due to its unusual appearance; and, indeed, it was described as such in the paper which he subsequently published. The new object was afterwards observed and followed by others, and Anders Lexell identified it as a planet – the first to be discovered, as the movements of the other 'wanderers' had been known since ancient times. Herschel named his new planet Georgium Sidus (George's Star, after King George III), but others suggested different names. The name Uranus (the father of Saturn) was proposed by Johannes Bode, but even that name was not generally accepted until the mid-nineteenth century.

Following Herschel's discovery, King George appointed him Royal Astronomer – a unique position (not associated with the official post of Astronomer Royal) – and awarded him a pension of £200 a year. A condition of this appointment was that Herschel should allow members of the royal family to look through his telescopes, and he therefore moved to Datchet, near Slough. Fortunately, however, royal visits were rare. Thus freed of the necessity to earn a living, Herschel was able to devote all of his time to astronomy.

Caroline continued recording William's observations by assisting him at his large telescope – where she sat in a hut and communicated with him via a speaking-tube – and by copying his drawings and preparing his papers and catalogues. For this she was entitled to the notable salary of £1 a week, and thus became one of the first women to 'make space pay'! One of her roles was to grind the manure used for making the moulds for the cast speculum-metal mirrors (glass mirrors were not used until the 1850s), but she was also a skilled observer and instrument-maker in her own right, and used the telescope that her brother had given her in 1782 to carry out her own observations. For this her time was limited, but on 1 August 1786 she discovered a comet, and over the ensuing decade she discovered another four comets and fourteen nebulae. Following the death of her brother in 1822, Caroline returned to Hanover.

Mary Somerville (1780–1872) was also highly respected for her knowledge of science, and published several popular books explaining and popularising some of the most important and difficult scientific concepts of the period – mostly in astronomy, physics and chemistry.[2] Both she and Caroline Herschel were so highly respected that they were elected Honorary Members of the Royal Astronomical Society. (Unlike the Royal Society, in which the title 'Fellow' is an awarded honour, Fellowship of the Royal Astronomical Society is a membership, requiring only the payment of a subscription. However, no women were allowed to join as Fellows until after the First World War – and even Caroline Herschel and Mary Somerville were Honorary Members, and not Honorary Fellows.)

Caroline Herschel was also awarded a gold medal by the King of Prussia and a gold medal by the King of Denmark for her contributions to astronomy, and was also an Honorary Member of the Royal Irish Academy. In her later years her interest in astronomy did not diminish, although failing eyesight was increasingly frustrating. In her letter of acceptance as an Honorary Member of the Royal Astronomical Society she was as modest as ever, and underplayed her important role in the advancement of astronomy: 'Regrettably, at the feeble age of 85, I have no hope of making myself deserving of the great honour.' But she lived on for another twelve years, and died, at the age of 97, on 9 January 1848.[3]

Maria Mitchell

On 1 October 1847, a young American woman, Maria Mitchell (1818–1889), proudly announced her discovery of a comet to her parents and their dinner guests. Maria had become fascinated with astronomy due to the influence of her father, an enthusiastic amateur astronomer. She was employed as a librarian at the Nantucket Athenaeum, and took full advantage by reading the astronomy books in the library during the day and turning her studies to practical observations by night, using the family telescope on the roof of her home. Her discovery of a comet in 1847 was rewarded by the presentation of a gold medal by the King of Denmark, and soon afterwards she became the first woman member of both the American Academy of Arts and Sciences and the newly created American Association for the Advancement of Science.

Mitchell was also a founder member of the Association for the Advancement of Women, and spent nineteen years working from home for the *American Ephemeris* and *Nautical Almanac*. In 1865 she was appointed Director of the observatory at the newly founded Vassar College (for women only) – a remarkable achievement given her lack of formal education – where, besides her observational work, she also taught and guided her students in preparing for their chosen degrees.

The Harvard computers

At the turn of the twentieth century, Edward C. Pickering, Director of Harvard College Observatory, brought together a group of women to work on the processing of astronomical observations and data. Prior to the First World War, they worked on the top floor of Building C, purpose-built to protect the astronomical files and glass negatives from the risk of fire. Pickering's offices were across the hall from the women's offices, and other male staff worked on the lower levels, with the archive at ground level.

These women became known as the 'Harvard computers' – and several of them would later follow distinguished careers. They included Margaret Harwood, a 1916 graduate of the University of California who later became Director of the Maria Mitchell Observatory – the first such honour bestowed on a woman by an independent observatory; Annie Jump Cannon (1863–1941), who carried out work in mass spectroscopy by individually identifying the spectral types of thousands of stars on photographic plates; Johanna Mackie, a recipient of a gold medal from the American Association of Variable Star Observers, for her discovery of a nova; Ida Woods, who was awarded a gold medal by the AAVSO in 1920, and later discovered several novae on photographic plates; Henrietta Swan Levitt (1858–1921), whose important work on cataloguing variable stars by studying photographs of the Magellanic Clouds led to her discovery of 1,777 new variable stars, including a type afterwards known as Cepheids – a fundamental key in determining distance scales throughout the Universe; and Cecilia Payne-Gasposchkin (*née* Payne, 1900–1979), who in 1925 was awarded a PhD for her work at Harvard, and later became a leading astrophysicist.

British amateur astronomers

Until the latter half of the nineteenth century, many universities barred women, and many societies and organisations did not allow women to become members. Another constraint, however, was the cost. Even in the first half of the nineteenth century, the subscription for the Royal Astronomical Society was £12 12s – which would have deterred most people, not just women, from joining, as it was the equivalent of several years' wages for those on a lower income. This did not, however, prevent their participation, and at public lectures on various fields of science – such as those presented at the Royal Institution and the Surrey Institution – the audiences often included many women.

After the Napoleanic Wars, European countries were keen to reaffirm their status – chiefly with the establishment of scientific institutions, including observatories run by professional astronomers. In nineteenth-century Britain, however, research in astronomy, and in other sciences, was pioneered by amateurs. (It should be remembered that despite the current derogatory use of the word, an 'amateur' is 'someone who loves', whereas a 'professional' is someone who is paid.) Some of these amateurs were very wealthy and could afford to build large telescopes, and were thus able to observe objects which no-one else could see, and make many new discoveries. There were only a few professional astronomers, and they were mostly ill-equipped and poorly paid. In the early 1860s, amateur astronomy and telescope-making became very popular among those of more modest means – including many women.

At that time, the Royal Astronomical Society did not admit women as Fellows, and this situation, together with the dramatic increase of popularity in astronomy, led to the founding of the British Astronomical Association in 1890. Furthermore, the RAS publications were beginning to contain an increasingly high proportion of papers on mathematical and theoretical astronomy, whereas the BAA – like the numerous provincial scientific and natural history societies founded in the latter half of the Victorian era – was primarily an organisation of practical scientists and observers from all walks of life and widely different social strata. Many women joined the BAA immediately – indeed, the door was willingly opened wide for them – and several of them served on the first Council. The gentlemen, of course, treated them as ladies, in keeping with the manners of the time – but they were also treated as fellow astronomers of equal merit. To name but a few: Annie S.D. Maunder (*née* Russell, 1868–1947) was a member of Council, and later married E. Walter Maunder (the founder of the BAA), whom she had met when they both worked at the Royal Observatory, Greenwich; Elizabeth Brown (1830–1899) had been Director of the Solar Section of the Liverpool Astronomical Society during the 1880s, and was Director of the BAA Solar Section and a member of Council from 1890 to 1899; Mary Acworth Orr (1867–1949), who served on the first Council, married the eminent amateur spectroscopist John Evershed (1864–1956) in 1906, and worked with him in solar research, cooperatively and independently, for the rest of her life; Agnes Clerke was a notable and respected author of several books on the history of astronomy and astrophysics; Lady Huggins – wife of Sir William Huggins, an amateur astronomer knighted for his work in astrophysics – worked both with her husband and as an independent researcher; and there were many others. The emancipationists were yet to appear on the scene, but in the British Astronomical Association, women were, and always have been, treated as equals with a common interest.

There was one BAA member, however, who was the daughter of an astronomer. During the 1890s, Gertrude Bacon – daughter of the Rev John Bacon – often accompanied her father on high-altitude balloon flights to carry out meteorological research. One one occasion, however, they ventured too high and were rendered unconscious. They were injured in the ensuing crash, but they recovered – and Gertrude was not deterred from further adventures.

From astronomy to space

In June 1985, a young female astronomer, Tamara Jernigan – aged 26, and working towards her PhD (awarded in 1988) – was selected as a Mission Specialist in the eleventh group of NASA astronauts. She went on to fly on five Shuttle missions that included, among other objectives and assignments, working with astronomical payloads. She was the first professional female astronomer to study the Cosmos from above the atmosphere. But the story of how Jernigan and other women from around the world became professional space explorers also has its foundations in the history of medicine, engineering, natural and physical sciences, exploration and, of course, aviation.

PIONEERING WOMEN AVIATORS

While women across Europe and America were striving for recognition in science, medicine and astronomy, there was also a rapidly developing field of technology that would change the world and lead, eventually, to the exploration of space: flight. Since ancient times, ways to permit humans to fly through the air have been imagined and recorded. Studies of birds, gliding, and man-powered machines have continued to the present day, but five key developments would allow humans to eventually leave the ground and travel into the 'wide blue yonder' and beyond: balloons, parachutes, gliders, engine-powered flying machines, and rockets. Remarkable advances in balloons, gliding and parachute technology began in the eighteenth century, joined by powered flight at the beginning of the twentieth century. Records show that the development of rockets is centuries old, but it was not until the twentieth century that the rocket would fulfil the age-old dream of flying in space. The rapid pace of development of the science of powered flight with aircraft (aeronautics) and spacecraft (astronautics) has been remarkable over the past hundred years; but to arrive at where we are today, we must thank not only Joseph and Étienne Montgolfier (balloonists), George Caley (gliders and aerodynamics), Wilbur and Orville Wright (powered flight), Charles Lindbergh (transatlantic crossing) and Chuck Yeager (breaking the sound barrier), but also the thousands of other male pilots who advanced aeronautics from a foolhardy and dangerous occupation to the international air travel network we take for granted, and the scores of pioneering women aviators. Together with the men, the women moved the 'ceiling' of flight to the very fringes of the atmosphere, allowing, in 1961, a human to break through the invisible barrier into orbit for the first time.

A sheep, a duck and a chicken

The competition to achieve the first manned balloon flight was between French physicist Jacque Alexandre César Charles – who had demonstrated the lifting potential of a helium-filled balloon – and the Montgolfier brothers – paper-makers who had succeeded with taffeta balloons at the French Academy of Sciences during the summer of 1783. History records that on 19 September 1783, the first land-bound creatures to ascend into the air were not humans, but a sheep, a duck and a chicken.

This three-creature 'prime crew' preceded human flight under a helium-filled balloon called *Martial*, which was launched from the courtyard of the Palace de Versailles in Paris, to the amazement of King Louis XVI, Queen Marie Antoinette and assembled courtiers. After a 'mission' of 2.5 miles, lasting eight minutes, the balloon landed in the Vaucresson Forest. It was reported by the 'recovery crew' of horsemen that the animals were dazed but had survived;[4] but when the animals were presented before the King, he ordered them to be cooked for his dinner! Ironically, 174 years later the first living creature in space was another animal (and a female): the Russian dog Laika. Unfortunately, she did not survive the ordeal, due to the limited technology and difficulties during the flight.[5]

The three animals preparing the way for human balloon flight were a logical stepping-stone to the honour of being the first human to do so. It had originally been suggested that perhaps condemned criminals should be the first human passengers of a balloon – in case anything should go wrong! But the young French professor of physics and chemistry, Jean Francois Pilâtre de Rozier, disagreed, and persuaded King Louis that a free man should take the honour of making the first flight. The King relented, and on 15 October 1783 Pilâtre de Rozier achieved his goal when he became the first person to fly in a balloon (though still tethered) some 50 feet into the sky. After a tethered flight to 340 feet on 19 October, the first free flight with human passengers in a balloon was completed on 21 November 1783. The age of air travel had arrived, and it was not long before women would join the exciting new adventure.

'She's actually been flying!'

On 20 May 1784, in Paris, four aristocratic ladies accompanied the Marquis d'Arlandes and Pilâtre de Rozier on a short captive flight close to the ground. Then, on the evening of 4 June 1784, Mme Elisabethe Dhible accompanied Mr Fleurant in the balloon *Gustav* to a dizzy height of 900 feet, travelling about two miles through the air in an event witnessed by King Gustav III of Sweden, who had given his name to the balloon. On landing after her historic flight, Mme Dhible admitted feeling a little giddy after viewing the countryside from up high. For such an historic event, the attire of the first female aeronaut was of course required to be fashionable: 'Mme Dhible was a charming young woman, for anyone less attractive would hardly have been suitable for the occasion. She wore a white and grey lace dress and a large hat with horizon-blue feathers ... but she had nearly frozen in her transparent layers of muslin.' Witnessing the event, her friends reportedly explained to those who had not witnessed the fever; 'My Dear! She's actually been flying! It's positively inconceivable!'[6]

The art of ballooning soon became an international endeavour. On 14 September 1784, on the other side of the English Channel, in London, the Italian aeronaut Lunardi was preparing to ascend with Mrs Letitia A. Sage, who was hoping to be the first English woman to ascend from the ground. Unfortunately, after some time spent sitting inside the basket, smiling obligingly for the growing crowds of onlookers, a problem that would befall several future space-travellers fell upon the Englishwoman: she was 'bumped' from the crew. It was found that the balloon could not rise with two human passengers, as the payload weight was too great, and so Mrs Sage was replaced by a pigeon, a cat and a dog. Undeterred, the next year she was a passenger with pilot George Biggin in an ascent from St George's Field in London on a flight to Harrow Common, and finally became the first Englishwoman to leave the ground. On 10 November 1798 – a few days before the fifth anniversary of the first untethered manned balloon ascent – the first all-female crew was launched over Paris: pilot Jeanne Geneviève Gamerin (Miss Labross), and co-pilot Miss Henry.

Marie Blanchard

On 7 January 1785, Jean-Pierre Blanchard and Dr John Jefferies piloted a balloon from Dover to a landing in the forest of Felmores, near Calais. Unfortunately, a few years later, and despite the fame of his aeronautical successes, Blanchard was facing bankruptcy. Unwilling to live in poverty, and knowing of the fortune that male aeronauts could expect to earn, his wife, Marie Madeleine Sophie Armant Blanchard, decided that she would attempt to earn a living by promoting herself as a professional female aeronaut. She certainly became famous in a career in which she was more than a mere passenger, unlike most of the earlier women that had made balloon ascents. In June 1810 she participated in a tethered ascent as part of a festival display by the Imperial Guard on the Champs de Mars during the wedding of Napoleon and Marie-Louise, and became a talking point of Paris. Unfortunately, she also had her share of mishaps during her short career. A celebrity across Europe, she crossed the Alps by balloon, but suffered a nose-bleed at 9,000 feet. She also nearly drowned, in 1817, when she mistakenly thought that a flooded field near Nantes was a safe grassy meadow; but fortunately, she was rescued by the horsemen tracking her flight. On 9 July 1819 she was to take part in a spectacular display by setting off a trailing line of fireworks from the balloon – which was a little reckless, considering that the balloon was filled with hydrogen. She ascended and duly set off the fireworks, with the balloon masked in smoke. Then suddenly, the horror of what had happened became clear to the thousands of onlookers across the Tivoli Park. A spark from the fireworks had set fire to the basket and the balloon itself, causing it to crash to the ground and killing Marie Blanchard – the first recorded female victim of an aviation accident.

The first giant leaps

One of the most important developments in aeronautics (and astronautics) has been the evolution of rescue systems. In the eighteenth century, the parachute became part of the progress of balloon ascents. Designs of parachutes had been studied by Leonardo da Vinci in the fifteenth century, and animals were dropped from a balloon in tests of such devices in 1785; but the first human trial was carried out by Jacques Garnerin, who decided to test his own parachute – a parasol-like device 40 feet across, fixed to the gondola under his hydrogen balloon. On 22 October 1797 at 2,300 feet, Garnerin severed the connections with the balloon and dropped. The parachute opened to suspend him in the gondola to a safe landing, and witnesses described the event as one of the greatest acts of heroism in human history.

Refinements to the design followed, including cutting a hole in the top of the device to relieve the air pressure under the parachute canopy, diminishing the oscillations, and reducing orientation sickness. On 25 August 1825, after safely completing a 'leap of faith' from 10,000 feet, Garnerin's wife Elisa became the first woman to parachute to the ground; and after Jacques Garnerin's death, his niece (also Elisa) became a favourite at public festivals, drawing larger crowds than her famous uncle ever had, and thrilling the audience with her 'pretty woman in peril' routine that, it was reported, had many women on the ground requiring a whiff of smelling salts to regain their senses after witnessing such daring feat. Of course, it helped to attract the crowds that Elisa was attractive, and she was also remarkably lucky in this fledgling endeavour. She retiring without injury, after thirty-nine successful jumps, in 1828.

Higher, further, faster

By the middle of the nineteenth century, enclosed envelope balloon flights were still being completed, but the strategic advantages of balloons were also being investigated at the same time as very early studies into winged apparatus,

On 18 October 1864, Mme Godard and her husband completed a 16-hour flight, but were almost killed upon landing. On 31 August 1876, the Duruofs were about to attempt the first crossing of the English Channel by a husband-and-wife team, when the weather closed in and they tried to abandon the attempt. Unfortunately, the crowd had paid a lot of money to see them leave, and were prepared to wreck the balloon and injure the occupants if they did not take off. Reluctantly – and not totally unscathed – the pair lifted off into the storm; and halfway across the Channel, the balloon came down into the sea. A vessel sighted them, but lost them again in the swell. The following morning, Mme Duruof was found unconscious in the water, but there was no sign of her husband. Not surprisingly, she promptly refused to ever leave the ground again! In 1886, American female balloonist Mary Myers set a new altitude record of four miles, but other events in the US less than twenty years later forever changed exploration of the air above us.

The success of the Wright brothers in December 1903 changed the face of aeronautics, but powered flight would need time to develop, and ballooning would continue for some years before becoming primarily a sporting pastime with the expansion of air travel in the 1960s and 1970s. Until then, there remained records to be claimed and mountains to be climbed – or at least flown over!

On 26 October 1910 – six years after the Wright brothers' first powered flight – Marie Marvingt completed the first female solo balloon crossing of the English Channel from Nancy in France to Southwold in Suffolk. The first woman to compete in an FAI classified balloon race, on 12 October 1913, was Rene Rumpelmayer, who piloted her balloon (co-piloted by Mme Gustavo Goldschmidt) to a remarkable sixth place in the eighth Gordon Bennet race at Jardin des Tuileries in Paris.

Over in the Soviet Union during the 1930s, parachute and balloon records by women were frequently pursued. In 1935 it was reported that a pair of Soviet women had parachuted together from 26,000 feet without oxygen equipment, passing through three layers of cloud and landing safely just 2.5 miles apart.[7] On 15 May 1939, A. Kondratyeva completed a record flight (for a woman) of 22 hrs 40 min in a gas balloon, flying from Moscow to Lukino Polie. This still stands as a record in the 400–600 m^3 Class A free balloons category AA-03, 65 years after it was established.

Then, on 5 August 1931, a remarkable lady, Kate Paulus, completed her final balloon flight at the age of 63. Born in 1868, she had logged more than 510 balloon flights and recorded more than 150 parachute jumps from balloons. She died at the age of 67, in 1935. A pioneer even by today's standards, a century after her exploits, she bridged the gap from women aeronauts and pioneering parachutists to the age of the aeroplane, stratospheric exploration, and the pre-dawn of the space age. However, the greatest of female aeronauts in the years up to the Second World War was Jeannette Piccard, the American wife of the famous Belgian stratospheric balloonist, Jean Piccard.

Into the stratosphere

If there was ever a dynasty of balloon-borne stratospheric explorers, it would be the Piccard family. Belgian Auguste Piccard (1884–1962) completed stratospheric flights by pressurised gondola in 1931 and 1932, pioneering the successful exploration of the region for later scientific and military purposes in the US and USSR. His twin brother Jean would also attempt to emulate and surpass his famous brother, and together with his wife, American-born Jeannette Ridlon Piccard (1895–1981), would set a record in 1934. Auguste was also the grandfather of the more recent record-setting balloonist, Bertrand Piccard, who participated in the first non-stop round-the-world balloon flight of twenty days in March 1999.

For Jeanette Piccard, fame reached a high point (literally) on 23 October 1934, with her ascent to 17,850 m (57,579 feet) – a new record, and a milestone in female aviation history. In several contemporary accounts of the exploit, Jeannette Piccard is described as 'the first woman in space'.

WITH WINGS AND ENGINES

The development of lightweight petrol-driven engines was a milestone in the development of transportation both on the ground and in the air. They were also adapted for fitting to balloons and refining in guidance, and produced a better-than-average chance of returning from the objective for which the balloonist set off! For some, it was also a way to advertise a film career. In 1914, Gaby Morlay (also known as Gabyde, Gaby de Morlaix and Gaby Morlaix) decided to promote her career by agreeing to undertake a petrol-driven ascent in a balloon – soon after she had won a camel race. The attentive media soon latched on to her aviation achievement, and her career was certainly not harmed. Reportedly, she later told newsmen that 'one has to get talked about somehow! At that time, film-producers in search of an actress weren't absolutely overwhelming me!' But the days of the balloon were numbered. There was a new machine called an 'aeroplane' – which would radically change the world forever.

The Wright stuff

The date 17 December 1903 is notably significant in aviation history. It was the day on which Orville Wright flew the *Flyer* on the first manned powered flight, covering 120 feet and lasting 12 seconds. Although arguments continue about whether this was the first powered flight, most acknowledge the event as the date that powered air travel began. Orville and Wilbur Wright are famous, but they had significant support from their sister Katharine – often referred to as the 'third brother'.

Bishop Milton Wright and Susan Catherine Wright had four sons – Reuchlin, Lorin, Wilbur and Orville – and a daughter, Katharine, born in 1874. It was a close family, but Susan Wright died of tuberculosis when Katharine was only 15 years old and, as was the practice in those days, the daughter became head of the household. Always close to her two famous brothers, she was well educated and became a teacher, using her skills to teach mathematics and aerodynamics to her brothers. Katharine also took on the duties of organising the family finances and managing the business finances, enabling the two brothers to concentrate on their experiments and test flights. With the success of their first flights, the brothers were tirelessly supported by Katharine, whom they called Sves (little sister). She not only designed and sewed the fabric that covered the wings of their aircraft, but also arranged cooperative work with the US Army, organised their social calendar to entertain royals, politicians and presidents to generate more funds, and acted as a combination of social secretary, marketing manager and public affairs officer. During the ensuing world tours, she accompanied her famous brothers and shared the limelight.

Over the next few years, Katharine immersed herself in her brothers' company as well as studying the emerging field of aviation science. She has been credited with designing the first flying suit for female pilots, in which ropes gathered the skirt at the ankles as a practical approach and to retaining the wearer's modesty. After the death of Wilbur in 1912, Orville sold the company in 1915, and the once close family drifted apart. By the late 1920s, Katharine was ill with pneumonia, and she died in 1929. Overshadowed by the fame of her brothers, it is clear that they would not have succeeded without the unswerving support of their sister, and many have suggested that she be recognised as the 'mother of flying'.[8]

Magnificent women in their flying machines

The pace of aviation development in the first decade after the Wright brothers' achievement was astounding. The image of dare-devil pilots taking rather flimsy aircraft aloft to increase height, speed, duration and distance was matched by those willing to make headlines, set records and stretch the limits of aerodynamics, setting the foundations for later and more lucrative adventures.

One of the first women to be credited as an aircraft designer and builder was E. Lillian Toddin but her design, in 1906, sadly never flew. Two years later, in 1908?, Thérèse Peltier, a sculptress, ascended to 600 feet in an aeroplane flown by her husband Léon Delagrange Peltier; and later that year she became the first woman to perform a solo flight. In 1910, Baroness Raymonde de la Roche became the first woman to earn an official piloting licence. That year also saw the first female solo flights by American women. On 2 September, Blanche Stuart Scott became the first

American woman to pilot an aircraft when, without the knowledge or permission of the owner and builder, Glenn Curtiss, she removed the chocks and took off without having had a single flying lesson! Later that same month, on 16 September, Bessica Raiche completed her first solo flight in an aircraft that she and her husband François had built. (Some sources argue that Raiche was the first American female pilot, and that it was not Scott, whose flight has been referred to as 'accidental'.)

In those days, aircraft were flimsy – made from canvas stretched over a bamboo framework – and apart from the danger of flying in them and trying to land them in one piece, it was sometimes dangerous even to try to get in or out of one. Bessica Raiche later suffered a serious fall from a cockpit and required a series of major operations, although she recovered, and resumed flying. She acknowledged that it was a potentially risky and dangerous profession, and often reminded her followers that she would only have herself to blame should she be killed. She was, like all early pioneering test-pilots, willing to accept the risks of taming these new machines, and on one ascent reached the height of 4,800 feet on one ascent – a remarkable record, considering that only a few years earlier, the first powered flight had barely lifted off the ground. In 1919 she predicted her own death from her flying exploits: 'I shall be the first woman to be killed flying. Well, I don't care, I'd rather finish that way than stay on the ground and be no good to anybody.' Unfortunately, her premonition proved correct, and she was killed in an aircraft accident later in 1919.

In 1911, Harriett Quimby, a talented journalist, was awarded the first American female pilot's licence and began performing at flying exhibitions. The following year, on 16 April 1912, she successfully became the first woman to fly an aircraft across the English Channel. She used an aircraft that she had not flown in before, and a compass that she had only just learned to use, and so many thought that she would fail. But 59 minutes later, after flying through fog, she landed near Hardeloit, France. She received a rapturous welcome from the locals and was carried shoulder high, but she would not receive the same adulation as the first man to cross the channel, Louis Blériot. Quimby's flight – historic as it was – took place the day after the loss of the *Titanic*. Although achievement and recognition were her primary objectives, she was also aware of the need to attract attention, and had clothes made both for flying and for 'post-flight activities'. She wore a purple satin costume with a blouse and knickerbockers, and a monk-style hood. The knickerbockers incorporated inside seams closed by rows of buttons which, when unfastened, converted into a walking skirt accompanied by high-top leather boots. Quimby's flying career was also tragically short: just 11 months. On 16 July 1912, during a demonstration of a new aeroplane at the third annual Boston Aviation Meet, near Quincy, Massachussetts, her monoplane pitched violently forward, causing her to lose control. Both Quimby and her colleague – the show's organiser, William A.P. Williard – fell from the aircraft and were killed. Ironically, Quimby had authored an article on the dangers of flying and the need for adequate safety precautions; but on this occasion was not wearing a seat belt.

One of the first female-owned flying businesses was established by Katherine Stinson and her mother in 1913. Two years later, Katherine would become the first woman pilot to accomplish a loop-the-loop – a stunt that would later be expanded upon to delight crowds in displays of dare-devil piloting skills. These skills would also be developed into combat skills for future aerial conflicts.

In Russia, two females were beginning to emerge in both the aircraft industry and in demonstrations of flying skills. One of the pioneers of Russian female aviation was Lidia V. Zvereva (1890–1916) who, with her husband Vladimir Slusarenko and fellow aviator A.A. Agfonov, regularly flew public demonstration flights, in their Farman-IV aircraft, around various Russian towns. To help attract additional funds, they sometime took fare-paying passengers for a 'thrill-ride', but regular support was uncertain. In 1912, prior to a show in Tiflis (now Tbilisi), Georgia, a storm destroyed the aircraft, and with no revenue they were forced to sign over the engine of the aircraft to the airfield owner in lieu of their takings. The following year, Zvereva and her husband secured a valuable military contract for an aircraft repair workshop and piloting school near Riga. Initially successful, the factory expanded and work improved, but by 1917 the business declined due to supply difficulties, shortages of manpower, and the Revolution. By then, Zvereva had died aged only 26. In recognition of her contribution to Russian aviation, her name was later given to a feature on the planet Venus.[9]

WOMEN AT WAR

Barely a decade after the Wright brothers achieved powered flight, the 'war to end all wars' began in 1914. For the next four years, the Great War raged on, and for the first time, in the sky above the battlefields, aeroplanes became the new tool of combat. Then, after a gap of two decades during which new records and headlines in aviation were achieved, the world was once again thrown into the darkness of global conflict for another six years. During these two major conflicts, as during the uncertain peace between the wars, aviation played an important, decisive and pioneering role. And women also played their part.

The Great War in the air

In 1915, Marjorie Stinson joined the family firm and created a flying school. During the First World War, she and her sister Katherine trained many of the first American and Canadian pilots to fly, and at the end of hostilities in 1918, the US Postmaster General approved Marjorie as the first female US Air Mail pilot.

During the First World War, aircraft were use for strategic reconnaissance and observation, bombing, and in aerial combat. Modern combat aircraft and 'dog-fight' techniques originated high above the bomb-cratered fields of France, and from those confrontations, heroes and legends emerged. Most of these were male, but there are reports of a few female pilots. With the outbreak of war, several famous female aviators volunteered for service, but very few were actually called up.

In spring 1914, Fraulin Riotte successfully qualified to become a Zeppelin airship pilot, but although she met the requirements she was not accepted. She was apparently offered 'honorary' status, but it is not known whether she accepted. Until the beginning of the First World War women played a traditional rather than a practical part in aviation. Although there were many who could be described as pioneering dare-devils, most women preferred to remain firmly on the ground. providing tea and cakes for the pilots and spectators, hosting garden parties to raise money for those brave men in their flying machines, and occasionally taking a picnic up in a balloon. As the war developed, these 'social clubs' disbanded, but a few of the leading characters tried to establish an air ambulance service – notably, Marie Marvingt and Jane Herveu, of the Stella aeronautical club in France. But the military hierarchy would not accept such a service, and although some male pilots performed medical evacuations during 1915, the idea and service was not developed after the First World War.

In 1914, Héléne Dutrrieu was accepted for air patrol duty in France, and apparently completed flights, from an airfield near Paris, to reconnoitre German troop movements. Marvingt also appears in official French government documents in a citation for the Legion of Honour. Mme Marvingt (1875–1963) began flying in 1908, and was officially qualified as an aviatrix in 1910. She was also an experienced parachutist and could fire a rifle, but was initially refused the French Legion of Honour because she was considered too young to have earned it. When it was pointed out that, in those days, pilots hardly had time to grow old, Marvingt received her honour. She also is credited with bombing missions over Germany during the First World War, and is often cited as the first female pilot to log combat missions.

Five distinguished Russian female pilots are mentioned in the record books. These and many unnamed others created the role model of heroic Russian fighting women that continued during the Second World War under the Soviet regime. During 1915 Nadeshda Degtereva flew reconnaissance missions along the Galician Front in Austria, and was the first woman pilot wounded in combat. Princess Sophia A. Dolgorukaya was a volunteer, and served as a pilot and observer for the 26th Corps Air Squadron for nine months. She had earned her pilot's licence in 1914, but because of her obvious connection to the imperial family she was officially demobilised following the October 1917 Bolshevik Revolution. Test-pilot Lyubov A. Golanchikova contributed her own aeroplane to the army of the Czar, and during the Russian civil war she returned to flying, apparently logging several sorties for the Red Air Fleet, probably flying Nieuports and SPAD VIIs. After serving as a member of the training squadron of the Red Air Fleet during the was, she fled to Germany. Reportedly, she eventually worked as a New York cab driver, and died in 1961. Helen P. Samonova served as a 5th Corp Air Squadron reconnaissance pilot, while Princess Eugenie M. Shakovskyaya served as an artillery and reconnaissance pilot in 1914 and is credited with being the first woman to become a military pilot. Unfortunately, she was later charged with treason and attempting to escape to the enemy lines. Her sentence of 'death by firing squad' was commuted to life imprisonment by the Tsar, but during the Revolution she was freed, and subsequently worked under General Tchecka. Tragically, she also became a drug addict, and while under their influence she shot one of her assistants, and was herself shot and killed.

EXPLORERS OF THE SUNLIT SKY

After the end of the First World War, tension was still high in the new Soviet Russia. The four long years of war had also accelerated the pace of aviation development, and had demonstrated that strategic superiority would in future require mastery of the air above the combat field and over enemy territory. The race to develop higher, faster, bigger and better aircraft began in the 1920s, and as military aviation developed, so did civilian use of aircraft for exploration, record-setting headlines, and the establishment of new trading and communication routes around the globe. Women aviators played just as important a role as their male colleagues, and also attracted some of the largest headlines. In 1916, two new American flying records had been set by Ruth Law, flying from Chicago to New York, and three years later she became the first person to deliver mail by air in the Philippines.

Barnstormers and wing-walkers

The 1920s was the era of the 'barnstormers' in the United States. These were war veterans or young daredevil pilots who performed dangerous stunts across the countryside of rural America, gathering paying crowds to watch them and giving rides and flying lessons to whoever wanted to pay for the experience ... or fright. They would often fly from relatively flat ploughed fields supplied by farmers with an interest in local aeronautical clubs, or to add revenue to the farm. These pilots would also provide a mean of transportation to those willing to risk the adventure of travelling between locations at a time when 'airports' were unheard of. From these demonstrations came cross-country air races, larger aviation 'shows' and the skills of aerobatics. During the prohibition years in the USA, an added danger was being shot at by moonshiners who must have thought the pilots were working on aerial reconnaissance for the government! Between the two World Wars, several American female pilots became famous barnstormers. Aircraft manufacturers apparently desired the publicity of glamorous women putting themselves in danger, and revelled in the records they attained.

One of these daredevil stunts was to have a woman stand on the top wing of a biplane as the pilot performed death-defying aerobatics. In the early 1920s, Florence Leach accompanied her husband (a former First World War pilot) around New Jersey as he performed stunt-flying at county fairs. Sometimes she would climb onboard the Curtiss Jenny and dance across the lower wing of the biplane, holding on to the struts for support. She attracted much attention for the feat, encouraging onlookers to pay $5 for a flight around the field – in the cockpit. Several years later her son asked her why she had stopped doing it in 1922. She told him that it was because she was pregnant; he was 'in the hangar'. He was born in March 1923, and was named Walter Marty Schirra (after his father): Wally Schirra who, thirty-six years later became one of the first American astronauts selected by NASA and the only astronaut to fly a mission in all three American pioneering space programmes (Mercury, Gemini and Apollo). Always known for his sense of humour and practical jokes, Schirra would often counter the claims of rocket research pilots Chuck Yeager and Scott Crossfield – who suggested that they were far more experienced than the early astronauts – by telling them that he had flown on the wing of a biplane before he was even born![10]

The passion for flying was not limited to older women, society women and adventurous women. In 1927, Dorothy Hesler's flying skills were recognised by one of the leading 'barnstormers' of the day, Tex Rankin. He gave her flying lessons and nurtured her talents so that by the age of just 17 she was already an aerobatics champion.

The Roaring Twenties

On 1 April 1921, Frenchwoman Adrienne Boland (1896–1975) became the first person to fly over the Andes mountain range in South America, in a Caudron G3 aircraft. The aircraft constructor realised that the aircraft was ungainly in appearance but was actually quite flexible and sturdy – so what better way to demonstrate how easy it was to control and fly than by having a woman fly it? Boland's first association with the aircraft, and indeed flying, was in 1919, just a few days after the death of Bessica Raiche. Boland approached officials at Caudron aircraft company for flying lessons at the apparently cut-down price of 2,000 francs for those first few to attain a flying certificate. Showing little concern for the recent news of the death of Raiche, she was determined that she should be taught to fly, and with funds available to support the venture she was accepted as a pilot candidate for the G3. The aircraft was noisy, unreliable, and prone to breakdowns, and the vibrations were so intense that they were likely to bounce the pilot out of the aircraft if they were not strapped into the open cockpit. The engine splattered oil across the pilot's clothing and face, and the smell from the aircraft's engine could be detected 100 metres away. The pilot sat behind a huge fuel-tank that blocked the vision, but at least protected the pilot against the wind stream. On her first ascent in the aircraft as a passenger, Boland was told that she would only feel the draft from the chin up, and would feel warmer if she pushed her hands up her sleeves, or wore a jacket with sleeve buttons.

Several years later she was interviewed on the experience of her first flight as a passenger, and was candid about her bluff of past experience in order to obtain a flying certificate: 'The most extraordinary part of it all was that I didn't know what I was up to. I hadn't the faintest idea what I was in for. When I impulsively put my name down as a candidate for a pilot's certificate it was a typical girlish frolic. I really didn't know what I was doing. When asked, 'Have you been up before?' I answered airily, 'Oh yes', but it wasn't true. Then they put me onboard that old G3 for a short flight, and I was frightened out of my wits. When the machine rolled to the left I leaned over to the right, which one should never do. One should participate in every movement of the aircraft. I couldn't have been more clumsy, more of a lame duck. But I'd paid my 2,000 francs and it was too late to be sorry. I just had to get that certificate.' When she was told that this was the aircraft she would be piloting, she boldly replied 'splendid,' but without confidence. In those days, aircraft manufacturers were worried about allowing women to fly difficult or troublesome aircraft. If they were killed in them, there would be a lot more fuss and attention than if a male pilot lost his life! It was also a familiar practice for students to be sent aloft alone to reduce weight if the engine was malfunctioning and could not support the mass of instructor and pilot. Fortunately for Boland, this had not been necessary.

By January 1920 she had logged about 12 hours' instruction, and was ready for her first solo flight. However, although instructed to fly only one circuit, she completed five circuits, touching the ground and immediately taking off again, and ignoring the instructor on the ground, who was shouting for her to stop. When she finally landed, the instructor expressed his opinion of female intelligence and independence.

Less than fifteen months later Boland was in South America preparing to fly the G3, with enough fuel for nine hours' flying and small adjustments to the wings to assist in lift. The main problem was that the ceiling for the aircraft was 13,000 feet, while the lowest peak in the range was 14,000 feet. Determined to succeed (and, to lighten the load, flying without a mechanic), she set off from Brazil after taking the advice of a local woman who told her to look out for an oyster-shaped lake as a landmark, and to head for a solid rock wall to the right and not the more inviting valley to the left. Flying at 2.5 miles above the ground in the thin air at that height, she was being bounced around the cockpit of the aircraft and buffeted by icy winds. The cold was biting at her hands and feet, she felt dizzy, and she also found it difficult to breathe. With her eyes stinging, she crouched down over the controls, feeling blood in her nose and mouth as the aircraft was jerked downwards in air pockets and stiff winds. Gripping the throttle, she searched the ground for the lake, and after a frightening ten minutes she spotted it far below. She then turned right and headed for the vertical cliff wall. The wind was being deflected upwards, and it took her aircraft with it, higher and higher. She was spitting blood as she climbed, until a break in the rock led to a peak, and suddenly all was peaceful. She had succeeded, and was soon flying above peaceful valleys and towards the Pacific coast. Upon landing at Santiago, however, she found that the French consul was not there to authenticate her achievement. Apparently, he thought that news of a woman pilot flying over the Andes was nothing but an April Fool joke!

In 1921, Bessie Coleman (1893–1926) became the first African-American to earn a pilot's license – in France. Her mother could neither read nor write, but encouraged her family to learn by borrowing books from a travelling library wagon that rolled into town twice a year. Bessie – the twelfth of thirteen children – was interested in learning to fly at an early age, but after leaving high school could not afford to go to college. Her attempt to find a school in America that also taught flying was marred by both her colour and her sex. She was advised to fulfil her ambition in Europe, and in particular, France, so she left America to learn French, and received her piloting licence from the Federation Aeronautique Internationale. A year later she was demonstrating her skills in flying exhibitions in Chicago, and became known countrywide for both her flying and her encouraging speeches to young African-American men and women. Her career, however, was short. Prior to realising her dream of creating a flying school for African-Americans, she was killed, while preparing for an air show, on 30 April 1926. Known as 'Brave Bessie' in her barnstorming days, her example in overcoming difficulties to achieve a dream continues to inspire African-Americans.

Bessie Coleman – the first African-American pilot.

Prejudice against female pilots in the early years of aviation was not solely through gender or race. Marvel Crosson was one of the first female pilots in the Alaskan territory, having accompanied her brother Joe to Alaska in 1927 after learning to fly in San Diego, California, and logging 200 hours solo to prove that she was as good as a man. In the same year she passed her flying examination, and became the first Alaskan female pilot licence-holder. She had previously found that the fellowship of male pilots was something that annoyed her, 'These good fellows never forgot I was a girl! There was a shade of condescension in their [friendship] ... They acted as though it was a pleasant thing for a girl to be interested in flying, but 'just among us men' it was of no importance. I could feel the sex line drawn against me.'[11] As a flyer she transported freight and delivered mail faster than the traditional dog-sled teams. But this was not appreciated, and all pilots were unpopular with dog-handlers. Businesses lost revenue, as dog-sleds and their mushers would refuse to stop in towns that accepted the pilots, and therefore did not spend their money there. This resulted in signs being displayed outside shops, stating, 'No dogs or pilots allowed.' Despite this, Marvel's career and skills improved, and in the spring of 1929 she set a new altitude record for women. In early August she entered the National Women's Air Derby, but this was her last main event. A few days afterwards, on 19 August, her aircraft suffered engine problems and crashed. She had jumped out, but her body was found entangled in a parachute that had failed to open correctly.

The first National Women's Air Derby

This national event, held in August 1929, brought together the twenty very best female pilots of the day in a race from Santa Monica, California, to Cleveland, Ohio. Despite rumours of sabotage, and thoughts that women would fail in their attempt to fly such a distance, fourteen pilots completed the race. It was won by Louise Thaden, while Gladys O'Donnell came second and Amelia Earhart came third.

Female distance and endurance flying was becoming more popular across the US and in Europe as the quest grew to fly across the Atlantic in the trail of Lindbergh, who in May 1927 had completed the feat solo, flying *Spirit of St Louis*. In 1922, Lillian Gatlin had become the first woman to fly across America as a passenger; but it was in the pilot seat that the fame and potential fortune lay. In 1928, Amelia Earhart became the first woman to fly across the Atlantic. But her co-pilots Lou Gorson and Wilmer Stultz had done most of the flying, and it would be another four years before she accomplished the feat herself.

In 1929, Florence Lowe Barnes became the first woman stunt pilot in motion pictures when she appeared in Howard Hughes' feature film *Hells Angels*. Barnes – better known as Pancho Barnes – went on to set speed records and establish a flying group that provided emergency assistance at disasters, after which she ran a farm and popular ranch with bar near a remote army airfield called Muroc, in the Mojave desert in California, where many of the X-plane pilots spent their time between test flights after the Second World War. This 'watering hole in the desert' became known as the Happy Bottom Riding Club, and is part of the legendary Right Stuff of America's pioneering rocket-aircraft test-pilots in the 1940s to 1950s. The site later became Edwards Air Force Base.

The decade closed with the creation of Ninety-Nines Inc International, formed by ninety-nine members of a new women's pilot association in 1929. The first secretary and host of the inaugural meeting of the now world-famous female piloting organisation was Fay Gillis Wells. Her career in journalism as well as aviation allowed her – while living in the Soviet Union during 1930–1934 with her engineer father, who worked there – to write for *The Herald* newspaper and arrange the landing rights and fuel supplies for the important Russian leg of Wiley Post's famous round-the-world flight in 1933. She is also credited as being the first Western woman to pilot a Soviet civilian aircraft. The first President of the '99s' was Amelia Earhart.

TROPHIES AND RACES

If the 1920s were the barnstorming years, then the 1930s can be classified as the years of trophy races and long-duration flying expeditions. Local air displays and aerobatics were becoming common, and something more was needed to demonstrate American aviation prowess in a rapidly changing political world. One of the first women to combine aeronautical dexterity and duration was Laura Ingalls, who in 1930 set a record for continuously looping her aircraft, without stopping, 980 times, and for the most barrel rolls, turning her aircraft over the longitudinal axis a staggering 714 times and enduring the then unfamiliar forces of vertigo, g forces and disorientation. Five years later, Ingalls became the first woman to fly non-stop from New York to California.

Anne Morrow Lindbergh was awarded the first female glider-pilot's licence in 1930 and a private pilot's licence the following year. With her husband, Charles Lindbergh she became one of the first female navigators, and they flew together around the world, mapping early air routes that are still used today, including a ground track from the US over Newfoundland and to Europe.

In 1931, Ruth Nichols surpassed the altitude record of 28,743 feet to add to more than thirty-five other records in altitude, distance and speed. Speed was one of the new targets in the early 1930s, but women were discouraged from competing in official races. It was seen as a male preserve, and many pilots openly disapproved of women competing. In 1931, Cliff Henderson approached businessmen to sponsor a cross-country air race to focus on reliability, endurance and speed, demonstrating American aviation skills. The sponsorship was taken up by Vincent Bendix, and the race was run, apart from the war years, until 1962. The heyday was in the 1930s, as in the post-war years it became a mainly military competition, and supersonic speeds limited the competitors to the armed forces or leading aviation contractors. (In 1961 it was won by future NASA astronaut Richard F. Gordon.)

In the 1930s there was more public appeal for the races, but there was little support for female competitors until 1936. This was not helped by the tragic death of Florence Klingsmith, who was killed while flying her Gee Bee racer during the Frank Phillips Trophy Race in Chicago. This even led Henderson to rule out women from competing in the 1934 Bendix final. Despite growing concerns that women could not endure the stresses and physical demands of flying such a race, pressure from leading female aviators led to the ban being lifted in 1935. Their determination was rewarded in the 1936 races, when Louise Thaden and co-pilot Blanche Noyse, flying a modified C17-R, beat the male competitors to become the first women to win the event. Second place was also taken by a woman that year, with Thaden's friend Laura Ingalls flying a Lockheed Orion to second place. Their time of 14 hrs 55 min from New York to Los Angeles contrasted sharply with the first race in 1931, when Jimmy Doolittle completed the course in a Laird Super Solution aircraft in 9 hrs 10 min. With advances in aviation, in 1962 Captain Robert Sawers won the race by completing the course in a USAF B-58 Hustler in just 2 hrs 56 min.

Another famous pilot of the era, Jacqueline Cochran, set several records in the late 1930s, including winning the 1938 Bendix Trophy Race. Born sometime between 1906 and 1910 (she was orphaned at the age of 4), she earned her pilot's licence in 1932 and entered her first long-distance race from England to Australia in 1934. Engine problems forced her to retire early from the 1935 Bendix race, but in 1937 she took first place in the women's category and third place overall, after completing the course in 10 hrs 19 min at an average speed of 194.74 mph. Her overall win in 1938 was in an elapsed time of 8 hrs 10 min 31 sec, averaging 249.77 mph. This remarkable pilot also went on to set new speed records in 1938, including a US women's national speed record of 203.895 mph, flying a Beech Staggerwing; a women's world speed record of 292.271 mph, flying a Seversky P-35, and the New York to Miami speed record of 4 hrs 12 min 37 sec, also flying a Seversky. In 1939 she set a women's national altitude record of 30,052 feet above Palm Springs, California, and an international speed record over a 1,000-mile course at an average speed of 305.926 mph. This was an amazing achievement in a short period of time; but she would also set other aviation records.

ACROSS THE ATLANTIC AND AROUND THE WORLD

Although flying across the US was a challenge, there was always somewhere to land. The next big challenge for female aviation was flying across the Atlantic Ocean or across continents alone. International interest in aviation was growing by the year. In 1931 Katherine Chung received the first pilot's licence awarded to a woman of Chinese origin, and in 1932 Ruthy Tu reportedly became the first woman pilot in the Chinese army. In 1931, Ruth Nichols set world distance records flying from California to Kentucky, but failed in her attempt to fly solo across the Atlantic. Achieving that record would not be easy. On the other side of the Atlantic, the challenge was to fly in stages from England to Australia – a distance of about 12,000 miles. It was a different type of challenge, but no less daunting.

Aviation academics and attainments

While there were headline-makers in female aviation, these were not the only notable achievements in flight. In 1932, Elsa Garner was one of only a small group of women honoured to join the American Society of Mechanical Engineers, having recently attended the Massachusetts Institute of Technology, where she studied aeronautical engineering. Two years later, the first female-directed US federal programme – the National Air Marking Program – was founded by Phoebe Omlie, and the senior staff were all women pilots. They organised the painting of town and city names on large barns and buildings as a visual navigation aid for pilots flying across the country. Omlie had been so impressed by her first aeroplane ride as a younger woman that she used the money inherited from her grandfather to purchase the same aircraft! Later, as a married woman, she and her husband operated an aerial circus, performing aerobatics and parachute descents. During the 1930s women were slowly becoming increasingly involved in all aspects of aviation, but in the cockpit they continued to demonstrate abilities equal to their male colleagues.

Helen Richey became the first woman pilot employed by a regularly scheduled American airline (Central Airlines) in 1934, while in 1938, as war clouds gathered over Europe, a young German pilot, Hanna Reitsch, became the first woman to fly a helicopter and be granted a licence for it. The following year, Willa Brown received a double honour as the first female African-American commercial pilot and the first African-American female officer in the US Civil Air Patrol. She was also instrumental in the foundation of the National Airmen's Association of America, which aimed to allow more African-American men to join the US armed forces.

Scientific research flights in aviation in the 1920s and 1930s were, as with the early space missions that followed, relatively few and far between. Apart from tests of the vehicle flown, the main objective was basic survival. The pure thrill of achieving solo flight developed into advanced aircraft-handling skills, both in normal flight and in difficult situations: recovering a faulty aircraft from in-flight difficulties, pushing the vehicle to the limit to determine its capabilites, supporting the primitive navigational equipment with visual observation – all these qualities evolved into test-pilot criteria for both men and women.

One of the first women to use aircraft as a platform for scientific information gathering was Osa Johnson (1884–1953), who, with her husband Martin, is credited as being the first aerial photographer. During the 1920s and '30s they documented several foreign expeditions with movie and still film, and in January 1934 became the first to fly over and film Mount Kenya and Mount Kilimanjaro. The footage was later used in their 1935 film *Baboona*, and their photographs were used in their book on their 1933–34 aerial safari, from Cape Town to Cairo and then on to London via North Africa and France.[12] From their various adventures, the Johnson's popularised the use of cameras on safaris and foreign travel, and generated a growing interest in wildlife and conservation. They also demonstrated the value of aerial photography for documentaries of some of the more remote parts of the Earth, adding to the value of photoreconnaissance for strategic objectives.

Winged sisters of the Soviet Union

In the Soviet Union of the 1930s there was a desire to demonstrate the 'power and superiority' of the Communist regime in the development of aviation. Several Soviet aeronautical spectaculars were the forerunners of space spectaculars three decades later, including record-setting flights by leading Soviet female aviators demonstrating the apparent devotion and dedication of all Soviets for the common good. Many of these pilots would serve alongside their male colleagues in the aerial front line during the Second World War.

During 22–25 May 1937, Polina Osipenko (1907–1939) set altitude records, in a passenger variant of the MP-Ibis hydroplane, of 8,864 m (no additional load), 7,605 m (carrying a 500-kg load) and 7,009 m (carrying a 1,0000kg load). Just over a year later, on 2 July 1938, Osipenko was joined by fellow pilots V. Lomako and Maria Raskova (1912–1943) in a record 2,416-km non-stop flight (Sevastopol–Kiev–Novgorod–Archangelsk) of 10 hrs 33 min, averaging a speed of 228 kmh. Then, just two months later, on 24–25 September 1938, Osipenko and Raskova joined Valentina Grizodubova (1910–1993), flying an ANT-37 on a non-stop 5,900-km world record flight from Moscow to Siberia. Previously, on 28 October 1937, Raskova and Grizodubova, onboard an AIR-37, had jointly set a world record for female long-distance non-stop flying. This flight was recreated by two American women during the 1990s.[13]

THE ROAD TO THE STRATOSPHERE

During the early 1930s, balloon flights into the stratosphere gave rise to competition to attain national headlines, political propaganda, new records and scientific achievements. In the midst of this 'race to the stratosphere', in the summer of 1934 Jeanette Piccard was preparing to accompany her husband Jean on a balloon ascent; but before they could do so Jeanette required many hours of training, and in order to assist in piloting the balloon she required a pilot's licence. At the same time, her husband was busy refurbishing the gondola (used for a previous ascent in November 1933) and convincing scientific investigators to supply suitable experiments for the flight – notably for the investigation of cosmic rays. They were determined to restore family pride by again placing the name of Piccard in the record books, but were frustrated with the success of their competitors, and found it difficult to obtain practical and financial support. The Detroit Aero Club, the People Outfitting Company of Detroit and the Grunow Radio Company were helpful, but their sponsorship was insufficient to fund the enterprise. Jeanette Piccard, however, fully appreciated the requirement to 'sell the idea' – not purely for setting a new altitude record, but primarily to obtain important scientific results. She worked to supplement the sponsorship funding by designing and selling series of commemorative stamps and souvenir programmes and folders. In addition, she approved official news releases by the North American Newspaper Alliance. She also commented that although record altitude flights were 'splendid achievements' and have a place in the development of aviation, to her and her husband this was only a means to an end. To them, the 'end' was scientific research.[14]

The element of danger and risk is evident in contemporary reports, but such talk only strengthen Jeannette Piccard's resolve to support her husband and participate in the flight. To her, the greater the difficulty, the more interesting the quest. Her real reasons for flying, as explained to her father, were to be reiterated by female explorers more than half a century later: 'There are many reasons, some of them so deep-seated emotionally as to be very difficult of expression. Possibly the simplest explanation is that we started along this road . . . and I cannot stop until I have won.'

Ascent into history

Throughout the summer the Piccards struggled not only with the preparation of hardware and experiments but also the funding. By September 1934, however, they were almost ready. The weight of the experiment payload would limit the maximum altitude, but the Piccards were adamant that the science was more important. With the hardware and experiments at the Ford Airport launch site, south-west of downtown Detroit, the funding issue still plagued the ascent almost to the last minute, even with free housing of the gondola at the airport, the installation of ground anchoring services, and the reduced costs of supplying the seven hundred cylinders of hydrogen. It was only due to the advance of several thousand dollars by Grunow Radio Company, against the expected revenue to be generated from the public exhibitions and interest after the flight, that the ascent was finally set for early October.

The first attempts were cancelled due to the weather, but shortly before daybreak on 23 October 1934 the Piccards lifted off . . . but not without difficulty, as not all of the ground control lines were released at the same time. As the balloon finally ascended, Jeanette Piccard could be seen through the still-open access hatch. The flight into the stratosphere took more than 8 hours, during which time the balloon and its occupants attained an altitude of 17,800 metres. Jean Piccard later claimed that he was not correctly informed about the weather conditions that day, that an erroneous report was the basis of their ascent on the 23rd, and that they encountered cloud cover and not clear skies. Fully aware of the dangers of being blown over the ocean, he adjusted the flight path and shortened the flight time. This, of course, reduced the amount of scientific work that could be carried out, but they still tried for the highest altitude. The landing – in a wood on a farm about 5 km south-west of Cadiz, Ohio – was not without danger. The balloon, descending through the trees, was ripped, and on impact the lower level of the gondola was crushed, although most of the instruments were undamaged. The occupants survived, although Jean suffered small fractures in his ribs and left foot and ankle.

With the flight successfully completed, the Piccard's embarked on a post-flight tour and lectures. In his book on American stratospheric balloon exploration (published in 1989), David DeVorkin summarised the scientific results which were to have been the major objective of the flight. The altered flight path changed the quantity of scientific data which could be gathered; and because of Jeanette's 'unplanned and impulsive manoeuvres', the record of their actions during the flight was incomplete. Despite the claim that they wanted the flight to be remembered for its scientific achievements, they had been promised a $1,000 bonus by the newspaper alliance sponsor should they surpass the altitude record. They had therefore jettisoned their last sand-bags to lighten the vehicle. In the end, not only did they not achieve the record, they also compromised the scientific data they had collected. Aware that the science return was not going to be as good as they hoped for, the Piccard's always felt that they had accomplished and learned a great deal. However, those who provided the experiments were not particularly happy with the lack of scientific results.

Jeanette Piccard continued to work with her husband on his balloon projects beyond the Second World War. Their experiments including research into sounding balloons and in the US Navy Helios programme. Helios was cancelled in May 1947, but it had prepared the way for Project Skyhook Man-High in the 1950s and 1960s. By then, America had placed men into space, and a President had committed America to the Moon. In the mid-1960s, while Jeannette Piccard was still talking romantically about manned high-altitude flight in balloons, Robert Gilruth, Director of the Manned Spacecraft Center in Houston, invited her to be a special consultant to support the effort to the Moon. But It would be another twenty years before American women would reach even greater heights than those attained by the Piccards.

AMY TO AUSTRALIA

When Amy Johnson arrived in Australia, after a solo flight from England, on 24 May 1930, she was surprised by the attention that such a flight generated, especially since she had not broken the flight record for that journey. She left England onboard her De Havilland Gypsy Moth I *Jason* (funded by her father and sponsored by Lord

Wakefield's oil company) on 5 May, and arrived at Port Darwin 19 days later. It was the ninth such trip, and not even the first solo flight; but it was the first by a woman. Johnson was largely unknown before her epic flight, but became a feted celebrity, and huge crowds gathered to witness her landing. But she was not like like the titled or privileged ladies of the era who took up flying and record-breaking attempts for fun and fame.

As details emerged it was revealed that Amy Johnson was an 'ordinary' girl with lots of charm and determination. Born in 1903, in 1925 she graduated from the University of Sheffield with a BA degree after studying economics, Latin and French, and later moved to London to work as a secretary. She joined the London Aeroplane Club in Edgware, and spent all her spare time learning how aircraft worked and learning to fly. In 1929, at the age of 26, she earned her pilot's licence and ground engineer's licence. She became the thirty-seventh Englishwoman to receive a licence to fly, just a year before her epic flight to Australia, but just as impressive was her achievement as the first woman in England to received a ground engineer's licence. In 1931 she flew a record-breaking trip from Siberia to Tokyo, and also set a new record for the return trip. The following year she married another record-breaking pilot Jim Mollison, and then the same year broke the record for a solo flight from England to Cape Town as well as for the return leg – both records formerly held by her new husband. In 1933 they made a non-stop flight to the US, and the following year broke the record for a flight to Karachi. In 1936 Amy recaptured the record for a solo flight not only to the Cape and back to England but for the overall time taken to make the return trip.

Her achievements in the 1930s were all the more remarkable considering that she had only experienced a short pleasure flight with her suitor in Hull in 1926, and only started flying after taking a bus ride to the local aerodrome to see the aircraft. After watching the activities she enrolled in the club. She enjoyed every minute of it, and became involved in the engineering aspect of preparing the aircraft for flight – which at that time was still rare for a woman. Her background was so interesting that the newspapers began to follow her progress in aviation and her desire to break the record of 15 days for a solo flight to Australia, and she used the publicity to help raise funds for the trip. In preparing for the adventure she flew short busts for an hour or two, and accrued more than 85 hours; but when she embarked for Australia the problems of long-duration flying become apparent. She felt nauseous due to the fumes in the cockpit, had little sleep, and conducted the repairs on the aircraft herself. (Similar challenges must be faced during short and long spaceflights, for which different skills are required.) She was two days ahead of the record when she reached India, but several problems, and exhaustion, affected the latter stages. Her endeavour was all the more remarkable because in those days aerodromes along the route were extremely basic: essentially dirt strips in the desert or small clearings in jungles, with almost no equipment nor even English-speaking helpers. On her desire to prove that women could endure arduous flights she was limited by the fact that most women had achieved several records in solo and distance flights, and the solo record to Australia was the last remaining long-distance record to be accomplished. Her reasoning was to inspire other young English girls to follow her and raise awareness and public support for the growing air travel business: 'I am certain a successful flight by an English girl, solo and in a light plane, would do much to engender confidence amongst the public in air travel', she wrote, during her request for sponsorship prior to the trip. 'Girls are shielded and sometimes helped so much they lose their initiative and begin to believe the signs 'girls don't' and 'girls can't', which marks their path.' Amy Johnson demonstrated on more than one occasion that in aviation 'girls did' and 'girl's can' achieve.

The fame excited and pleased her, and songs were written recalling the exploits of the 'Queen of the Air', 'Aeroplane Girl', and even 'Johnnie, Heroine of the Air'. From 1934 to 1937 she served as President of the Women's Engineering Society, and won many international awards and honours. In 1936, however, her marriage was dissolved.

She often voiced annoyance that male pilots were being paid as much as twice the rate offered to women pilots, and that it was difficult to secure a permanent job in aviation. She tried to gather support for equal pay, and encouraged other women to 'take the role of a woman pilot and try to persuade your male boss that you should have the same pay. Get a friend to be the boss. Once you have argued as a woman pilot, switch roles and take the role of the male boss'. Although she loved the adulation and fame, she did not wish to make headlines as a woman doing unusual things – rather, that aviation should be considered usual for a woman.

Although she was disappointed at not being able to secure regular flying work, she was at her happiest when serving as a pilot for the Air Transport Auxiliary during the Second World War. She ferried all types of aircraft across England, flying all hours of the day and night, and conscious of the dangers, including potential attack by enemy aircraft.

Amy Johnson, like many of her contemporaries in aviation, said that her love of flying would kill her, and on 5 January 1941 this proved true. While ferrying an aircraft for the Air Ministry she was lost over the Thames estuary, Her body was never found, and it was almost three years before she was officially listed as 'missing, presumed dead'. She was 38 when she disappeared.[15]

AMELIA ACROSS THE ATLANTIC

During May 1932, American aviatrix Amelia Earhart became the first woman to fly the Atlantic solo. On 20 May she took off from Newfoundland in her Lockheed Vega (bought in July 1929), and landed in Ireland the next morning. Born in 1897 in Kansas, Earhart realised at an early age that she was meant to have an adventurous life. During 1919 she studied medicine at Columbia University in New York, at a time when most women wanted nothing more than to finish college, get married, and raise a family. She afterwards gave up her studies to join her family in California, and there she attended several air shows at which former First World War pilots used their flying skills to entertain crowds and earn money. Such shows impressed her, and a ten-minute ride thrilled her so much that she announced to her family that she wanted to be a pilot. She worked long hours in various odd jobs to pay for flying

Amelia Earhart stands in front of the Lockheed Electra which she was flying when she disappeared in July 1937.

lessons, and after two years she was awarded a licence. For her 24th birthday she treated herself by buying a bright yellow Kinner Airster for the huge sum of $2,000. In this aircraft – which she called *Amelia* – she set numerous records up to 14,000 feet. The problem with the aircraft was that it had the tendency to stall, and even landing was not without its hazards. On one occasion it flipped on its back and threw Amelia from the cockpit.

Financial strains forced her to eventually sell the aircraft and try to secure a mudane job; but the draw of aviation was too strong, and she soon joined the National Aeronautics Association, flying wherever she could. She became so well known that in 1928 she was elected to join the crew of the large Fokker *Friendship* in an attempt to fly across the Atlantic Ocean. It was not a solo flight, nor did she get to handle the controls; but she was the first woman to fly across the Atlantic Ocean – albeit only as a passenger. This was an important milestone in proving that a large aircraft could fly across from America to Europe, supporting the growing interest in setting up commercial flights between the continents. Earhart proved that a woman could endure a long, cold and difficult flight across the Atlantic, and the resulting press coverage made her more determined to make flying her life.

Over the next decade she achieved many records in various aircraft, including an Avro Avian, and was the first woman to fly across the continental US and back again. She was placed third in the 1929 Women's Air Derby, and demonstrated her flying skills by overcoming the loss of her altimeter on her record-breaking solo flight over the Atlantic in 1932. She was the first pilot to fly solo from the Hawaiian

Amelia Earhart with VIPs on the steps of the Langley Research Building in 1928. During the tour, part of Earhart's fur coat was sucked into the High Speed Wind Tunnel.

Islands over the Pacific to California (and therefore the first woman to fly over both the Antlantic and the Pacific); the first to fly solo from Los Angeles to Mexico City; and beat her own record by flying across the US in 17 hrs 7 min. Following the 1929 Women's Air Race Derby, Earhart described the event as being the beginning of concerted activity amongst female fliers. She was instrumental, with other competitors, in forming an association of women pilots, called the Ninety-Nines (from the number of founding members), dedicated to the improvement of opportunities in aviation for women, but called the 'Petticoat Pilots' or the 'Ladybirds' by the press. Today this association still exists, and includes Shuttle Commander Eileen Collins as a member.

In 1937 Earhart and her husband George Palmer Putnam planned a brand new adventure: a round-the-world flight, during which she would be accompanied by experience navigator Fred Noonan. The route would take them from Florida, east over the Atlantic, across Africa, on to India and Australia, and back across the Pacific to the America, The plans called for a number of refuelling and rest stops in a flight that would take more than a month. To circumnavigate the globe (which would later take astronauts about 90 minutes), the Lockheed Electra was loaded with extra fuel, and took off from Miami on 1 June 1937. For the next four weeks they progressed across the globe, with huge crowds welcoming them at each stop. The leg across the Pacific was by far the most challenging, and great navigational skills were required to locate Howland Island to refuel before flying on to Hawaii and then back to the US. Unfortunately, on 2 July 1937, whilst approaching Howland Island, radio contact with Earhart and Noonan was lost. Apparently flying through rain clouds, they had possibly become disorientated after many hours in the air, and missed the island. The Electra continued flying until it ran out of fuel, and finally crashed in the Pacific, taking the lives of both aviators. Despite numerous searches, no trace of the aircraft or the bodies of the flyers were ever found. In addition to her achievements in aviation, Amelia Earhart is also remembered for championing equality between men and women, using aviation as a vehicle of similar pioneering sprit, technical skills and bravery. Many continued to protest that women had no place as pilots and that flying was the domain of men, but Earhart, Johnson and scores of other women pilots proved this not to be the case. They were determined to change that attitude – even if it cost them their lives. The challenge to their bravery was soon to include new dangers in a world war, and – in order to gain technical superiority in the work to break the sound barrier – a major step from atmospheric flight to spaceflight.

A SHRINKING WORLD AND A NEW WAR

Alongside the record-breaking flights ever high, faster and further, and the numerous air shows, a new business of aviation was developing, delivering mail and cargo and eventually passengers. From 1919 the first civilian airline companies began operating from Britain using stripped-down former First World War bombers, reconfigured with seats for the passengers and the loan of flying clothes to help keep them warm. It was not long before these early passenger aircraft were replaced by custom-built aircraft designed to be more comfortable. The longer the flights, the more the requests for refreshments and assistance, generating the need for aircraft crew-members to look after the passengers en route. Now, a hundred years after the first powered flight, millions of passengers cross the world in thousands of airliners every day. However, in the 1920s and 1930s they numbered only tens of thousands, and this, for some women, offered the opportunity to work as 'air hostesses', flying when they could, or working as secretaries for the airline. But female airline pilots were few and far between. In 1930, eight female nurses were employed by one American airline as the world's first stewardesses (the nursing skills were thought to be a useful additional skill). There was an early selection criteria: under 1.64 metres (5 feet 4 inches) tall, under 25 years old, and able to walk through the aircraft compartment comfortably. One of the pioneer women airline pilots was Amelia Earhart, who helped create the first regular passenger route flying from New York and Washington.

In Europe the dark clouds of war were looming, and a significant change was on the horizon. During the First World War, with men being shipped to the front in their thousands, there was a significant drain on man-power in the factories and institutes at home. The employment of women was at first opposed; but then the reality and seriousness of the situation forced reluctant acceptance of women in the workplace, and gradual support of the armed forces as nurses, maintenance workers and drivers. In April 1918 the Royal Flying Corps and the Royal Naval Air Service were incorporated into the Royal Air Force, and women were urged to volunteer for what eventually became the Women's Royal Air Force. But although supporting the construction, maintenance and operational activities of the Air Force, women were not allowed to fly. The WRAF was disbanded in April 1920, and during the 1920s and 1930s the role of women in society changed. The world was going through a depression, and in aviation there was a small move towards equality.

By the late 1930s, several volunteer services were formed in England to mobilise the workforce and population as the threat of a new world war loomed. One of these was the Woman's Auxiliary Air Force (WAAF), formed in June 1939.[16]

The Air Transport Auxiliary

One role not afforded to the early members of the WAAF was service as a pilot. Initially, the attitude of the Air Member for Personnel was that a female flying section of the WAAF was not wanted. But, as during the First World War, necessity and escalation of hostilities changed that attitude, as did the realisation that when the women were given a task they adapted quickly and were very successful.[17]

In 1939 it became evident that because of a shortage of pilots, aircraft could be moved around the country only by fully trained RAF pilots taken of active service. This problem was solved by allowing civilians with pilot's licences to ferry new aircraft from factories to airfields and military sites across the country, deliver and return aircraft for maintenance, and flying the small Anson passenger aircraft as 'taxis', delivering pilots and personnel from base to base. As a unit in the four years of operational activity, the Air Transport Auxiliary moved 309,011 aircraft of 140 different types to where they were needed. Of more than 750 pilots assigned to the ATA during the war, 150 were women. They each held an 'A' licence and a logbook showing at least 250 flying hours, but their background was irrelevant. It was flying skills that mattered most, with 'rank' structured on experience: those who were limited to single-engine aircraft (Second Officer rank), and those with more than 500 flying hours and could pilot twin- or multi-engine aircraft (First Officer).

A large number of qualified female pilots initially applied, but not all were accepted, and only eight were included in the first training squad, focusing on light trainers including the Tiger Moth and Magister. Even in the dark days of the war, the popular aviation magazine *The Aeroplane* questioned why was such an encroachment on men's jobs was threatened! On 1 January 1940 the first group of women pilots entered the service. These 'First Eight' were Winifred Crossley, Margaret Cunnison, Margaret Fairweather, Mona Frielander, Joan Hughes (who at 21 was the youngest of the group), Gabrielle Patterson, Rosemary Rees and Marion Wilberforce. Each was highly experienced, with more than 600 hours of flying time, and seven were also rated as flying instructors. In charge of these pilots was Pauline Gower, who eventually became the Deputy Commandant of the ATA.

In May 1940 Amy Johnson joined the ATA, and in her first year logged more than 275 flying hours moving aircraft across the country. During the first years of the war, significant obstacles had to be overcome, including flying in bad weather, navigation, and dealing with a lack of female facilities at most RAF bases. In addition, because of transport irregularities and a shortage of fuel, the return to home base was not always achieved on time. By the end of 1944, the ATA network had increased to twenty-two ferry pools around the country, including three crewed entirely by female pilots at Hamble (Southampton), Cosford (near Wolverhampton) and Hatfield.

In addition to those in the RAF and ATA, females served in the Fleet Air Arm of the Royal Navy. Their duties varied considerably, and some of the first to fly as part of their normal duties were women radio mechanics. Like their ATA and Air Force colleagues they participated in almost every job on several stations except actual combat flying. Their duties included aerial photography, training operations, servicing armaments, constructing and repairing aircraft, and moving aircraft of the Fleet Air Arm between stations.[18]

By the summer of 1941 the demand for more and larger aircraft increased dramatically. On 19 July 1941 Winifred Crossley, Margaret Fairweather, Joan Hughes and Rosemary Rees became the first of the eight to fly Hurricanes, and soon afterward Lysander, Walrus, Spitfire and Mosquito aircraft. Training on a new aircraft type was via a small folder, called the 'ferry pilot notes', that included the required technical data, weight ratios, starting and stopping speeds, and stalling speeds. As the types of aircraft increased, so the need for training required attendance at a specialised school at which students were instructed to fly various categories of aircraft from Class 1, single-engine light aircraft (trainers); Class 2, single-engine operational aircraft (mostly fighters); Class 3, twin-engine light aircraft; Class 4, twin-engine operational aircraft (medium bombers, with Class 4A assigned to tricycle undercarriage types of aircraft) ;Class 5, four-engine aircraft (bombers, including Avro Lancasters, Short Sterlings, and the American Boeing B-17 Flying Fortress and B-24 Liberators); and Class 6, flying boats (Catalinas and Sunderlands).

The first woman to fly a four-engine bomber was Lettice Curtis. RAF records reveal that in a single day she flew two Class 1 aircraft, a Class 2 Spitfire, a Class 3 Mitchell, a Class 3 Mosquito, and a Class 5 Sterling bomber. Twelve women attained Class 5 ranking, although no-one reached Class 6. The skills required to fly such different aircraft on the same day demonstrated their amazing flying skills and ability to endure sudden changes in handling skills and flying techniques.

When America entered the war, Pauline Gower approached Jacqueline Cochran to select about two hundred American female pilots whose skills and experience would help in transporting aircraft from the US to England to support the war effort. At the height of the war, female pilots in the ATA came from Great Britain, American, Australia, Canada, South Africa, New Zealand, Poland and Chile. By the summer of 1943 the women were finally given equal pay for equal work, but were not officially allowed to fly aircraft to the Continent. However, after Diana Barnato Walker flew a Spitfire to Brussels (with special permission, and whilst on leave), female ATA pilots were cleared to fly to Europe – and a few flew to Berlin after the end of the war. This also gave them the opportunity to ferry the Meteor – the first jet fighter to enter service with the RAF. With the end of the war the ATA was disbanded, although several women continued flying in the RAF. In 1963, Barnato Walker became the first British woman to break the sound barrier, flying a Lightning T4 jet fighter.

After the war the WAAF was not disbanded, but declined over the following five years. In May 1948, fourteen former ATA female pilots underwent refresher courses and were allowed to wear a flying badge on their uniforms, and on 1 February 1949 the WAAF was renamed the Women's Royal Air Force (WRAF.) During the 1950s it was still difficult to be accepted as a female pilot in the RAF, as it was argued that a woman's body could not endure the sudden acceleration of more modern supersonic aircraft, or the cost of training would be wasted when they left the service to start a family. Both of these arguments, however, were eventually proven false. Although piloting qualifications were to take much longer, the RAF offerred other roles to women: stewardesses on Transport Command, from 1959; air crew status, from 1962; pilots, during the 1960s; and air electric operators, air engineers and air navigators, during the 1970s and 1980s. In 1990, Flying Officers Jackie Chapman and Christine Martin became the first student pilots; and by 1994. when the WRAF was merged into the RAF, women were flying all types of helicopters, tankers, transports, trainers and jets.

Women's Air Service Pilots

In the United States, during 1941, there was a growing awareness that America would enter the war in Europe, and the then US Army Air force (USAAF) faced a shortage of men to fill both combat and civilian piloting roles. Many allied forces faced the same problem. In England the ATA was being expanded, with many female pilots moving aircraft around the country, and in the Soviet Union, women were flying ground support combat missions. In order to ease the situation General Hap Arnold approached the famous female aviator Jackie Cochran for ideas. The suggestion was to review the medical and flying records of all the women listed under the Civil Aeronautics Administration, after which the suitable women would be recruited for civilian flying work with additional training through the AAF. Despite approval from the commanding office of the Ferry Command of the AAF, which would control the movement of the aircraft, the plan was rejected. At the same time, however, Pauline Gower asked Cochran to recruit and train a cadre of American women pilots for ferrying duty. As a result, in 1942 twenty-five American women pilots moved to England to serve in a uniformed civilian capacity as part of the British ATA.

In September of the same year, another US aviatrix, Nancy Harkness Love, formed an experimental women's squadron called the Women's Auxiliary Flying Squadron (WAFS), for ferry duty. These pilots were given only four to six weeks training to familiarise them with military procedures, and the success of the group prompted new efforts to create a US female pilot training programme. The criteria were limited to high-school educated 21–25-year-old US citizens with cross-country flying experience and at least 200 hours in the log book. As more women went through the training, the pool became smaller for later groups and so the minimum flying experience was reduced to 100 hours and then 35 hours.[19]

The selection began with a letter explaining the plan to each potential candidate, followed by a personal interview and medical by a flight surgeon. All of them were then screened by Cochran (serving as Director of Women's Flying Training, Flying Training Command General Staff, Fort Worth, Texas). The first recruits trained at Municipal Airport in Houston, in cramped conditions on a collection of very old or surplus stock, and graduated in the first class of Women's Air Service Pilots (WASPs) on 28 April 1943 (Class 43-W-1). The training then moved to larger premises at Avenger Field, Sweetwater, Texas, later in the year. The trainig course was divided into three sections – military, ground school, and flying – and was completed over twenty-three weeks. This included 115 hours of flying and 180 hours in ground instruction, but as the experience of applicants reduced, so the course was lengthened to thirty weeks, encompassing 210 hours of flying and 393 hours on ground schooling. The trainees dressed in military uniforms, lived in military barracks, and took their meals in mess halls – and the training was difficult. The group was combined with the WAF's, headed by Nancy Harkness Love, on 4 August 1943, and formally called the Women's Air Service Pilots (WASPs).

In addition to the ferrying of aircraft, other duties included towing of targets, searchlight and tracking missions, simulated strafing, layering of smoke shields, engineering test flying, and administrative flying. Assigned to a variety of bases across the US, the women flew a wide range of aircraft including the B-17 Flying Fortress, the B-26 Marauder, and the B-29 Super Fortress. They also served in various Air Commands, including Air Transport (ferrying duties), 3rd Air Force (towing targets, radio-controlled target flying, and personnel transport), Material Command (assisting in the development of personal flying equipment and flying experimental jet aircraft), the Weather Wing (personnel transport), and Flying Training Command (bombardier, pilot and navigational training).

In *Test Flying at Old Wright Field*, one of the WASPs, Ann Baumgartner, recalled her experiences of flying a variety of old aircraft 'to train artillery men' including 'small cubs, old B-34 bombers, ancient SBD dive bombers, C-45s, tired old fabric covered C-78s and heavy SB2C dive bombers'. She longed to fly newer and sleeker aircraft, and according to the Edwards AFB History website she apparently received her chance (possibly becoming the first woman to fly a turbo-powered fighter) testing the experimental YP-59. Baumgartner also recalled her experiences in testing (with another female pilot colleague) high-altitude equipment and clothing in the nose of B-17 up to 12,192 metres (around 40,000 feet – 7.6 miles).

With the invasion of Europe in June 1944, the male pilots began to return home and become more available for home duty, resulting in less demand for female pilots. The order for deactivating the WASPs was activated on 3 October 1944 and came into effect on 20 December, and though some WASPs left the service, most of them stayed until the very last duty. The final graduation class of WASP pilots took place on 7 December 1944. Although some stayed in Air Force reserve, many returned to civilian life. In 1948, former WASP personnel could apply to the USAF Reserve and include their service in the WASPs as commissioned service; but many found 'flying a desk' in peacetime not so challenging or exciting as their time in the WASPs, and very few progressed to a full military career. (This is also the case in the space programme, and many astronauts and cosmonauts, after retiring from active spaceflight, return to a military career or pursue opportunities in the private sector.) It was well into the 1970s that the USAF finally announced its plans to train its first female military pilots.

Jackie Cochran

Jacqueline Cochran (*c.*1906–1980) was orphaned at birth. She had very little formal education, but managed to build an impressive career as the owner of a salon with a line of cosmetics. After meeting her future husband, the millionaire businessman Floyd Odlum (whom she married in 1938), she learned to fly to help develop her business and use her time in travelling to greater effect A natural pilot, in 1932 she flew solo just two days after her first flight, gained her licence 18 days later, and in a few weeks was proficient enough to fly her own aircraft. Her flying career was remarkable.

1934	Flew and tested the first turbo-supercharger to be installed on an aircraft; entered her first long-distance race – the MacRoberston London-to-Australia Race – with Wesley Smith, flying a Gee Bee QED. She led the race for a time, but was forced to retire due to aircraft faults.
1934–36	First person to fly and test development models of Pratt and Whitney engines
1935	First woman to enter the Bendix Trophy Race, flying a Northrop Gamma. Due to engine problems, she did not finish.
1935–42	Experimental test-pilot for Sperry Corp, testing many new aircraft and gyro instrumentation. She also worked with Dr Randolph Lovelace in the field of aviation medicine, helping to design the first oxygen mask, and became the first person to wear one while flying at over 20,000 feet.
1937	First in the Women's Division and third overall in the Bendix Trophy Race, flying from Los Angeles, California, to Cleveland, Ohio, in 10 hrs 19 min, averaging 194.74 mph.
1938	Won the Bendix Trophy Race, flying a Seversky P-35 in an elapsed time of 8 hrs 10 min 31 sec, averaging 249.774 mph; flew the first 'wet wing' (fuel tanks inside the wings), and set a world altitude record of 33,000 feet.
1940	Made the first flight on the Republic P-43, recommending longer tail wheel installation later incorporated on all P-47 aircraft; first female pilot to ferry a bomber across the Atlantic to Great Britain.
1941–45	Helped create the WASPs, involved in the recruitment and training, and later worked as a press correspondent; present at the surrender of Japan, and visited China, Russia and Germany, where she attended the Nuremberg trials.
1946	Finished second in the first resumed Bendix Trophy Race, flying a P-51B with an average speed of 420.925 mph.
1948	Flew her final Bendix Trophy Race, achieving third place in a P-51B at an average speed of 445.847 mph.
1953	First woman to break the sound barrier, in a F-86 Sabre jet, with the help of her friend Air Force General Chuck Yeager (the first man to break the sound barrier) in her late 40s.
1958–59	First woman to hold the position of President of the Federation Aeronautique Internationale; re-elected for a second term.
1962	Set sixty-nine intercity and straight-line distance records for Lockheed, flying a Lockheed Jet Star, and became the first woman to fly a jet across the Atlantic; also set nine international speed, distance and altitude records in the same year, flying a Northrop T-38.

Between April 1963 and May 1964, Cochran set a series of records – all in excess of 1,200 mph – flying a Lockheed F-104G Starfighter, including setting a world speed record of 1,429 mph when in her late 50s. In addition, she also won the Clifford Burke Harmon trophy three times, set three speed records, all before 1940, and in 1948 saved the life of Lyndon B. Johnson (future US President) when she flew him to hospital for emergency kidney surgery. Flying was a passion, and she became a company pilot for Lockheed, Canadair and Northrop aircraft companies. In the 1970s she developed heart problems, and required a pacemaker. This ended her 'fast-flying' career, but she afterwards began flying gliders. She held more speed and altitude records than anyone else in the world.[20]

The Edwards AFB website article on Cochran refers to an article published in the *Smithsonian* magazine in August 1994. Cochran possibly contributed to the grounding of the early Mercury female astronaut applicants (the Mercury 13) when she testified before the US House of Representatives Science and Astronauts Committee in the early 1960s. She apparently warned NASA that a large group of women should not be selected, as the money spent on their training would be wasted when they left to get married!

Soviet female combat pilots

On 21 June 1941, Russia was invaded by the Germans, who planned to crush the Soviet Union in ten weeks. By November the Germans were just 19 miles from Moscow, and Leningrad was under siege. With large units of the Red Army destroyed and the Air Force grounded, the Russian winter would be a valuable ally. However, the surviving members of the Soviet Air Force (VVS) included a number of women who served in the air squadron and conducted combat missions. This squadron was formed by Soviet aviatrix Maria Mikahailovna Raskova, who had urged members of the Political Bureau of the Central Committee of the Communist Party to allow her to do so. On 8 October 1941 the Peoples Defence Committee issued a decree (number 0099) 'On the Activation of Female Air Regiments for the Air Force of the Red Army'[21]

During the 1920s and 1930s, arguments about female involvement – not only in aviation, but in all aspects of the relatively new Soviet lifestyle and implementation of the Communist ideology – resulted in several Soviet women achieving notable records in aviation. These included Marina Roskova, Valentina Grizodubova and Polina Osipenko, who in 1938 set the world endurance record for a direct non-stop 6,000-km flight by women across eleven time-zones of the Soviet Union. It was natural for these women, and dozens like them, to volunteer to help their motherland at the time of greatest need, in what became known as the Great Patriotic War – or, in the West, the Second World War.

Raskova organised the 588th night-bomber squadron, consisting of a complete cadre of females, from mechanics and navigators to pilots and officers – many of them around 20 years old – from the Air Force, civil aviation and the Ossoaviakhim military assistance organisation. It was a propaganda coup for Joseph Stalin, who approved the plan for three air regiments: the 586th IAP, 587th BAP and 588th NBAP. The plan was to train the pilots at Zhukovsky Aviation Engineering

Academy, but the siege of Moscow changed this plan, and the first of them commenced training in Engels, in Saratov Raion, north of Stalingrad (Volgograd). The original candidates numbered 450, but few were professional, and most of them were former glider pilots. In the first training, beginning on 16 October 1941, Major Raskova and her deputy, Major Yevdokia Bershanskaya, condensed two years' flight training into six months. On 8 June 1942 the first combat mission, by three aircraft, was flown on a raid on the German headquarters. It was a successful mission, but one of the aircraft was lost. In all, VVS female pilots flew 24,000 sorties, from the siege of Stalingrad to the fall of Berlin, with sixty-eight pilots, were awarded the honour of Hero of the Soviet Union. Raskova was later killed in action, and is buried in the Kremlin Wall.

Some of the achievements of the women included the following:

586th Women's Fighter Regiment (Commander Tamara Kazarinova) flew the first operations at the front, in Yak 7B and Yak 1 aircraft, totalling 4,419 operational sorties with 125 combat missions, with thirty-eight enemy aircraft shot down. Squadron Commander Olga Yamshchikova flew 93 sorties and shot down three enemy aircraft. Lilya Litvyak and Ekaterina Budanova were pilots with the 588th, and were so proficient that they were transferred to the male unit of the 73rd Fighter Regiment, took part in aerial combat over Stalingrad, and became fighter aces. Budanova was credited with eleven victories, and Litvyak with twelve official and three shared before she was lost on 1 August 1943. At first, the men found it difficult to accept the women and refused to fly with them, even after they proved their worth in the air, and so the women flew together; eleven female pilots each flew more than a hundred sorties during the war. Lieut Maria Mikhailovna Kuznetsova flew 204.

588th Women's Night Bomber Regiment (Commander Yevdokia Bershanskaya) Called the 'Night Witches', they completed 24,000 combat missions, dropping 23,000 tons of bombs from ancient PO-2 biplanes from May 1942 to the end of the war. Twenty-three pilots and navigators became Heroes of the Soviet Union.

587th Dive Bomber Regiment The members of this regiment did not enter combat service until January 1943, due to the differences between the aircraft they were to fly compared with those in which they trained. Later in the war the regiment used replacement male pilots, as there were not enough qualified women left to fill the vacant positions.

On 20 July 1945 the demobilisation order was received, but not all the women stopped flying, and many of them afterwards flew for several leading design bureaus. They were famous in the Soviet Union for their courage and patriotism. Seventeen years later, another small group of Soviet women were selected for a new advance in aviation – to be the first women in space.

PANCHO BARNES AND THE HAPPY BOTTOM RIDING CLUB

Florence Lowe 'Pancho' Barnes was a contestant in the 1929 Women's Air Derby, alongside Amelia Earhart. A legendary pilot and character in the aviation circles of the 1920s to 1950s, she later owned a ranch near Muroc army airfield, in the hot, windy, dusty and dry lake-bed deserts of California. The 368-acre ranch was called Rancho Oro Verde Fly Inn Dude Ranch, and became known to the young test-pilots at Muroc as the Happy Bottom Riding Club; and although throughout a decade it was given a number of names, it was always known locally as Pancho's. Muroc, of course, became Edwards Air Force Base – the home of the X-series of research aircraft that broke the sound barrier and created the image of the test-pilots with the Right Stuff.[22]

Born into a wealthy family in 1901, Florence Barnes was, during her early life, inspired by two men: her father – an avid outdoorsman – and her grandfather, Professor Thaddeus Lowe – a founder of the California Institute of Technology, and a balloonist in the American Civil War (he had spied on the Confederate lines). When Florence was nine years old, her grandfather took her to her first air show, and it was not long before aeroplanes ranked with horses as the major passions of her life. She later went to Mexico, where she was given the name Pancho due to her 'revolutionary' appearance. Back in the United States she was inspired by the exploits of Charles Lindbergh. She therefore took flying lessons, and was able to go solo after only six hours. Totally dedicated to flying, she wore men's clothes (often oil-stained) and smoked cigars – which did not reflect her genteel upbringing, and placed her differently from other more refined early female aviators.

She was one of nineteen women in the 1929 Women's Air Derby (also known as the Powder Puff Derby) transcontinental air race from Santa Monica to Cleveland, and reached Pecos, Texas, before having to pull out of the race after she collided with a truck which was being driven along the runway when she was trying to take off. She then followed a career of demonstration flights and air races – in 1930, taking the world speed record of 196.19 mph away from Earhart – flew as a stunt pilot in several Hollywood silent movies of the era, and helped form her own company for guaranteed film work called the Association of Motion Picture Pilots. In 1935 she bought a small ranch in the desert, and promptly set up a business supplying pork and milk to the growing number of military personnel close by. Her business prospered, and the ranch grew from 80 acres to 368 acres. With the precipitation of the war in Europe and the requirement to test new aircraft, Muroc expanded, and she opened her ranch to off-duty fliers, who rode horses, stayed for dinner, and talked 'planes' into the small hours.

The ranch expanded to include a restaurant, bar and coffee shop, and Barnes soon became acquainted with many of the leading figures in American aviation – the test-pilots out at Muroc, who were pushing the sound barrier. She organised hunting and fishing trips in Mexico, and promised to give a free steak dinner to the pilot who broke the sound barrier for the first time. Chuck Yeager received that first steak after his historic flight on 14 October 1947.

Towards the end of the 1950s, plans to expand the base into what became

Edwards Air Force Base, with a 15,000-foot runway, signalled the end of the Happy Bottom Riding Club. Pancho went to court to fight the plan, but before it could be settled a mysterious fire destroyed most of the ranch. Although she was awarded a large sum of money by the courts, and bought a new property, the costs prevented setting up a new Pancho's. After a serious illness, she died in 1975,

In her later years she often claimed that she had more fun in one week than most people do in a lifetime. Although she was not a member of the Air Force or even a resident of the Muroc facility, Pancho Barnes is as much a part of the history and character of breaking the sound barrier as the pilots, the aircraft and the airbase.

From Aglaonike to Pancho Barnes: pioneers of the skies

By the 1960s the first men were flying into space – military and civilian pilots with a background in test flying and engineering – and the first professional scientists were being selected for space missions. In the Soviet Union a few women pilots and parachutists were being prepared for a pioneering space flight, whilst in the United States a similar group of female aviators was battling with bureaucracy and a male-dominated world to try to fly into orbit.

Historically, men have led the way in inventions, science, technology, engineering and medicine, as well as in exploration, politics and combat. For women to enter this domain they have had to demonstrate the same skills. With experiences in pioneering speed, distance and altitude – often at great cost to themselves – women, like men, pushed the limits to the fringes of the atmosphere. It was time to take the extra step, and journey beyond the atmosphere into the Cosmos. In June 1963 that journey was begun by a young former textile worker, Valentina Tereshkova, and over the next forty years the journey would be continued by women from many other countries.

REFERENCES

1 College Board of Advanced Placement examination paper, European History, Section II, Women in Science, Historical Background (undated, *c.* 2000); also Londa Schiebinger, 'Maria Winkelman at the Berlin Academy: A Turning Point for Science', *ISIS*, 1987.
2 Allan Chapman, *Mary Somerville*, Canopus, 2004.
3 Carole Stott, 'Caroline Herschel', in *Into the Unknown: Women History Makers*, Macdonald, 1988, pp. 10–19.
4 Richard P. Hallion, *Test Pilots: The Frontiersmen of Flight*, Doubleday, 1981, pp. 6–19.
5 Rex D. Hall and David J. Shayler, *The Rocket Men*, Praxis–Springer, 2001, pp. 64–67.
6 Hervé Lauwick, *Heroines of the Sky*, Frederick Muller, 1960, pp. 17–29, English translation from French original.
7 *Flight International*, **27**, No. 1393, 5 September 1935, p. 247.
8 Jim McBeth, 'The third Wright 'brother'', *The Scotsman*, 18 January 2003.
9 Russell Naughton, 'The Pioneers: Celebrating the Bi-centennial of Aviation, 1804–2004', 2002: www.ctie.monash.cdi/hargrave/pioneer.html.

10 Walter M. Schirra Jr with Richard N. Billings, *Schirra's Space*, Quinan Press, Boston, 1988, p. 10.
11 Elizabeth Beckett and Sarah Teel, *Women in Alaska's History*, 1997.
12 Martin Johnson, *African Jungles*, 1935.
13 Hargrave, *The Pioneers*, www.ctie.monash.edu/hargrave/pioneer.html; also, information taken by D.J. Shayler from displays at the Air Force Museum, Monino, Moscow, in June 2003.
14 De Vorkin, *Race to the Stratosphere*, Springer, 1989, pp. 108–130.
15 Carole Stott, *Into the Unknown: Women History Makers*, Macdonald, 1988, pp. 22–29.
16 Cathy M. Morgan, 'The WRAF: Women in Blue', RAF History website, http://www.raf.mod.uk/history/wraf.html.
17 Cathy M. Morgan, 'The ATA: Women with Wings, RAF History website, http://www.raf.mod.uk/history/ata.html.
18 http://www.fleetairarmarchive.net/RollofHonour/Women.html.
19 'Women's Air Service Pilots', in 'Notable fly girls', Edwards Air Force Base history website:
http://www.edwards.af.mil/articles98/docs_html/splash/feb98/cover/wasp.html;
http://www.edwards.af.mil/articles98/docs_html/splash/feb98/cover/wasp.html.
20 'Jackie Cochran', in 'Notable Fly Girls', Edwards Air Force Base history website: http://www.edwards.af.mil/articles98/doc_html/splash/feb98/cover/cochran.html; and Stagger wing website biography of Jacqueline Cochran, http://www.staggerwing.co.uk/J_Cochran.htm.
21 Extended details of the activities of Soviet female pilots during the Second World War can be found on several websites: Vitaly Gorbach, Yekatrinia Poliunina and Dmitri Khazanov, 'The Night Witches: Female Faces of the Air War', http://www.samolet.co.uk/femalefaces.html; 'Soviet Women Pilots of WWII', http://www.elknet.pl; http://airsports.fai.org/dec98/dec9824.html. Books: Bruce Myles, *Night Witches: The Untold Story of Soviet Women of Combat*, Academy Books, Chicago, 1990; Anne Noggle, *A Dance with Death: Soviet Airwomen of World War II*, Texas A&M Press, 1995; Henry Sakaida, *Heroines of the Soviet Union, 1941–1945*, Elite Books, 2003.
22 'Florence 'Pancho' Barnes, Aviation Companion', Edwards Air Force Base history website:
http://www.edwards.af.mil/history/docs_html/people/pancho_barns_biography.html; also Chuck Yeager *et al.*, 'The Happy Bottom Riding Club', in *The Quest for Mach One: First-Person Accounts of Breaking the Sound Barrier*, pp. 57–79, Penguin Putnam Books, 1997.

A seagull in orbit

The announcement of a second Soviet cosmonaut joining Valeri Bykovsky's Vostok 5 in orbit was no great surprise, as there had already been a dual flight of Vostok 3 and Vostok 4 in August 1962. What *was* a surprise was the identity of Bykovsky's colleague. Vostok 6 confirmed the many rumours that Russia had a cadre of female, as well as male, Vostok cosmonauts in training – but America's Mercury programme did not officially have any female astronauts. The reasons for flying a female cosmonaut at this time, and why no other women followed Valentina Tereshkova for another twenty years, is a story of the Soviet regime, its systems, the opportunities it presented, and the restrictions it imposed.

A SOVIET WOMAN IN SPACE

On 12 April 1961 a young Soviet Air Force pilot, with a fresh face and an engaging smile, made history by becoming the first person to fly in space. His historic flight began a new facet of human exploration, and the dreams and plans of hundreds of thousands of others soared when he proved that a human could make the rapid rocket ride above the atmosphere, survive in space, and endure the dangerous re-entry and landing. Yuri Gagarin became one of the most famous and recognised people in world history, and pioneered the trail for others to follow. Weeks later, Alan Shepard became the first American in space in a brief, 15-minute sub-orbital flight. Two months later he was followed by Gus Grissom on another sub-orbital mission, after which the Soviets again upstaged the US with a day-long flight by Gherman Titov. In February 1962, John Glenn became the first American to orbit Earth, and was followed by Scott Carpenter in May 1962. But again the Americans were upstaged – this time with the dual flight of Vostok 3 and Vostok 4 in August 1962, with Andrian Nikolayev on Vostok 3 setting a new endurance record of more than 94 hours and 64 orbits. The Americans planned a day-long flight for Mercury in 1963 and a possible three-day mission was hinted at; but in October 1962 America's response was a short 8-hour flight by Walter Schirra.

By December 1962 Mercury was being phased out to make way for the two-man

Gemini, which sought to develop techniques and gain experience for the three-man Apollo lunar landing programme. But America's entire human space programme was in effect a 'man in space' programme using military jet pilots, and even the Soviets had accumulated biomedical and operational data on male pilots only. The question remained about how a woman might cope with a flight into space. It was a question that America was not pursuing – at least not before Apollo – but it was being investigated in Russia.

A female cosmonaut

The Soviet Vostok spacecraft – developed in the late 1950s – was designed to orbit a single cosmonaut for a few days; but the pilot had little control of the spacecraft, there was barely room for movement inside the crew compartment, and only a few simple experiments and observations were possible. Therefore, Sergei Korolyov – Chief Designer at the OKB-1 bureau that had developed the launch system and spacecraft – wanted to replace the pioneering spacecraft with a more advanced vehicle that could perform orbital rendezvous and docking, help to create a space platform, and support manned flights to the Moon. Frustrated with delays to his longer-term plan, Korolyov sought the support of the military for a multi-person version of Vostok. In the meantime, despite a series of delays and changes, new flights by Vostok were authorised, although each manned mission required approval by the Presidium of the Central Committee and a confirmation by decree from the Council of Ministers, adding to the political and bureaucratic hurdles.[1]

After the success of Gagarin and Titov, hundreds of letters were written by Soviet citizens (including a large number by women) expressing a desire to join the cosmonaut team. At the time, the thirteen American female pilots were in the news as they attempted to be accepted for training for a flight on Project Mercury. Not wishing to be beaten by the Americans, Nikolai Kamanin – Director of Cosmonaut Training, and Deputy Chief of the Air Force – realised that flying the first woman in space would be as big a political coup for Communism as Gagarin had been. In October 1961 he wrote in his diary: 'We cannot allow that the first woman in space will be American. This would be an insult to the patriotic feelings of Soviet women.' Kamanin began arguing for the inclusion of a small group of female cosmonauts in the programme, but it would take six months to convince several influential and key figures in the Soviet space programme that it would be an advantage to orbit one or perhaps two women before the Americans did so.

Two months later, on 30 December 1961, the Central Committee of the Communist Party gave their approval for new cosmonauts, though only twenty would be finally authorised, and of these, five would be women. Selection of the fifteen male pilots would be delayed until 1963, but the process to select the first female cosmonaut began immediately. The general idea of putting a Soviet female into space appealed to Premier Nikita Khrushchev, who was convinced that it would be an important demonstration of the Communist system for an 'above average' worker to be trained for such a demanding role. It would be another technological blow to the Western world. Of course, the reliability and simplicity of the Soviet spacecraft would help, so that an abbreviated training programme would enable them to achieve a flight possibly within the next year. Selection of the candidate was restricted to pilots, but this was a limited resource at the time, and so the criteria were extended to include experienced active sport parachutists – a leading pastime in the Soviet Union. Due to the shortened training programme this skill would prove very useful, as at the end of each Vostok mission the pilot was required to eject from the capsule and land by separate parachute.

The first female selection

The responsibility of finding suitable candidates fell to the All-Union Voluntary Society for Assistance to the Army, Air Force and Navy (in Russian, DOSAAF). The criteria encompassed unmarried women aged between eighteen and thirty with some aviation experience, though this was not solely restricted to pilots. In the search for suitable candidates, the records of parachutists, sports pilots, aerobatic pilots and military trained pilots were reviewed, and the entire corps of the Soviet National Sky Diving Team was nominated. Spotters were dispatched to aero clubs to personally interview possible candidates (presumably without informing the 'candidate' about the reasons), and by the middle of January 1962 a list of more than four hundred names had been compiled. Kamanin hoped that at least a hundred candidates could be found, but only fifty-eight met the minimum requirements. This was not helped by the weight restriction (in part limited by the lift capability of the launch vehicle) and the height limit (determined by the size of the Vostok suit and spacecraft). The shortage of suitable candidates was also due to some of them being married and some being over thirty years old.

The list of candidates was agreed on 15 January 1962, and the plan was then to review all them and select thirty to forty to go to Moscow for medical and related tests. In reviewing each file, however, Kamanin was disappointed with what he found. In his diary entry of 19 January he wrote: 'Yesterday I considered the files of fifty-eight female candidates. Generally disappointed and dissatisfied. The majority are not suitable for our requirements and have been rejected. Only twenty-three will be brought to Moscow for medical tests because DOSAAF did not examine their credentials correctly. I told them I needed girls who were young, brave, physically strong and with experience of aviation, who we can prepare for spaceflight in no more than six months. The central objective of this accelerated preparation is to ensure that the Americans do not beat us to place the first woman into space.'

Each of these women underwent extensive medical tests. Four or five apparently failed the early medicals, and only seventeen (or eighteen) went on for further consideration as candidates. Many of these women have remained largely unidentified, but those who have been named include the following:[2]

Pilot applicants

Galena Korchuganova (*b.*1935), an aerobatic pilot and a champion of the USSR, who later became a world aerobatic champion.

Marina Popovicha (*b.*1931), wife of cosmonaut Pavel Popovich, and a famous Soviet military test-pilot with more logged flying hours than her husband. After leaving school in 1947 she joined the Antov design bureau, and later joined the Air Force,

where she rose to the rank of Colonel. When she met her future husband in 1955 she already held thirteen world aviation records, and was a Major in the Air Force and a test-pilot. She and Popovich later divorced, and after retiring from the Air Force she became known as a UFOlogist and lecturer around the world.
Rosalie Shikina (*b*.1928), married (married name Zanozina) with one child. In 1964 she became a USSR and World aerobatic champion, but was killed on 8 September 1965 in an aircraft crash while working for the Yakolev aircraft design bureau as a civilian test-pilot.
Vera Kvasova – nothing known.
Marina Sokolova, an instructor pilot with DOSAAF who set a world speed record for a female pilot in a Mig 21 in 1965.
Ludmilla Solovyeva, a sports pilot with DOSAAF.
(?) Yefremova – nothing known.

Parachutist applicants
(?) Borzenkova, Valentina Daritcheva and Svetlana Ivleva – nothing known.
Natalya Maslova – the youngest finalist, born in 1943; a member of the USSR parachuting team.

The five known finalists
Irina Solovyova, a member of the USSR parachuting team.
Tatyana Kuznetsova, a member of the USSR parachuting team.
Valentina Ponomaryova, also a recreational DOSAAF pilot.
Valentina Tereshkova.
Zhanna Yorkina.

Other candidates identified but not included in the final twenty-three
Galina Koroikova (*b*.1933), a graduate of the Moscow Aviation Institute and possibly a hobby pilot.
Tatyana Morozitchevo, a colleague of Tereshkova from Yaroslavl who became a parachuting champion of the USSR in 1968. She may also be the same person often mentioned as Tatyana Torchillova.
Natalya Prokhanova (*b*.1940), a sports pilot with DOSAAF.
Svetlana Vaslova, a member of the USSR parachuting team.
Lydia Zaitseva, a sports pilot with DOSAAF.
Vera Zoubova, also a member of the USSR parachuting team.

On 22 February, Kamanin wrote that nine girls 'from the first batch' had passed the tests and were to be interviewed, and that from these, four or five would be selected. He also noted that to achieve a female spaceflight by August 1962 (his original aim), they had to begin training on 1 April. In his diary entry of 28 February he noted seven candidates that were interviewed by the selection Commission the previous day: Yefremova, Kvasova, Kuznetsova, Sokolova, Solovyeva, Solovyova and Tereshkova, all of whom passed the medical tests. It became clear that from these seven only three or four could be nominated for training, with the most probable being Solovyova, Tereshkova and Kuznetsova. Of the other four, Sokolova was listed as least likely to be selected, while the others were possibilities. Kamanin wrote of the three probable candidates that all were strong. He believed that one of them would be the first Soviet woman in space, 'and hopefully the world's first, too.'

The medical commission authorised the appointment of Solovyova, Tereshkova and Kuznetsova on 3 March (to report on 12 March), but a decision on Kvasova and Yefremova would be deferred until the end of the month, after the second group of finalists had been interviewed. From this group, Borzenkova and Yorkina were the most noteworthy candidates. No other details were forthcoming on the selection of Ponomaryova or Yorkina, who were approved on 3 April. All five finalists were in their 20s: Kuznetsova was 20; Ponomaryova, 28; Solovyova and Tereshkova, both 24; and Yorkina, 22. Each of them was a very experienced parachutist, with Solovyova being a world champion member of the Soviet national parachuting team. There was only one pilot, Ponomaryova, who had learned to fly as a member of a DOSAAF air club. As she was also married and a mother, there was some reluctance by the Commission to select her, but Mistislav Keldysh, President of the Academy of Sciences, supported her, and persuaded the rest of the Commission to accept her for training. All five were civilians, but due to the organisational structure of the cosmonaut team at that time, they were enrolled in the Air Force.

Tatyana D. Kuznetsova (née Pitskelauiri) (*b*.1941) was a qualified parachutist with a number of world records to her credit. At the age of 20 she became the youngest person ever selected for space training. She served in a support role for Vostok 6, and it is possible that her decision to marry after agreeing to remain single until completing at least one spaceflight affected her chances to fly; but she remained a member of the team. When the group stood down in 1969 she remained at the training centre and assisted cosmonauts with geophysical experiments. She rose to serve as a Lieut-Colonel in the Air Force Reserves.

Valentina L. Ponomaryova (*b*.1933) was, as a young girl, interested in flying. She was a 1959 graduate of the Moscow Aviation Institute, where she learned to fly and became a parachutist. She was marred to fellow student Yuri Ponomarev, who was himself selected for cosmonaut training as part of the NPO Energiya team, although he never flew in space. When selected for cosmonaut training she was working at the Institute of Applied Mathematics. She served as second back-up for Vostok 6, and continued training for a flight as Commander of Voskhod 4, which was subsequently cancelled. She continued working at the cosmonaut training centre until 1988, and then worked in an institute affiliated to the Academy of Science. She also became a Colonel in the Air Force Reserves.

Irina B. Solovyova (*b*.1937) was a 1957 graduate of the Ural Polytechnic Institute, where she joined a local air club and learned parachuting. She became a member of the Soviet national team, becoming a Master of Sport and setting a number of world records while completing more than 2,200 jumps. She was first back-up for Vostok 6, and would have flown as a Pilot on an all-female 'endurance' Voskhod mission that would have included her conducting an EVA, but it was cancelled in 1966. After the

Three of the first female cosmonauts – Valentina Ponomaryova, Irina Solovyova and Valentina Tereshkova – prior to the launch of Vostok 6 in June 1963.

female cosmonaut group disbanded she continued working at TsPK as a scientist and psychologist, and also participated in an expedition to Antarctica. She continued her parachuting career for many years, setting new records and evaluating new equipment. She rose to serve as a Colonel in the Air Force Reserves.

Valentina V. Tereshkova (*b*.1937) worked in a textile mill after leaving school, and later joined a parachute club, making more than a hundred jumps. As the Pilot of Vostok 6 she became the first woman to fly in space, and became as famous as Yuri Gagarin. After the flight she married fellow cosmonaut Andrian Nikolayev in November 1963, and the couple had a daughter. She also continued her cosmonaut career throughout the 1960s, participating in training for Voskhod and Soyuz missions, but there is no evidence that she actually trained for a second mission. She remained in the cosmonaut team until March 1997, and attained the rank of Major General in the Air Force. Since her mission she has held a number of responsible positions in the Soviet and Russian governments, focusing on cultural and international movements and societies.

Zhanna D. Yorkina (née Sergeuchik) (*b*.1939) earned a degree in English, and took up parachuting as a hobby. After supporting Vostok 6 and working on the Voskhod 2 mission in 1965, she was assigned as back-up Commander of Voskhod 4. After the cancellation of the mission in 1966 she trained on Soyuz until the female cosmonaut team was disbanded. She continued working at the training centre, and has risen to the rank of Lieut-Colonel in the Air Force Reserves.

Training the cosmonauts

Upon selection for cosmonaut training, all five members were initially enrolled into the Air Force as Privates attached to the Cosmonaut Team. The successful candidates would be commissioned as Junior Lieutenants after completion of basic training. Very few detailed accounts have been released about their training programme, but it is often reported that this was similar to that of the first male group selected in 1960. The sequence and priorities of this programme are unknown.

All the candidates began a qualification programme to fly as passengers in Mig 25 UTI jet trainers, including a programme of five flights on successive days. Their preparations included weightless parabolic flights in Tu-104 aircraft, more than eighty parachute jumps – including some while wearing the Vostok pressure garment for drops onto land (as would be planned for the flight) or into the Black Sea (to practice for emergency water recovery) – a programme of isolation chamber tests simulating long solo flights in space, and centrifuge runs up to 10 g. Their academic courses covered the theory of rocketry, radio communications, spacecraft engineering, navigation (including astronomy courses), and simulations inside mock-ups of the Vostok spacecraft. They also completed a series of survival training courses to cover emergencies during the flight or immediately afterwards, and participated in a physical training programme that included various sports and integration into various men's teams.

During their training there was no guaranteed seat into space, as all five were competing for just one mission. The group became very competitive, and were being observed for their personalities as well as their technical skills. As each was vying for the position of 'first woman in space', the image of a 'good Soviet citizen' was as important as it had been for the selection of Gagarin. Essentially unknown prior to the flight, the chosen candidate would become an international celebrity before return to Earth, and the Soviets needed a candidate who could cope with the additional strain that this historic mission would bring. Gagarin was well chosen both for the flight and for the iconic role thrust upon him, and each of the five women was being evaluated for their public skills and the persona that they would need immediately after the mission.

In the late summer of 1962, an informal assessment of the team rated Tereshkova, Solovyova and Ponomaryova as the most likely candidates to fly in space. According to Kamanin, Yorkina was 'too fond of chocolate and cakes', and had missed some of the training – partly as a result of an ankle injury from a bad parachute landing; while Kuznetsova was 'still very young, and her character is not fully developed.' At the start of the training programme, Kuznetsova had apparently been a strong contender for the flight, but by the late summer she was less well favoured. During September she became ill, and eventually stood down.

Basic training was completed in October 1962, and during the following month examinations were held on all except Kuznetsova, who had been ill and had had lower performance levels on the centrifuge. Due to her combined health problems, she fell behind in her studies and did not receive her official cosmonaut designation until January 1965, more than two years after the other four received them.

In theory and practical tests, Ponomaryova was clearly in the lead with the highest overall scores and with better skills than some of her male counterparts. But Kamanin was concerned about her personality, and worried about her independence, self-assertion and over-confidence. She also displayed the 'unsteady morals' of swearing, smoking, and leaving the base without permission. Tereshkova, according to Kamanin, was feminine and charming, and had a more appealing personality; and though Solovyova was displaying both acceptable moral and physical skills – being the toughest, both physically and emotionally – her personality made her a loner. She did not actively participate in social engagements and became separated from the other team members, and was less of a 'team player'. The 'weakest' of the four was Yorkina, but though she was lacking in professional skills she was perceived as to be improving as the programme proceeded, and would probably eventually become a suitable cosmonaut. None of the team knew which one of them would be finally chosen to make the flight, and they would have to wait for several more months. In November 1962 the four remaining candidates were commissioned as Junior Lieutenants in the Soviet Air Force.

On 29 November 1962 Kamanin noted in his diary each woman's strengths and weaknesses; but he still favoured Tereshkova for the first flight: 'We must send Tereshkova into space first, and her back-up will be Solovyova. Tereshkova, she is a Gagarin in a skirt.'

By December there was no possibility of a female flight within the ensuing few months, and after an intensive eight-month training programme the female group was granted an extended leave of six weeks, being told not to report back until early 1963. During this time, Solovyova was given permission to rejoin her colleagues in the national parachuting team. By the time they reported back in the middle of January the plans for the flight were being refined, but exactly what the flight programme was and who should fly was still far from settled. Though impatience was evident in the group in early February, by March, confirmation of a June flight had left them much happier, with the clear objective of a seat into space – for one of them.

PLANNING FOR FLIGHT

By the beginning of 1963, mission planners were evaluating their options for the mission and others in the Vostok series, and the four women were formed into a training group to prepare for their specific mission. Funding for additional Vostok missions was not immediately forthcoming in 1962, and though plans were made for several additional flights later in 1963 and 1964 – crewed from a training group of male cosmonauts – when it became clear that no funds existed beyond a sixth Vostok mission, the plans for the female flight were also affected. There were plans in progress to modify some of the Vostok hardware for longer missions, with multi-person crews and early spacewalking experiments, but the limitations of hardware reliability and the constant desire to 'beat the Americans' heaped additional pressures on the programme and its resources.

A female pressure garment

One of the early factors affecting the female flight was the adaptation of the pressure garment for the female cosmonauts. The SK-1 pressure garment was developed by the Zvezda research development and production enterprise for the

Vostok programme, utilising experience with high-altitude pressure garments.[3] This suit provided adequate conditions for the cosmonaut for twelve days inside a pressurised cabin: five hours in orbit (just over three orbits), or, in the case of a depressurised cabin (at an operating pressure of 270–300 hPa, corresponding to an 'altitude' of 10 km), safe occupation inside the Descent Module for twenty-five minutes during re-entry and landing. It also provided protection during ejection from Vostok at altitudes of up to 8 km, and an oxygen supply for use during the parachute descent. If the cosmonaut landed in water, the suit had thermal support for twelve hours in water or three days in a rescue dinghy (or after landing) in temperatures of 15° C.

When the flight of a female cosmonaut was authorised, production of the SK-1 suit was amended during 1962 to incorporate female-specific features, and was redesignated as SK-2. This new design incorporated reduced shoulder breadth, with amendments to the shoulder harness system for more arm movement, plus an increased hip girth and a reduced neck opening. The cord used to retain the helmet at the front of the SK-1 unit was located at the breast area, and for the female version this was lowered. Amendments were also required for the gloves, which had reduced thermal layering and more mobility in the thumb digit. The respiratory valves and helmet visor handles also required modification to render them more accessible and easier to use. The waste management receptacle was also modified to fit the female form.

The suit was not taken off during the mission, and one of the cosmonaut's tasks was to evaluate the comfort levels during flight. Some improvements, based on the first four manned flights of the system, had been incorporated, but the flight of the first woman in space was also the first opportunity to test the new design operationally. To assist in evaluating the design prior to authorising the suit for flight, a female Zvezda engineer, G.I. Viskovskaya, was assigned as suit tester. She also assisted in the preparation of the flight and the suiting of Tereshkova for the mission. An even smaller production line for this suit resulted in just four test models, two training models and two flight models of the SK-2 design. The changes to, and production of, this suit affected plans for the mission.

Progress towards launch

During a goodwill visit to the United States in May 1962, Kamanin and cosmonaut Titov met with a number of astronauts and NASA officials. During a barbecue party attended by the Soviet delegation at the home of John Glenn, the astronaut informed Kamanin that it was possible that an American woman might make a three-orbit flight before the end of 1962. It is debatable whether or not this was US propaganda, but the news spurred Kamanin into telling Soviet mission planners that it was imperative that a Soviet female cosmonaut be orbited by the late summer. However, the next two missions (Vostok 3 and Vostok 4) were slipping towards August, and delays with various hardware elements for the proposed female flight plagued this plan.

In June 1962 Kamanin personally approached the Chief Designer at OKB-Zvezda, Semyon Alekseyev, to determine whether the production of the suit could be

Vostok women cosmonauts during training. (Courtesy Tony Quine.)

hastened – and even took each woman's measurements with him to hand over personally. But he was told that production of the female version of the suit would not be completed before the end of 1962, which delayed the flight until at least the spring of 1963, near the end of the production lifetime of the spacecraft. In his diary entry for 27 August 1962, Kamanin noted that the plan was for two female cosmonauts to fly on the final two Vostok spacecraft; but after a meeting in November, several options were considered for a flight profile that included the simultaneous flight of two female cosmonauts, or for one female on one mission and a male colleague for a record five to seven days, with a launch still planned for March 1963. In January 1963 there were three profiles for the proposed mission: a two- to three-day solo flight by one female cosmonaut; two cosmonauts launched a day apart in a group flight, but landing on the same day; or what Kamanin described as a 'ridiculous option' – flying a female cosmonaut for a three-day solo flight, and then, two days after her return, flying a male cosmonaut on a five- to seven-day mission. The planners chose the two-female option, as it was a repeat of the flights of Vostok 3 and Vostok 4, and was therefore much easier to prepare.

On 1 February the four female cosmonauts were placed in a training group to prepare for the missions, but the flight was challenged by other programme leaders and slipped beyond March. Caution suggested sending one woman up in spacecraft 007 (designated Vostok 5), while spacecraft 008 was held in reserve. If this flight was a success, spacecraft 008 would not be used; but if Vostok 5 was not successful, spacecraft 008 would fly as Vostok 6 in a second attempt.[4]

The suggestion that a second flight might be flown after a 'failed' mission (which may have included the loss of a cosmonaut) is interesting. The reality following the accidents encountered with Soyuz 1 in 1967, Soyuz 11 in 1971, the American Apollo 1 pad fire in 1967, Apollo 13 in 1970, *Challenger* in 1986 and *Columbia* in 2003 included the long delays while the accidents were investigated and safety recommendations incorporated to prevent similar accidents, as well as allowing for the inevitable grief of the loss. With the Vostok operational lifetime approaching by the middle of June, the mission would have to be flown by then; and if only one mission was attempted and it failed, it would certainly signal the end of the Vostok programme.

On 21 March 1963, a meeting of the Presidium of the Communist Party decided that there would only be two flights in 1963 – one female and one male – using the two remaining craft and rescheduling them for early June, as the 'shelf life' of Vostok could not be extended into August. On 13 April, Korolyov and Kamanin agreed on a plan to fly a man on an eight-day mission onboard 'Vostok 5' and a woman on a two- to three-day mission onboard 'Vostok 6'. On 29 April, the Central Committee approved the plan.

Mission training

During April 1963, all four women completed a three-day simulation in the Vostok simulator, and all passed the test. Yorkina, however, ended the test very weak, and fainted, as she had eaten only a third of her rations, and had removed her boots during the first day. This deviation from procedures effectively eliminated her from consideration for the Vostok 6 assignment.[5]

During considerations for flight assignments in May, many of the technical instructors supported Kamanin's nomination of Tereshkova for the flight; but Keldysh led a strong lobby for Ponomaryova (who had worked for Keldysh at the Academy of Sciences!), and even tried to sway Gagarin, who had a vote in the decision but remained uncommitted either way. When lobbied, Gagarin tired of the politics and supported Tereshkova for the flight. This was a decisive move, which resulted in the Commission nominating Tereshkova to fly.

Who would fly?

With the question of the flight and hardware settled, there remained the question of which cosmonauts would fly the mission. The original male training group consisted of Valeri Bykovsky, Boris Volynov and Vladimir Komarov, who were to have flown two or three long-duration solo missions of five to seven days each. With the changes to the female mission and the launch time restrictions of the hardware, the male and female training groups were reassigned in April for the dual flight in June. In May, Komarov was temporarily grounded due to a medical condition, and left the group, to be replaced by Alexei Leonov and Yevgenei Khrunov. On 10 May 1963, a month before the planned flight, the final assignments were announced. Bykovsky would fly on Vostok 5, backed up by Volynov and with Leonov in support. Tereshkova would make history as the first female in space, with Solovyova as back-up and Ponomaryova as second back-up. The selection of two back-ups reflected concern that the timing of the flight might eliminate one of the three due to their menstrual cycle. Some years later, Tereshkova recalled the issue of menstruation during flight: 'Doctors and biologists carefully monitored our progress to see if the female body was in any way different to the male in undergoing tests and training exercises.

Valentina Tereshkova shortly before her mission.

Menstruation in space was an obvious problem, and we were not required to undertake centrifuge training at this time of the month. My spaceflight took place between periods.'[6]

The selection of Tereshkova reflected observations of her character and personality, but as Ponomaryova was a leading candidate for the flight, it is strange that her assignment was as second back-up and not first back-up. In reviewing Vostok 6 crewing assignments, researcher Tony Quine put forward his theory of how this could have come about. 'It is my belief that training pairings were created with Tereshkova/Solovyova and Ponomaryova/Yorkina in anticipation of two female flights. Assuming that Tereshkova had been chosen to make the first female flight, her back-up would have been Solovyova, leaving Ponomaryova to prepare for the second flight the following day. When the plans changed to a single female flight, the pairing stayed, leaving Solovyova as Tereshkova's back-up even though she does not appear (from Kamanin's diaries) to have been seriously considered for the prime slot.'[4]

At first the news was not passed to the women, as it depended upon medical checks and the completion of a final parachute training session. On 14 May, Tereshkova and Solovyova completed their parachute training with seven jumps of varying difficulty. Kamanin met Tereshkova, who was visibly happy, leading to him to suspect that someone had leaked the news to her. The news was officially given to the cosmonauts on 21 May at a meeting of the State Commission. Dressed in their Air Force uniforms, they waited in turn to be given the news. Tereshkova was the first to be told, and she accepted the assignment, assuring the Commission that she would do her best to complete the flight programme. The launch was dependent on the Vostok 5 mission, but she had to be ready to fly by 7 June – just seventeen days away.

The other two were also told of their roles in supporting the flight, and though not totally unexpected, Ponomaryova was very disappointed and was not afraid to let it show. Both women were assured by Korolyov that future flights with female cosmonaut crew-members were planned, and that both would fly in space. Tereshkova, on the other hand, was elated and happy that all the work had paid off, and showed no fear of what lay ahead.

Over the next few days, a light training programme and briefings were conducted while the final plans for departure to the cosmodrome were prepared. After discussion of when they should leave, how many people from the training centre were required, and whether all should travel in a single aircraft, they planned to be at Baikonur before the end of May. Unfortunately, on 28 May Chief Parachute Instructor and one of Tereshkova's supporters, Nikolai Nikitin, was killed in a parachute accident. A delay to attend the funeral on 30 May was a concern to Kamanin, who was worried about what effect the loss of a close member of the team would have on the women's morale. However, the preparations for the flight were in motion, and on 1 June the team flew to the Baikonur cosmodrome for final preparations.

As the time of the flight approached, the Western press again conjured up a mixture of fact and fiction, with rumours about a one- or two-shot mission, or that a version like America's two-man Gemini spacecraft was under development for the Soviet programme. There were also suggestions of a two-person Vostok spacecraft.

Irina Solovyova (Tereshkova's back-up) on the transfer bus, 16 June 1963.

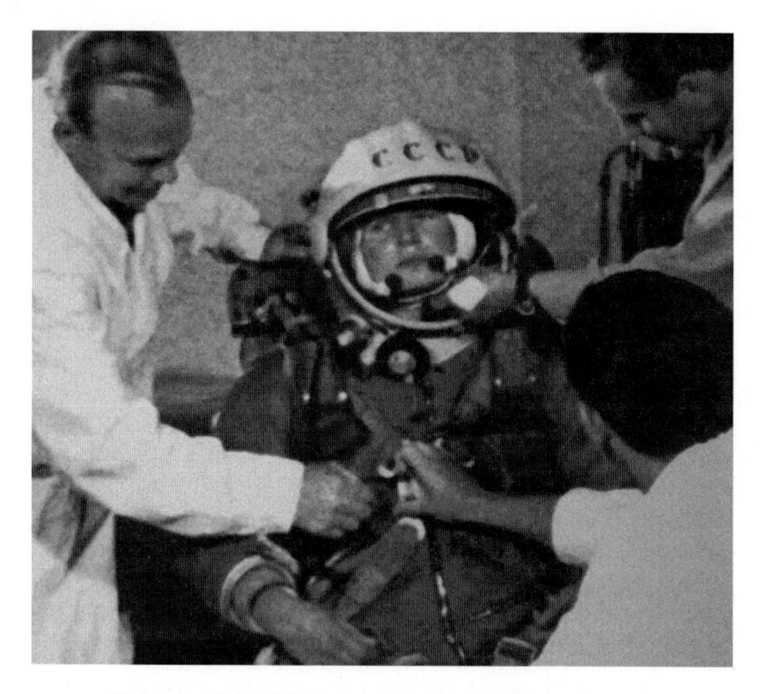

Valentina Tereshkova undergoes pre-flight suit tests.

Newspapers recalled films of female cosmonauts participating in parabolic flights in aircraft and apparently in training. They also repeated erroneous reports of a female space fatality that supposedly had occurred on 17 May 1961, in which a female cosmonaut was one of two killed in a launch accident. In one report of the time,[7] a newspaper actually proposed the identity of the first woman in space as Anne Massevitch, Vice President of the Aeronautical Academy of Sciences, but then revealed that no date was set for the expected flight. Reminiscent of reports of several 'phantom' cosmonaut accidents prior to the launch of Gagarin, and the identification of the son of leading aircraft designer Vladimir Ilyushin just two days before the flight of Vostok, these latest speculations were exaggerations by the newspapers, rather than factual reports.

With the launch of Bykovsky on 14 June 1963, these rumours were heightened with the expectation of a space link-up and, again, exaggeration of information from 'official' sources'. Supposedly, a 25-year-old girl named Ludmilla was ready for take-off, with news agencies in Moscow having prepared biographical details ready for the announcement of the launch.[8] With Bykovsky in orbit, the world would not have to wait long before the news was confirmed, but it would be a 26-year-old called Valentina, not Ludmilla.

Technical problems delayed the launch of Vostok 5 from 7 June to 11 June, but an increase in solar activity delayed this further to 14 June. On that day, after being strapped in the capsule for more than six hours, due to several problems, the cosmonaut was finally launched on his record-breaking solo mission. With

Bykovsky in orbit, it was time to prepare Tereshkova for her flight. His planned orbit had not been as precise as expected, and the eight-day mission was cut to five days; and as a result, this affected the launch and landing dates for Vostok 6.

The clock is counting

The female cosmonauts had been at Baikonur cosmodrome since early June, and had spent the following two weeks completing final preparations and working with the launch crews in reviewing the status of the hardware and the weather forecasts. The day before launch the rocket and spacecraft were moved to the launch pad – the same pad that had supported the launch of Vostok 5 the previous day. Tereshkova woke at 8 am local time and had breakfast with the other cosmonauts, and Kamanin told her that her launch was planned for the next day. She was a favourite of Kamanin, and he took a paternal interest in her well-being and in preparing her – as had Koroylov with Gagarin. In good spirits, she was in good physical shape and had been eating well according to medical reports. Kamanin notes in his diary that she has also put on a little weight since her arrival at Baikonur, is pale in complexion and naturally is feeling a little worried about what she is about to do.

After a three-hour update on the status of the R-7 launch vehicle and her spacecraft by two of the leading designers (Konstantin Feoktistov and Boris Raushenbakh), Tereshkova visited the launch pad to follow the tradition of accepting her spacecraft from the ground crew. After the delays to Bykovsky's launch they were eager to reassure her, and were impressed by her apparent calmness. Asked if she is afraid, she replied with a smile, 'Of course not'.[9] Later in the day she reviewed her flight logs with her back-ups, and also finalised the

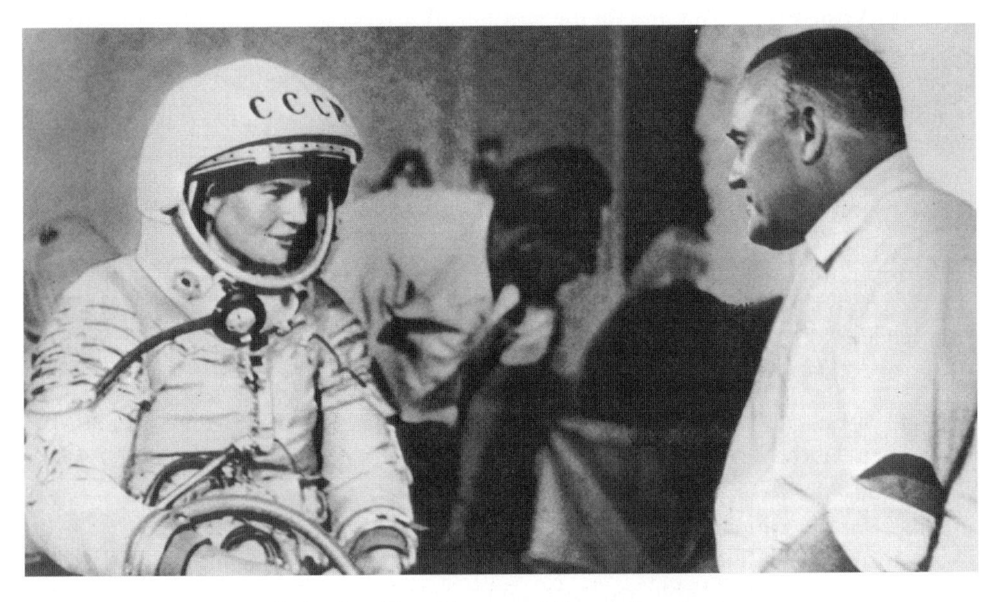

Valentina Tereshkova and Sergei Korolyov.

messages of goodwill and propaganda messages that she was to broadcast during the flight. The ceremony at the launch pad later that day, to accept the readiness of the vehicle, was to have been in full military uniform, but a late message insisted that women should be filmed in civilian cloths so as not to identify their military status. Tereshkova delivered a thank-you speech, and dedicated her forthcoming flight to the Soviet people. Ponomaryova also commented, but with obvious reluctance. The ceremony, though running late, went well, with Tereshkova receiving an enthusiastic reception – especially from the pad workers, who were not all aware that they were preparing the spacecraft for a woman. Changing to tracksuits, the cosmonauts ascended the launch tower to try out the inside of the capsule (called 'Sharik', or 'little sphere', referring to the shape of the Descent Module). Both Tereshkova and Solovyova sat inside Vostok 6 for a few minutes, but Ponomaryova, aware that her chance of making a spaceflight had effectively been lost, declined the chance to sit inside the spacecraft that could have propelled her not only into space but also into history. With the pre-launch preparations completed, Ponomaryova was stood down from her role on the mission, but Tereshkova and Solovyova moved to the cosmonaut lodge with the medical team, and after a brief meeting with Korolyov and a meal they retired for a few hours' rest. Tereshkova slept in the bunk that Gagarin had slept in on the eve of his historic flight.

A new start: 16 June 1963

At 7.30 am local time (5.30 Moscow Time), Tereshkova and Solovyova were awoken to prepare for the launch of Vostok 6. After a short exercise period, and breakfast with the other cosmonauts and a number of space officials, they received an update on the condition of the vehicle (launch was scheduled for 14.30), on Bykovsky onboard Vostok 5 in orbit and on the weather. It was a clear day at the cosmodrome, with low winds and no cloud, but it was very hot (which Tereshkova vividly recalled more than forty years later).[10]

Four hours prior to launch, the cosmonauts arrived at the preparation area for suiting. Solovyova suited first, followed by Tereshkova, who remained calm and cheerful. Kamanin recorded that both cosmonauts seemed relaxed and ready to fulfil their assignments. The group left for the pad at noon, and as she approached the bus, onlookers saw her smiling and waving, but seemingly a little apprehensive. During the ten-minute bus-ride Tereshkova and her three colleagues chatted and sang songs. At the pad she bade farewell to Solovyova, Ponomaryova and Yorkina, and proceeded to conduct the brief formal ceremony in which she reported: 'Commander of the spaceship Vostok 6, Jr Lieut Tereshkova, is ready for the flight.'

Seagull ascending

With the ceremonies over, Tereshkova climbed the steps to the elevator that would take her to the spacecraft. Awaiting her was Chief Vostok Engineer Frelov and members of the launch team. Anticipation in what she was about to do was evident in a recorded pulse level of 140, but once inside the capsule she calmed down, began the pre-flight checks, established radio communications, and reassured the doctors. She entered the spacecraft at 12.15 pm local time, and over the next two hours went through the routine of pre-flight preparations, which included long periods of inactivity. During these phases she talked over the radio to Kamanin, Korolyov and Gagarin, and listened to music piped into the spacecraft. The pad was then cleared of personnel and the support gantries were retracted as the clock approached 14.30.

Fifteen minutes prior to launch, Tereshkova was instructed by Gagarin to close her helmet and put on her gloves before pressurising her suit for flight. Remaining calm, and chatting to Gagarin, she stated that she was ready to go. Twenty seconds before lift-off she reported that the booster was coming to life as she heard the rumble of opening valves and the running of the fuel pumps in the launch vehicle below her. Although calm, her pulse quickened to 156 as she braced herself for ignition and lift-off. At 14.30 pm local time – exactly on time – Vostok 6 was commanded for launch from the pad at Baikonur cosmodrome, and moments later the journey of the first woman into space had begun. In her memoirs, published in 1964, Tereshkova recalled the event: 'The music of launch begins with low sounds. I hear the roar that reminds me of the sound of thunder. The rocket is swaying like a thin tree under the wind. The roar grows, becomes wider, the upper notes were distinguished in it. The spaceship is shivering ... Unexpectedly I say to myself, 'I'm flying!"[11]

As the R-7 accelerated, the ground-based telemetry revealed that the cosmonaut's pulse was increasing and her breathing was difficult and fast. On the whole, however, she handled the ascent very well – in biomedical respects, better than Nikolayev on Vostok 3 and Popovich on Vostok 4. 'I feel my heavy hands and feet, the hidden weights that were shaking my chest. The weight grows. It becomes hard to breathe; I can't move a single finger. The [other cosmonauts] told me that's how it's going to be. So everything is OK then ... I feel like the weight has reached its limit, but it still grows. How much time has passed since the start? A minute? An hour? A day? I cannot collect my thoughts; I know I have to but I cannot.'[11]

About two minutes into the flight, the four strap-on stages separated, followed about 30 seconds later by the jettisoning of the launch shroud. The core stage of the launch vehicle was separated 5 minutes into the ascent, leaving the final stage to take Vostok 6 into orbit, shutting down about 8 minutes into the mission. A couple of minutes later the spacecraft entered orbit, with the initial orbital parameters of 180.9 × 231.1 km × 64°.95. In less than ten minutes, Tereshkova – a former mill-worker and sports parachutist – was elevated from relative obscurity to become one of the most famous women in the world. And Kamanin was delighted.

The voice of Gagarin took Tereshkova with surprise, as though he was sitting next to here inside the spacecraft, when he reported that everything appeared to be excellent with the ascent and the performance of the spacecraft. Tereshkova did not respond immediately, but collected her thoughts on the ascent. 'The flight into orbit is over, and the pressure disappeared as it melted under the warm wave spreading over my body. Breathing becomes easier. I open my eyes and look out of the [porthole]. In a loud noise I comment about what I see: 'I am Chaika. I see the horizon. There is a blue stripe. This is the Earth. How beautiful it is! Everything is going well. Hello Universe."

In her official onboard journal she recorded that the ascent phase was completed without much agitation, and that the transfer from boosted flight to weightlessness was very smooth, with no sharp deviations – probably because she did not notice, as she was busy observing the small dust particles that had begun to float all around her. She also recorded: 'In the state of weightlessness you feel a lightness, workability not impeded, mood always cheerful.'[12]

The flight programme

Vostok 6 completed forty-eight orbits of the Earth over a period of more than 70 hours. Tereshkova's flight was aimed mostly at biomedical objectives, and described as a 'Comparison of the effects of spaceflight on men and woman; a continued study of the effects of spaceflight on the human organism; new medical and biological research studies, and the flight of improved systems of manned spacecraft under conditions of a dual flight.' Data were recorded during the flight, both onboard and on the ground, and during post-flight debriefings Tereshkova reported on her activities and condition. Official reports during and immediately following the flight reflected the official line that all had proceeded according to the flight plan, but over the years there has been speculation that the flight plan did not proceed as smoothly as suggested by the formal (and propaganda-orientated) press releases at the time of the mission.[13]

Dual flight

The flights of Vostok 5 and Vostok 6 had been planned so that they would be more of a propaganda coup than a significant development in orbital rendezvous, although they offered the second opportunity to simultaneously monitor two separate manned spacecraft in orbit. Alhough the two cosmonauts thought that they spotted each other's spacecraft, there were no confirmations, though Tereshkova clearly stated that she saw the spent third stage of the launch vehicle shortly after she entered orbit. During the flight, the two cosmonauts conducted joint communications sessions and joined each other in singing patriotic songs. At one point in the mission, Bykovsky had to relay information to and from Tereshkova when direct communication with Vostok 6 seemed difficult.

Observations

More than forty years later, one of Tereshkova's most memorable recollections of her flight was the view of Earth passing below her. At first, as to many Earthlings, the planet seemed huge; but after she had orbited it several times every 90 minutes she realised how small and fragile it was – an impression of many explorers have who have benefited from looking upon the Earth from space:[10] 'From space, the beauty of Earth was overwhelming . . . I realised how small Earth is, and how fragile, so that it can be destroyed very quickly.'[14] During her flight she filmed and photographed cities, clouds, the Earth and the Moon, and observed weather patterns, though at times she found it difficult to change film cassettes. She recorded images of the Earth with a Konvass camera – used to study weather phenomena and the structure of the atmosphere, including the twilight corona.

Habitability

Vostok 6 internal parameters were recorded at a pressure of 754–777 mmHg, with 34% humidity, a temperature varying between 18° and 22°.6, and 20% oxygen in the atmosphere. Tereshkova did not experience any unpleasant feeling during weightless flight, but the constant wearing of the pressure suit throughout the flight did trouble her as the mission progressed. During the first day the suit felt comfortable, but during the second day the right knee area was becoming uncomfortable, and by the next day the discomfort had considerably increased. The constant wearing of the helmet (in the event of cabin decompression) was also a bothersome requirement , as it pressed against the shoulder and the ears. Also, although the medical harness was not a problem, its sensors caused itching and headaches. Many years later, Tereshkova said that she was young and it was her dream to fly in space, and that she 'did not care about the discomforts ... It was nothing to do with being a woman ... The difficulties did not hinder me ... This is the same for women as for men.'

Personal hygiene

Tereshkova reported that it would have been nice to have something with which to clean her teeth during her flight, as the sanitary napkins provided were not moist enough, and were too small. The flights of Vostok – as with all early manned spacecraft – were pioneering missions, and personal comfort was, essentially, basic. There were hardly any provisions for crew comfort, as the primary aim was to sustain life and to safely achieve the objective of the mission. It was only as spacecraft became larger and space stations allowed longer missions that home comforts were incorporated.

Medical

Medical tests were performed using electrocardiography, pnuemography, electroculography, kinetocardiography, electroencephalography and skin galvanic responses. Tereshkova at first reported that she was feeling well on the flight, and the flight period was increased from an initially planned 24 hours to a full three days. Discussions were held with the cosmonaut to determine whether she was capable of completing the mission, and she assured the ground that she would carry out the whole flight programme. But there were reports of her illness in space – including rumours of her being carried from the spacecraft. She reported that some of the food made her ill and caused at least one instance of vomiting, but it is also more widely suspected that she was suffering from space adaptation syndrome (first experienced by Gherman Titov two years previously), which affects one in three space explorers in their first 48 hours in space. She was exposed to 25 mrads of radiation during the mission, but this was not significant, as higher radiation levels had been encountered in early missions. Her reproductive system was not affected by the spaceflight – as proven by the birth of her only child, a daughter, a year after the mission. And her daughter later married, and gave birth to a healthy son. Tereshkova's heart-rate during the time she was awake was recorded at 64–82 beats per minute.

Exercise

Exercise was emphasised, to prepare her for the stress of landing.[15] This took the form of stretching on the first and second day of flight, and was increased on the third day. Her perspiration was handled by the onboard ventilation systems.

Sleep

The planning allowed for six-hour sleep periods during the mission. Sleeping in the seat and wearing the suit was the only option, and Tereshkova reported, during and shortly after the flight, that she slept well and did not oversleep. However, many years later she revealed that she had slept a little longer on one occasion, and had missed a call from the control centre (possibly adding to the concerns for her health and the communications systems). Her pulse rate during sleep was recorded at 52–60 beats per minutes.

Food and drink

Tereshkova was provided with four Earth-like meals and 1.5 litres of water for each flight-day, and her rations amounted to 2,529 kcal. She admitted that she was not fond of the sweet dishes, which apparently caused a vomiting session during the flight. She found the bread very dry, but enjoyed the meat dish and the juices, although she longed for the more traditional Russian dishes. She apparently did not eat all of her prescribed meals, and reportedly gave away some of the food upon landing – which must have affected the accuracy of some of the medical results from the flight.

Experiments and results

Although the flight of Vostok 6 lasted three days, there were few onboard experiments. The investigation of the human body's adaptation to launch, orbital flight and re-entry was the most important consideration, with Tereshkova providing the first set of biomedical data on a female. In addition, observation of the Earth was a primary objective, together with photographic studies of the Earth's horizon and early studies of the ozone layer in the atmosphere. In her discussions with A. Lothian, Tereshkova said that on later missions, aerosol layers had been found at altitudes between 11.5 and 19.5, ±1 km, and that scientists had referred to the photographs taken on her flight to collate the findings with balloon and aircraft studies. The study of photographs taken during the early years of the programme (such as on Vostok 6 in 1963) were used to understand ozone layering and depletion over the next few decades of the space age. Although Tereshkova was able to observe the Moon on several occasions, she had difficulty in identifying the solar corona, and was unable to observe it as planned. Due to the restrictions in the pressure suit and crew compartment she was also unable to activate the package of biological experiments (including *drosophila* flies) during the flight.

The seagull lands

Vostok 6 was programmed to re-enter and land before Vostok 5. During the entire re-entry phase on 19 June, Tereshkova remained silent, causing some concern on the ground, although all was well. Ejecting from the spacecraft at an altitude of 6.5 km, she completed a safe parachute landing 620 km north-east of Karaganda, Kazakhstan, at 11.20 Moscow Time, completing her forty-eight-orbit mission in 2 days 22 hrs 50 min. Bykovsky landed safely, 800 km from Tereshkova's landing site, less than three hours later.

During the descent Tereshkova noticed that she was heading towards a lake, and though she had trained for a splash-down she was not looking forward to it. Fortunately, gusts of wind carried her over the lake towards a field. With winds at 17 m/s, the landing was not gentle, and she suffered some bruising from the helmet as she landed, in addition to those she had incurred during descent when a piece of metal cut her on her nose as she looked up into the canopy of the parachute. After landing she briefly enjoyed the warm sunny day, quickly removed her pressure suit, and opened a nearby container (that had landed with her) to change into a tracksuit, to allow more movement when collecting the equipment.[16]

Celebrations in Moscow following Tereshkova's flight.

With the help of local farm-workers she gathered the suit, parachute and ejection seat close to the capsule – which had landed 400 m away – and then asked to be taken to the nearest village to telephone officials and inform them of her safe landing. 'As you can see, the final descent stage of the flight was arduous, physically and mentally, and nobody helped me. I did it all myself. Obviously I could not have done it if I was not feeling strong', she told A. Lothian in a later interview.

Back at the landing site she awaited the rescue team, who took her to Karaganda for an overnight stop before flying on to Kuibishev for a series of medical tests and an official address to the State Commission to report on the successful conclusion of her mission. The official return to Moscow, to deliver the report to the Soviet leaders, took place on 22 June. Wearing make-up to hide the bruises on her face, she was reunited not only with Party officials but also with her family. During selection and training, none of the women could reveal the true nature of their assignment, and her mother, in tears, repeatedly said that Valentina had deceived her. She thought that her daughter was undergoing special parachute training, and she did not know about the flight until it happened. Despite the success of the flight, and Tereshkova's subsequent fame, it was a long time before her mother forgave her.

A second spaceflight?

Following the mission, some officials reported that Tereshkova performed poorly, and this contributed to her not making a second flight. Tereshkova has always denied this, and during an interview conducted during the fortieth anniversary of Gagarin's flight she refuted the reports as absolute nonsense,[17] although she did say that the flight was very difficult and a terrible strain on her body, especially during landing. Her flight and her performance were, however, judged against the performance of her male colleagues. When she was asked during the interview (in April 2001) if she wanted a second flight, she replied that she 'wanted very badly to be in space once. You might say I was eager to go again. In 1965 they discussed a possible second flight, but it did not happen.' As the years slipped by, her role on international women's committees and in public relations work expanded and became a new 'career', of which she was particularly proud.

Whether this discussion of a second spaceflight for Tereshkova related to her role as training manager for an all-women Voskhod mission, including an EVA, was not made clear. A photograph of Tereshkova in a Voskhod EVA suit has been released, but it is probably a posed portrait rather than a photograph taken during a training session. For publicity purposes Tereshkova has posed in a Soyuz Sokol suit (used since 1973), but although completing some work on Voskhod and during the early Soyuz programme during the period 1964–69, she did not train specially for a flight as a prime crew-member. All the women of the 1962 group attended and graduated from the Zhukovsky Air Force Engineering Academy prior to the group being disbanded on 1 October 1969. It would be almost a decade before any other women would be selected for spaceflight training, though for a while it seemed that a second spaceflight with female crew-members was a strong possibility.

Voskhod 5 and a female EVA?

With the first flight of the more advanced replacement spacecraft (Soyuz) delayed until late 1966 at the earliest, the Soviets evolved a series of flights, using modified Vostok hardware, to compete with America's Gemini series of missions involving space rendezvous and docking, long-duration flights, and spacewalking. Called Voskhod, the first mission with three crew-members flew in October 1964, and a second in March 1965 saw the world's first EVA only a few days before the first manned flight of the Gemini programme. Although several other missions were planned, none flew. One of these was a flight by two women which included an EVA, and possibly including the flight test of an early Manned Manoeuvring Unit to increases the propaganda effect.

Called 'Voskhod 5' in the West, this mission originated in early 1965, with crews assigned on 17 April. Ponomaryova was assigned as Commander, with Solovyova as the Pilot who would complete the EVA – essentially repeating the mission which Voskhod 2 had completed the previous month. Four male cosmonauts were assigned as the first back-up crew (Zaikin and Khrunov) and second back-up crew (Shonin and Gorbatko). However, opposition arose from the male cosmonauts – who considered that the four men were far more qualified to fly the mission, especially since Khrunov had backed up Leonov for his spacewalk – and from the other three women, who thought that Ponomaryova was not suitable for assignment as a Commander for the same reasons that she was not assigned to Vostok 6. These

Zhanna Yorkina.

objections were overruled, but conditions for flying an all-women crew were not favourable.

Kamanin has also stated that he did not think that either Kuznetsova or Yorkina would be ready to fly before 1970 at the earliest. One of the more serious difficulties for the female EVA was that Zvezda (the suit manufacturer) was opposed to an all-female flight, and refused to fabricate a special EVA suit. As the months progressed there was an increase in support for longer flights, or cancellation in favour of moving on to Soyuz. By early 1966 the State Commission had indicated its support for a 15–20 day mission by the two female cosmonauts, but both Kuznetsova and Yorkina were now assigned as back-ups on Voskhod 4, replacing the male cosmonauts now assigned to a later two-week EVA mission (often termed 'Voskhod 6'), and probably the result of discussions in letting the all-woman crew still fly. Apparently, little training was carried out, although Ponomaryova would have used the call sign 'Silver Birch' had she flown. Without warning, the group was sent on holiday and the mission was cancelled and forgotten.

Shortly after her flight, Tereshkova was shown a copy of Jerrie Cobb's book about her flying career and her exploits in trying to fly on a Mercury mission a few years earlier. Tereshkova was aware of Cobb's talents as a pilot, and found it amusing that Cobb had authored a book about being a 'woman in space' without actually making a flight herself.[18] The two women meet briefly during the 56th International Astronautical Federation Congress held in Mexico City during October 1963. During a post-flight tour, when she was accompanied by Bykovsky and Yuri Gagarin,[19] she responded in typical Party fashion to a question on the influence of her mission upon women of the world: 'Since 1917, Soviet women have had the same prerogatives and rights as men ... They are workers, navigators, chemists, aviators, engineers, and now the nation has selected me for the honour of being a cosmonaut ... As you can see, on Earth, at sea, and in the sky, Soviet women are the equal of men.' This equality, however, apparently did not extend to further consideration for females to be selected for spaceflight for the next two decades.

FOLLOWING VALENTINA

Irrespective of the political intentions of the mission, the lack of science, and the absence of major advances in the techniques of human spaceflight, the flight of Valentina Tereshkova was nevertheless a major milestone in pioneering space exploration and a landmark for women's future participation in space. But it would be another twenty years before the role of women in supporting the space programmes of several nations was assured – with their increasing skills in medicine, science, engineering and technology, as well as in the more traditional roles of administration and media work.

Tereshkova's flight proved that a woman could, as well as a man, survive the stress of launch, sustained orbital flight, and violent re-entry and landing. Although the barriers of politics, qualifications and opportunity would still prevent women from training for spaceflight until the advent of the Space Shuttle and the growth of space stations, many would eventually follow Valentina on the road to orbit. But no matter how many take the journey, Valentina Tereshkova will forever be known as the *first* woman to leave Earth and fly into space.

REFERENCES

1 David J. Shayler and Rex D. Hall, *The Rocket Men*, Springer–Praxis, 2001.
2 http://www.astronaut.ru; http://www.spacefact.de; also, information supplied by Tony Quine.
3 Isaak P. Abramov and A. Ingemar Skoog, *Russian Spacesuits*, Springer–Praxis, 2003, pp. 31–58.
4 Correspondence from Tony Quine, 25 October 2004.
5 Personal research on the preparations for the Vostok 6 female cosmonauts has been compiled from official sources – notably Kamanin's diaries – by Tony Quine, and is cited with permission.
6 A. Lothian, *Valentina: First Woman in Space*, Pentland Press, 1993, pp. 224–248.
7 'Russia First?' and 'Spaceflight by Russian Woman', *The Sun* (Australian newspaper), undated. (Courtesy Colin Burgess.)
8 'Will Space Girl be Next?', UPI/AAP reports, Australian news reports, 15 June 1963. (Courtesy Colin Burgess.)
9 Private research notes based on a variety of sources including Kamanin's notes, by Tony Quine, regarding the launch of Vostok 6.
10 David J. Shayler, conversation with Valentina Tereshkova, Coventry, 10 October 2004.
11 Valentina Tereshkova, *The Stars are Calling*, Moscow 1964, English translation; 'The First Lady of Space Remembers', *Quest*, **10**, 2, 2003, pp. 6–21.
12 F. Lushnikov, 'An Exploit Which Humanity Will Not Forget', *Red Star* (newspaper), 14 December 1963, trans. USAF Wright–Patterson Air Force Base, Dayton, Ohio.
13 Ref. 1, pp.206–214; Asif Siddiqi, *Challenge to Apollo*, NASA SP-2000-4408, 2000. pp. 361–373.
14 Ref. 6, p.232.
15 Nicholas Johnson, 'Summary of Vostok Flight Experiments, Handbook of Soviet Manned Space Flight', *American Astronautical Society*, **48**, Science and Technology Series, 1980.
16 Ref. 6, pp. 234–238.
17 Marina Uvariva, 'On the Heels of Gagarin', *Moscow Times*, 26 April 2001; http://www.themoscowtimes.com/stores/2001/04/06/102-print.html.
18 Ian Moule, conversation with Valentina Tereshkova, Coventry, 9 October 2004.
19 Mitchell Sharpe, *It is I, Seagull*, 1975, pp. 132–133 and 145–146.

The Right Stuff, the wrong sex

In the heat and humidity of a July night, a specially invited audience patiently waited, for the third time that week, to witness a night launch of a Space Shuttle from NASA's John F. Kennedy Space Center. But this would be no ordinary launch, for sitting in the prestigious left-hand seat of the spacecraft was USAF Colonel Eileen Collins – the first female to be appointed Commander of a Space Shuttle mission. Originally scheduled to coincide with the thirtieth anniversary of the first landing on the Moon, fate had intervened yet again to frustrate the progress of women into space.

Among the spectators waiting to witness this historic event was a lean, elderly woman with honey-blond hair: Jerrie Cobb. If events had worked out differently forty years earlier she would have been the first woman to travel into space onboard a 'pilotable' spacecraft; but she had had first-hand experience of the fickle nature of fate.

THE SEVEN MERCURY ASTRONAUTS

On 4 October 1957 the Soviet Union amazed the world (especially the US) by placing the first man-made object, Sputnik, into space. Coming as it did at the height of the Cold War, emotions were heightened, and the American public was very much afraid that the orbiting Russian satellite could easily become a new and terrifying weapon of war. Control of the skies was of paramount importance, as it was widely believed that whoever controlled the high ground of space would control the world. Consequently, the US and the Soviet Union became locked in a 'space race', with the initial goal of sending the first human into space.

In America the task of trying to ensure that that first human was an American fell to the country's fledgling civilian space agency, NASA (National Aeronautical and Space Administration), which officially began operating on 1 October 1958. To assist in its quest to find potential occupants of 'tin can' capsules, NASA was loaned a number of eminent scientists from the military, who had been working on the human aspects of high-altitude research.

Dr Randy Lovelace was the senior Life Sciences adviser to the Administrator of the newly formed NASA, and was therefore involved in establishing the medical, psychological, physiological and behavioural criteria that were in the initial selection of what were later called 'astronauts'. Brigadier-General Don Flickinger was the senior Air Force representative on Dr Lovelace's Advisory Committee for Life Sciences, and together they brought to NASA the initial Life Sciences staff on loan from the military organisations, including Drs Augerson, Voas and White, within the first month of its existence.[1]

On 17 December 1958, NASA Administrator T. Keith Glennan publicly announced that the US programme to place a human into space would be known as Project Mercury. One month later, on 5 January 1959, NASA finalised its criteria for the initial selection of the first US astronauts, which were 'a direct extension of the selection criteria that had been used in the screening of aircrews for the high-altitude research and operational flights using special aircraft.'

To qualify as a potential Mercury astronaut, in addition to serving in the military, a candidate would have to meet certain criteria. (The decision to use military test-pilots came from wanting to select from a group whose profession required dangerous flight. The fact that they were still alive at selection time testified to their ability to survive in this hazardous profession). The criteria were as follows:[2]

- Less than 40 years of age.
- Less than 5 feet 11 inches tall (to fit within the dimensions of the spacecraft).
- In excellent physical condition
- Holding at least a bachelor's degree or the equivalent.
- A graduate of test-pilot school.
- A qualified jet pilot with at least 1,500 hours total flying time.

These criteria eliminated many groups which had been initially considered as potential sources of crew – such as commercial pilots, general aviation crew-members, divers, racing-car drivers, and many others with hazardous occupations. The first three of the above criteria proved to be the most difficult to meet in the same person. For example: test-pilots at that time tended to be tall, and over 35 years of age. Reducing the age limit to 30 reduced the available pool of more than four hundred people to less than fifty. The experience level in this small group was limited, but raising the age cut-off to 40 increased the pool to more than over 150.

President Eisenhower's decision that NASA should use only military test-pilots also meant that women were eliminated as a potential source of astronaut candidates.

In January 1959, 110 men were chosen from the 508 service records provided by the US military. Five were from the US Marine Corps, forty-seven were from the US Navy, and fifty-eight were from the US Air Force. The 110 chosen were then divided into three groups, for briefings and interviews by White, Voas and Augerson. However, having screened the first two groups, and based upon the high rate of volunteering, NASA realised that it had more than enough potential astronaut candidates, and therefore cancelled invitations to the third group.

By March 1959, NASA's pool of suitable candidates had been reduced to thirty-six men, with thirty-two of them volunteering to undergo the detailed physical examinations ro be conducted (under a NASA contract) by Randy Lovelace's Lovelace Clinic in Albuquerque, New Mexico. The Lovelace Foundation was chosen for the initial aspects of the physical examination of candidates because this was a civilian programme, and NASA wanted to be sure that it would not be misinterpreted by other nations as simply an extension of the previous military programmes. The Lovelace Foundation also had an excellent reputation for conducting medical examinations for special civilian test and high-altitude record flights.

During their stay at the Lovelace Clinic, the volunteers were subjected to a battery of tests aimed at assessing their general health and fitness. Only one of the volunteers failed the tests, and the remaining thirty-one were then moved to the Wright Air Development Center (WADC), where they underwent stress testing. This was one of the first instances of using treadmills for testing by loading on the lungs to determine their capacity for handling stress. The tests also involved the centrifuge, thermal loads, and psychological tests. As the volunteers completed one cycle, the medical staff increased its complexity in order determine not only their capability, but also how they solved problems, and the reasons for failure.[3]

At the end of the WADC testing, a NASA selection committee (which included Dr White) reviewed all of the test results in an attempt to select the required six astronauts for Project Mercury. However, although, in agreement, they nominated five of them, and were supposedly limited to six, there were two others who were equally as good, and so they increased the first group to seven.[4] These seven, introduced to the press on 9 April 1959, were US Navy Lieutenant Malcolm Scott Carpenter (35); USAF Captain Leroy Gordon Cooper (32); US Marine Corps John Herschel Glenn (37); USAF Captain Virgil Ivan 'Gus' Grissom (33); US Navy Lieutenant Commander Walter Marty Schirra (36); US Navy Lieutenant Commander Alan Bartlett Shepard (35); and USAF Captain Donald Kent 'Deke' Slayton (34).

The Mercury astronauts subsequently flew two sub-orbital missions in 1961 and four orbital missions in 1962–63, and were the first to manually control their spacecraft.[5]

Dee O'Hara: nurse to the astronauts

Lieut Dolores 'Dee' O'Hara, USAF – the Mercury 7 astronauts' personal nurse – exemplified the accepted role of women during the Mercury flights. At that time (1959–60) a woman usually became a secretary, a nurse or a teacher. There was the rare female engineer, but women were not really guided or encouraged to go into other careers.[6]

Dee O'Hara was born in Nampa, Idaho, on 9 August 1935. Although her family were in the low income bracket, and she never really had many special opportunities, Dee had a very normal upbringing. However, tragedy struck two weeks after she graduated from Lebanon High School, Oregon, when her father was killed in an accident at work. Being strong-willed, and having a lot of self-pride, Dee did not want her mother (who had worked very hard to put her through school) to support her. Consequently, she went into nursing. 'I had always planned on being a teacher or social worker, but I gained my inspiration to go into nursing during my senior year at Lebanon High. A Career Day was held with people from different professions, and one of the speakers was a very elegant lady from the Providence Hospital School of Nursing in Portland, Oregon ... Nursing had never occurred to me, as I would feel faint if I had to go into hospital, and became ill at the site of an accident. But she looked so pristine in her white uniform and nurse's cap, and she really made an impression on me. She handed out brochures of the nursing school and I thought the school was impressive. Then, as a clincher, I noticed a classmate at the back, and I thought, 'Well, if Luanne can do this, so can I ... I was selected for entrance, and the minute I walked into the school I knew I had made the right decision – one of the best of my life.'[7]

O'Hara duly graduated from the nursing school. Then one day, not long after graduating, her room-mate, Jackie McMahan, suggested that they both 'join the Armed Forces and see the world.' Her initial response was, 'Nice girls don't do that!' (she was 22 or 23 at the time); but she later went to the recruiting station and said: 'Well, here we are. Where do we sign? Of course, the recruiting sergeant met his quota for the month when we walked in, and he was thrilled.' Jackie McMahan went on to Mobile, Alabama, and in May 1959 Dee O'Hara went to Patrick Air Force Base (PAFB). Having completed the USAF flight nurses' course, O'Hara was called in by the hospital commander, Col George M. Knauf, MD, to talk about NASA and Project Mercury. She had no idea what they were talking about. They mentioned astronauts, but she did not know what that was, and when they said 'Project Mercury', she thought: 'Now, there's a planet named Mercury, and there's mercury in a thermometer' – and that was the extent of her knowledge. When they asked if she wanted the job of astronaut nurse, she accepted it without really knowing what it was.

O'Hara was assigned to NASA in November 1959. Her duties included setting up and coordinating an eight-room Aeromed Laboratory at Cape Canaveral, which would serve as an examination area for the Mercury 7 astronauts, and she also assisted the astronauts' physician, Dr William Douglas; but she was mostly with the astronauts as 'their nurse'. However, she was not the only USAF nurse assigned to work with the Mercury 7 astronauts. Lieut Shirley Sineath worked on the 'recovery team', and during a launch Sineath acted as the surgical nurse and O'Hara served as the intensive care nurse at the Cape's medical station.[8]

On the day of launch, O'Hara – who assisted with the pre-flight physicals – had mixed emotions. Despite her 'faith and confidence' in both the hardware and the astronauts, she was 'always quite afraid when they launched.' She was a devout Catholic, and would always say the Rosary during a mission. 'It was like putting one of my best friends on a Roman Candle. When they returned from a mission it was the best time for me. I knew they were back safely.'

In 1962 NASA began construction work on its new Manned Spacecraft Center in Houston, Texas. With the personnel working on Project Mercury making arrangements to move to Houston, O'Hara was asked by a couple of the Mercury astronauts to go with them. This placed her in something of a dilemma, as she was still a USAF nurse. But having thought about it, and wanting to remain with the manned space programme, she resigned her commission. She then spent a couple of months out of work until the Manned Spacecraft Center was completed in March 1964, and then received a telephone call inviting her to set up the Flight Medicine Clinic.

Deemed by many people to have had one of the most enviable jobs in the world, O'Hara felt fortunate to have been a part of a unique and exciting time in space history. However, she did have one regret. 'As I look back, I wish I had been more interested in school. If I had studied harder in maths and science it would have made it all easier. If I had studied the right courses I would have had an easier time once I got to nursing school, although I never felt that I was discriminated against. As to being selected to work with the first astronauts and the space programme, I just happened to be in the right place at the right time!'

A 'GIRL ASTRONAUT' PROGRAMME

Much has been written about the early testing of women as potential astronauts for Project Mercury, but is was not until recently that the real sequence of events, and the rationale behind them, was known. Between 1998 and 1999, Dr Margaret Weitekamp spent a year in residence at the NASA History Office in Washington, as the NASA/American Historical Association Aerospace History Fellow. During her time with NASA, she was able to piece together the story of the early, abortive, tests on women as possible astronauts.

There was a programme called Project WISE (Women in Space Earliest), or, as it was sometimes known, WISS (Women In Space Soonest), instigated by USAF Brigadier-General Donald D. Flickinger, MD, (who simply referred to it as the Girl Astronaut programme).[9]

Having been involved in the selection of the Mercury 7 astronauts, Dr W. Randolf 'Randy' Lovelace II wanted to conduct a similar study on women to establish whether they could pass the same selection tests without the need to modify them. Sharing Lovelace's enthusiasm for conducting 'fitness for space' tests on women was Brigadier-General Flickinger, who, in addition to being the senior USAF representative on Lovelace's Advisory Committee for Life Sciences (which established the selection criteria for the Mercury 7 astronauts), was Chief of Human Factors research at the Air Research and Development Command (ARDC) headquarters in Baltimore. As one of the few advocates within the USAF for 'spaceflight' research, he decided to set up a women's astronaut testing programme under the auspices of the ARDC. What was required was a pool of suitable women pilots and, at the Air Force Association meeting at Miami Beach in September 1959, they found, by chance, a suitable 'guinea pig.' Walking towards them as they emerged from a pre-breakfast swim, was the accomplished female aviator, Geraldyn M. 'Jerrie' Cobb.

Jerrie Cobb was born in Norman, Oklahoma, on 5 March 1931, and began her flying career at the age of twelve when she was taught to fly by her Air Force father, Lieut-Col William Harvey Cobb, in his open-cockpit Waco biplane. Soloing on her sixteenth birthday (the minimum legal age), while still a student at Classen High School in Oklahoma City, she made the conscious decision to go into professional aviation as a pilot, which at that time was not in keeping with the gender-assigned roles for women. She gained her Commercial Pilot's Licence on her eighteenth birthday, and went on to become a Certified Flight Instructor by the time she was twenty-one. During this time she started work as an international ferry pilot – a job which would see her flying USAF military aircraft, such as the T-6 'Texan' and the B-17 bomber, to foreign governments around the world. However, by the time of her meeting with Flickinger and Lovelace in 1959, she had changed jobs, and was working as part of the management team for the small aircraft manufacturing firm Aero Commander. She was also the holder of four world aviation records – one for speed, one for distance and two for altitude, all of which had been achieved in a twin-engine, propeller-driven, Aero Commander – and had logged more than 7,000 hours flying time. By the end of 1959 she had also broken the sound barrier in a USAF TF 102 Delta Dagger at Tyndall Air Force Base in Panama City, Florida.[10, 11]

During her preliminary conversation with Flickinger and Lovelace, Cobb learned that they had only just returned from Moscow, where they had attended a meeting of space scientists. Then, upon hearing about her accomplishments as a pilot, Flickinger informed Cobb that he was very interested in female aviation achievements, and that the USAF had recently constructed a pressure suit for the famed French female pilot, Jacqueline Auriol. Cobb's companion, Tom Harris, Sales Manager for Aero Commander, responded by telling Flickinger that he should make one for *her*, as she was 'liable to try for a record in space.' Lovelace then told Cobb that at the meeting in Moscow, the Russians had indicated that they planned to send women into space. At this point the conversation paused, and the three of them agreed to meet again and continue their discussions. Reconvening in the public rooms of the Fontainbleau Hotel, Flickinger and Lovelace began questioning Cobb about female pilots in the US. They then informed her that 'medical and psychological investigations had shown that women were more capable than men of withstanding pain, heat, cold, loneliness and monotony' and, as a result, they were looking for a pool of suitable female pilots to undergo the same selection tests as the Mercury 7 astronauts.[12] Cobb was then asked if she would be interested in being a test subject. With her excitement growing, and feeling that she was in the right place at the right time, Cobb confirmed her interest. Then, after telling her that they would contact her after they had checked that her aviation and medical records were satisfactory, Flickinger and Lovelace left.

Three months later, Cobb received a letter from Flickinger (dated 7 December 1959), but its content was not what she was expecting. In the letter – one of two key documents uncovered by Dr Weitekamp – Flickinger informed Cobb that the ARDC had withdrawn support for the 'Girl in Space' programme. Expressing his own deep disappointment in the change of events, Flickinger offered the following reason for the ARDC's decision: 'The unfortunate 'Nichols' release did much to 'turn the tide' against Air Force medical sponsorship of the programme, and to this day I cannot find out the individual responsible for approving the release.'[13]

Flickinger was referring to the 'physical testing for space' of the famed female pilot and aviation pioneer, Ruth Nichols, at the USAF's Wright Air Development Center (WADC).

Ruth Nichols and the WADC 'astronaut tests'

Ruth Nichols was born in New York City in 1901, and made her first flight (as a passenger) in 1919. Although this was a present from her parents for graduating from High School, she was very scared of flying, and therefore viewed her $10 flight in a JN-4 'Jenney' First World War trainer aircraft as a challenge. However, having experienced flying – her flight included one loop, during which she kept her eyes very tightly closed – she realised that there was nothing to fear.[14] Soon after her graduation flight, she and her parents visited her two brothers, who were at school in Palm Beach, Florida. They asked her if she would like to go for ride in a flying boat, to which she agreed, and it was as a result of this particular flight that she decided that she wanted to learn to fly.

Nichols undertook her first training flight in 1922, and first flew solo in 1924. Then, in 1929 she spent six months publicising the Aviation Country Clubs, during which time she flew 12,000 miles and landed in forty-six American States. In 1931 she followed this achievement by setting three women's world flying records for altitude (nearly 29,000 feet), speed and distance in a Lockheed Vega. Her next goal was to become the first woman to fly solo across the North Atlantic, and her record-setting flights had been a means of gaining publicity to help finance it. Fate intervened, however, and as a result of an injury to her back, sustained during a crash, followed by problems with the weather, she was beaten to the record by Amelia Earhart. Undeterred, and against her doctor's orders (and with the aid of a 'steel corset' to support her healing back), Nichols set a cross-country record by flying from Oakland, California, to Louisville, Kentucky. In 1932 she became the first woman pilot to fly a passenger aircraft, and in 1935 she was once again injured in a crash at Troy, New York. Nichols continued to play an active part in flying – having undergone a 'flying at altitude' lecture course at Mitchell Air Force Base, and a 'reactions' test in an altitude chamber – and achieved a women's altitude record of 51,000 feet in a supersonic Air Force jet in 1958. Towards the end of 1959, when she was 58 (and 21 years older than John Glenn, the oldest Mercury 7 astronaut), she sampled some of the astronaut tests at the USAF's WADC.

Nichols considered that of the astronaut tests which she undertook, three of them were of particular interest: weightlessness, isolation, and the centrifuge. The original plan had been for her to experience weightlessness in a KC-135 aircraft, but when the time for the test arrived, the aircraft was out of commission. Consequently, she had to use a simulator, consisting of a platform, suspended off the floor by jets of steam, and steered by means of a hand-held gyroscope. 'They handed me this very heavy – I should think it weighed nearly 50 lbs – gyroscope, and you were supposed to steer yourself somewhat by how you angled it.'

Nichols also noted that the WADC scientists and technicians appeared far more concerned about the safety of the gyroscope than her own safety. However, despite her previous crash injuries (a broken back and a broken leg), she was able to overcome the strain of holding the 'heavy' gyroscope, and found that she could easily control her movement in any direction. She also alluded to having tried to simulate tightening a screw in space – a feat that she could achieve only by means of a 'hand hold' on the outside of the so-called 'space-ship'.

The isolation test was a sensory deprivation experiment which involved her being placed inside a completely darkened room. To help her endure this psychological test (being cut off from the rest of the world), she employed the same mental processes as those she used to survive a ditching in the Irish Sea.

The centrifuge test was a capsule attached to the end of a long girder-like arm, used to mimic the g forces that an astronaut would be subjected to during the launch and landing phases of a space mission. This was achieved by spinning the capsule and its occupant around in a circle at varying speeds. Nichols had no problems with this test, and likened it to the sensations she had experienced during the execution of the loop on her graduation flight.

Having successfully completed the first female 'fitness for space' tests under scientific scrutiny, Nichols urged that women be used in space flights. But the scientists at WADC said 'no' – on the basis that they did not have the prerequisite physiological data. Nichols considered this to be an extraordinary statement: 'I therefore suggested a crash programme to determine how a female reacted, because women are organically meant to withstand a crisis in childbirth, and a woman is more passive than a man, and could therefore endure long isolations. From every viewpoint a woman could hold her own in a space situation and be of tremendous service.'

Unfortunately, the results of Nichols' tests were released prematurely, with respect to Flickinger's Girl Astronaut programme, and implied that the Air Force was interested in promoting a woman astronaut, when they had no intention of doing so. However, at about the same time as Nichols was carrying out Air Force-sanctioned astronaut tests, another woman, Betty Skelton, was carrying out NASA-sanctioned tests.

About four or five months after her tests, Nichols discovered that a very attractive picture of Betty Skelton had appeared on the cover of *LOOK* magazine: 'They must have picked up the idea, and decided that it wasn't such a bad one. It was supposed to be a hush-hush policy to look for some guinea pigs, but they set the age so very low that it would leave me out. That was interesting too, because I believe that one should consider physiological rather than chronological age, and I am very blessed with having a good constitution, so that my cholesterol count is low in my bloodstream, as well as blood pressure and all the other physical conditions that are necessary to meet a commercial pilot's requirements. So I may not be the first one, but I hope that I will some day have the opportunity of being in space.'

Interestingly, Nichols' remarks appear to indicate that she was aware of Flickinger's programme, although it had no connection with Betty Skelton's tests. Sadly, however, Nichols would not be able to champion the cause for women astronauts, as she died (perhaps by suicide) in September 1960.

Betty Skelton and the 'astronaut tests' for *LOOK* magazine

The appearance of Betty Skelton – a pilot and aerobatic champion – on the cover of *LOOK* was regarded as a publicity stunt. The magazine had contacted her to enquire if she would be interested in undergoing fitness for space training using the same facilities that NASA was using to train its Mercury 7 astronauts. Skelton had seen a picture of the Mercury 7 astronauts in a newspaper the day after they were introduced to the press, and had thought: 'Wow, wouldn't that be wonderful to be one of those people'. She therefore agreed to the offer. As a result, the front cover of the February 1960 edition of the magazine featured a colour photograph, taken at the McDonnell Aircraft Corporation in St Louis, which showed her wearing a prototype silver Mercury spacesuit in front of a mock-up of the Mercury capsule. Accompanying the picture was a caption that posed the question: 'Should a Girl Be First in Space?'[15]

Betty Skelton was born in Pensacola, Florida, in 1926, and, like Jerrie Cobb, had started flying at an early age, although in Skelton's case she had begun pushing the limits for flying while still a schoolgirl. 'I soloed when I was twelve, and it wasn't quite legal then, so I couldn't tell anybody. But I figure now, about fifty years later, nobody will bother me about it. I fibbed to Eastern Airlines about my age, and got a job about three weeks before I graduated from high school. Then I would work from midnight to eight so that I could fly during the daytime. I became a flight instructor, and got commercial and other ratings.'

In addition to becoming a professional aerobatic pilot (which happened purely by chance), Skelton also owned the second Pitts Special aeroplane ever built, which she named Little Stinker. (This aircraft now resides in the National Air and Space Museum.) She also became the first woman to cut a ribbon with the propeller of an aeroplane while flying upside down ten feet above the ground. In 1948, 1949 and 1950 Skelton held the title of International Feminine Aerobatic Champion. However, pushing the limits for flying was not her only interest, as she was doing the same on the ground by racing cars and in the water by jumping boats. One of her ground-based achievements was in becoming the first woman to achieve a land speed of more than 300 mph.

It was this experience of pushing the limits (coupled with her mental discipline and fast reflexes) that made Skelton the ideal candidate for *LOOK* magazine. Her mental aptitude and drive can be seen in her response to a question in her NASA Oral History interview regarding practice for aerobatic flying: 'I would go up high, to be safe, and would do maybe the same manoeuvre a hundred times in one practice setting, and I'd practice two or three times a day, not too long at a time, because you get very tired.'

According to the feature in *LOOK* – entitled 'The Lady Wants To Orbit' – Skelton spent four months acquainting herself with the Mercury 7 astronauts, undergoing tests at the various astronaut training facilities, and visiting the Cape. In the article, a large photograph taken at Cape Canaveral shows a hard-hatted Betty standing in front of an Atlas booster.

Skelton's first visit was to NASA's Langley Field in Virginia, where she met the Mercury 7 astronauts in their ready room. At 33 years of age she was the same age as some of the astronauts, whom she found to be both very kind and extremely helpful. She also felt that they were not resentful, but qualified this by saying that they did not have any reason to be, because they knew that she did not have a chance.

While Skelton was with the astronauts, the *LOOK* team took several photographs of her: standing in the cockpit of an F-102 supersonic military aircraft while astronaut Wally Schirra explained its instrumentation; looking at a 'returned from space' capsule with astronauts Wally Schirra and Al Shepard; flipping a coin with Slayton, Schirra, Carpenter, Glenn, Grissom and Shepard for morning coffee (which she was reported to have won on the final turn of the coin and, as a result, was given the label of 'No 7½' by the astronauts); flying the orbital (air-bearing) flight simulator under the guidance of astronauts Wally Schirra and Al Shepard (which Betty likened to a Link trainer), in a less than flattering pose; and preparing for underwater training to simulate aspects of weightlessness, which involved, among others, astronauts Scott Carpenter and Al Shepard, and the astronauts' doctor, William Douglas.[16]

The underwater training exemplified Skelton's mental determination and drive, as, unknown to her male companions, she could not swim: 'I don't swim. I never have. I am not particularly afraid of water, but I don't swim. But I didn't dare tell them. We all went underwater, and they were very nice to hang around and help me when I wanted a little help. I don't think they ever knew I couldn't swim.'

At the USAF's School of Aviation Medicine in San Antonio, Texas, Skelton was subjected to some of the astronauts' physical 'fitness for space' tests. However, as she prepared to undertake the 'tilt-table' test she soon became aware of the lack of preparation for female testing: 'When I arrived, they had a test that stands you up on a platform and your feet are up in the air, and they couldn't figure out what I should wear, so they decided to put me in a hospital gown, because they'd never tested women before. Then somebody said, 'Oh, I don't think that would work when she gets on this platform thing that goes upside down.' So they put me in a pair of overalls that were far too large. And I had forgotten to take shoes with me, so I wandered around the whole place in high-heeled shoes.'

At the US Navy's Aviation Medical Acceleration Laboratory in Johnsville, Pennsylvania, Skelton rode the human centrifuge. However, like Ruth Nichols, g loadings were not a new phenomenon for Betty: 'The centrifuge goes round and round very, very fast, and creates g loadings on the body, and each g loading carries up your body weight. I was used to g loadings because I'd had a number of them in Little Stinker.'

However, an indication of the unusual nature of Skelton's centrifuge test, and how the role of women was perceived at that time, can be gauged by the comment made by one of the technicians as he was preparing her for her test. 'As the two technicians were putting me in the capsule, one of them said – and this was an advertising campaign that was going on back in those days – 'Golly, wouldn't this make a great ad? I dreamed I rode the centrifuge in my Maidenform bra!'' In addition to meeting the Mercury 7 astronauts and experiencing some of their testing and training first-hand, Skelton also had the opportunity to meet a team of Russian space scientists at an American Rocket Society 'show', in Washington. *LOOK* magazine reported that she had learned that the Russians had overcome their bias against giving women hazardous tasks, and that one of the Russian scientists had stated that 'there was no objection to using women' on Russian space missions.

LOOK therefore speculated as to what America's first female astronaut would look like, based upon discussions with Generals, doctors, psychologists and engineers directly involved in the US manned space programme. 'Our first girl in space will probably be a flat-chested lightweight under 35 years of age, and married. Though not an outstanding athlete, she will have extraordinarily precise coordination. She will be a pilot. Her interests will tend toward swimming and skiing, rather than a more muscular sport like wrestling. She will adjust well to isolation and be able to 'hibernate', but also to snap into immediate alertness. Her personality will both soothe and stimulate others on her space team. Her first chance in space may be as the scientist-wife of a pilot-engineer. Her specialities will range from astronomy to zoology. She will not be bosomy because of the problems of designing pressure suits. She will not smoke or have a history of major surgery or mental disorder. Her menstruation will be eliminated by inhibiting medication. She will be willing to risk sterility from possible radiation exposure.'

Although conforming to some of the above, Skelton was under no illusion that she would be America's first female astronaut: 'I knew at the time that they were not considering a woman, really. In a way, I had to agree with them, although I'm always gung-ho about things like this; but they were working on such a small budget, and the equipment was really not totally developed. I felt that then was not really the time for them to uproot everything they were doing, and the progress they had made, to try to put a woman into the programme. I figured it would probably take twenty or twenty-five years due to the feeling about women ... But what little time there was associated with the NASA test and the astronauts, I did everything I could. I felt it was an opportunity to try to convince them that a woman could do this type of thing and could do it well. I think my entire association for even that brief period of time was probably the most soul-searching thing I've ever been involved with – to suddenly walk into the NASA compound, so to speak, and have them explain what it was they were trying to do, how they were going about it, and what the problems were. They had a tremendous job to do, and I had a great respect for all their efforts.'

But the issue of women astronauts did have *some* support within NASA – as Skelton discovered during her conversations with Dr William Douglas. She recalled that he was a little less negative than most about a woman astronaut, and that women might be better adapted than men to being less restless about the monotony of space travel. He also felt that a woman's reproductive organs were better protected than men's, and that physically it might be a little safer for women than men.

Concerning her experiences, and the ability of women to perform as potential astronauts, *LOOK* merely concluded that Betty Skelton (who was 5 feet 3 inches tall and weighed about 100–105 lbs) was 'just a petite example of the anatomical fact that women have more brains and stamina than men.' The magazine therefore failed to answer the question that it posed on its cover: 'Should a Girl Be First in Space?' But perhaps of greater significance is the fact that during the *LOOK* assignment, Skelton met Dr Donald Flickinger, who, it was reported, believed that women would be seriously considered for space missions after the development of semi-manouverable, orbiting space vehicles that could carry three people.

Jerrie Cobb and the Lovelace 'astronaut tests'

As a result of the Nichols' release and the adverse reaction of the Air Force to sponsoring female fitness-for-space tests, Flickinger hoped that Lovelace would be able to privately carry out his ARDC Girl Astronaut programme, and decided to transfer the programme to the Lovelace Foundation. His reasoning for this decision is given in an 'action memorandum' dated 20 December 1969, which he sent to Lovelace – the second key document uncovered by Dr Weitekamp: 'We in the ARDC have officially terminated our plans for pursuing this study any further with a wire so stating to both WADC [Wright Air Development Center] and SAM [School of Aviation Medicine]. The consensus of opinion was that there was too little to learn of value to Air Force medical interests, and too big a chance of adverse publicity to warrant continuation of the project. Since there was such great unanimity of opposition I did not see fit to overrule it, and do not plan on reopening the issue with anyone at SAM or at Air Force level.'[17]

It would also appear that ARDC scientists, in addition to public relations concerns, were unwilling to fund modification of the partial pressure suits (PPS) that were necessary for female testing because they expected to learn little of value. This issue seemed somewhat perverse, as the PPS in use at that time was designed and manufactured by the David Clark Company (DCC), whose core business was the design and manufacture of brassières and girdles. Furthermore, DCC had made a PPS for the famed French female pilot, Jacqueline Auriol.

Flickinger concluded his memo to Lovelace by saying: 'Please let me know how you proceed with this project, since I continue to have a keen personal interest in it and believe it should be done on as scientifically sound a basis as possible. I feel – by instinct perhaps – that if carefully done with a large enough series, there would be some interesting differences between male and female responses noted.'

Lovelace duly took over the research project – which he called the 'Women in Space' programme. By Christmas 1959 Jerrie Cobb had received another letter informing her that her aviation and medical records were satisfactory, and that she would be informed of the date of her tests early in the new year. Arriving at the Lovelace Foundation on 14 February 1960, Cobb was handed an itinerary that listed the Phase I tests that she would be going through over the ensuing few days. These were physiological tests, and were designed to determine whether the human body could withstand what the aerospace doctors and scientists of the day expected astronauts to experience in space. 'One of the things that they thought was perhaps when the capsule ejected from the booster in space it would start tumbling very fast, and this would give the human being inside vertigo, which would destroy the sense of balance. To test this, they took super-cool water and squirted it into your ears to freeze the inner ear drum [to induce vertigo]. They would then see how you could cope with it, and how long it would take you to recover from it. If it showed them anything about how to survive in outer space, we were glad to do it.'[18]

Another unpleasant test that Cobb recalled involved swallowing three foot of rubber hose: 'Every time I gagged a little bit they would shove a few more inches down my throat until you'd got three feet down, all the way into your stomach. They weren't pleasant tests, but they were necessary, and I was just so glad that I could pass them.'

Cobb – who had been named 'Woman of the Year in Aviation for 1959' – performed well, and on the basis of the results of the seventy-five physical tests she was recommended to progress to the next phase of testing, although she was told not to discuss the astronaut programme.

Before Cobb undertook the Phase II tests, however, she was given the opportunity to ride NASA's Multiple Axis Space Test Inertia Facility (MASTIF), was used to simulate a capsule tumbling in space. Located in the Altitude Wind Tunnel (AWT) at the Lewis Research Center in Cleveland, Ohio (now the John H. Glenn Research Center), MASTIF was a 21-foot-diameter tubular rig, with three metal 'cages' which allowed movement in pitch, yaw and roll. At the centre of MASTIF was a skeletal replica of a Mercury space capsule, complete with a contoured astronaut couch and hand-controller. As MASTIF tumbled in all three directions at a speed of 30 revolutions per minute, the astronauts had to bring themselves to a stop by means of the hand-controller (which in space would operate the capsule's thrusters), while the three cages continued to spin.

Cobb arrived at NASA Lewis in early April 1960. Although not strictly part of Lovelace's programme, she was keen to add another test to her list of achievements. Wearing a standard Air Force orange flight suit and helmet, she was strapped into MASTIF's contoured couch. The 'beast' was then unleashed – but Cobb soon had everything under control. At the end of her 45-minute 'ride' she was informed by the technicians operating MASTIF that her response was exceptionally quick. Only two months earlier, in February 1960, Al Shepard had taken his first ride, but when MASTIF finally began to spin he turned green and pressed the red 'chicken switch' to sound a claxon as a signal to stop.

On 19 August 1960, at the Space and Naval Medical Congress in Stockholm, Lovelace made public Cobb's Phase I test results, declaring that: 'Jerrie demonstrated a point that many scientists have long believed: that women may be better equipped than men for existing in space. Women have lower body mass and need significantly less oxygen and less food, hence may be able to go up in lighter capsules or exist longer than men on the same supplies. Since women's reproductive organs are internally located they should be able to tolerate higher radiations without sustaining harm.'[19]

As a result of Lovelace's announcement, Cobb became headline news, and was identified as the first successful American female astronaut. The following week (29 August 1960), *LIFE* magazine ran an exclusive photo-feature on her entitled 'A Lady Proves She's Fit for Space Flight', which was similar in lay-out to the earlier Betty Skelton photo-feature in *LOOK*.

One month later, in September 1960, Cobb began the week of Phase II tests at the Veterans' Administration (VA) Hospital in Oklahoma City. These were psychological and psychiatric tests, and the most demanding of them involved the isolation tank. Designed to simulate the weightless confinement of a space-suited astronaut in orbit, a test subject was required to float in a 10-foot-diameter tank of warm, body-temperature water, 8½ feet deep, in a pitch-black, sound-proof room. Cobb duly entered the tank, and the VA medical staff waited for her to begin hallucinating – and after 9 hrs 40 min of sensory deprivation, and no hallucinations, she ended the

Seven women of the Mercury 13 attend the launch of STS-93 in 1999. (Courtesy Al Hallonquist.)

test. Unbeknown to her at the time, she had just set a record for isolation tank testing, as no previous test-subject (male or female) had exceeded 6½ hours without hallucinating. The results of her Phase II tests had shown her to possess several exceptional, if not unique, qualities and capabilities for serving on special missions in astronautics. None of the Mercury 7 astronauts had been tested in the isolation tank.

The following year, in May 1961, Cobb was invited to take the Phase III tests at the Navy's School of Aviation Medicine (SAM), in Pensacola, Florida. These were stress tests, and they marked the final stage of Lovelace's research project. One of them was an altitude chamber test to 60,000 feet (higher than the 47,000 feet that Cobb had attained during her supersonic flight in the Air Force's F-102 Delta Dagger), which involved her donning the Navy's smallest full pressure suit. This procedure – including lacing, fastening the gloves and boots, and donning and sealing the helmet – occupied about 1½ hours. At the end of the test she was rather sorry to take off the suit, and hoped that she would soon have a 'space suit' of her own!

Another test she underwent was the electroencephalogram (EEG), which measured brain activity (via eighteen needles stuck into the subject's head), and was used to ensure that astronauts could cope with the high g-forces at lift-off and landing and the zero g of space. As this test involved Cobb flying as a passenger in the Navy's EEG-instrumented, fully aerobatic jet aircraft, the SAM had to obtain permission from Navy headquarters at the Pentagon. Stating that the purpose of the test was to ascertain the differences between male and female astronauts, the Pentagon's humorous but stereotypical response was: 'If you don't know the difference already, we refuse to put money into the project.' However, the two tests which Cobb found of most interest were the 'Dilbert Dunker' – an aircraft cockpit that was shot down a 45-degree railed ramp and then turned upside down in a pool of water to simulate an emergency egress in the event of a crash into the sea – and the 'Slow Rotating Room' – a full-size, 15-foot-diameter round room attached to a centrifuge, designed to study the ability of a test subject to function in a 'disoriented environment'.

Cobb successfully passed all of the Phase III tests, and in doing so paved the way for the other female pilots who were undergoing the Phase I tests at the Lovelace Foundation. 'I was sort of the guinea pig that went through the test first, and if I did all right they would bring the others in.'

A few days after Cobb completed the Phase III tests, NASA Administrator James E. Webb appointed her as a consultant to NASA.

The Lovelace class of 1961

Towards the latter part of 1960, Lovelace had begun to develop Flickinger's ARDC Girl Astronaut programme into an independent Women in Space programme. He had already inherited a list of potential candidates from the female pilots that Jerrie Cobb had suggested, and which Flickinger's ARDC researchers, having checked their credentials, had reduced to a group of eight: Frances Bera, Geraldyn Cobb, Barbara Erickson, Marilyn Link, Betty Skelton, Geraldine Sloan, Marian Petty and Jane White.

Furthermore, as a result of the media coverage of Jerrie Cobb's tests there were now other potential candidates interested in the programme. Lovelace's most pressing problem, however, was how to initiate and maintain the programme. He needed financial support. The female pilots had, in general, little spare money to cover travel and accommodation costs, the tests cost money, and the personnel conducting the tests and analysing the results had to be paid. Lovelace therefore sought assistance from his Foundation's financial benefactors – the famed female pilot and aviation pioneer Jacqueline 'Jackie' Cochran, and her husband, Floyd Odlum.

Jackie Cochran was America's premiere female pilot. She responded positively to Lovelace's enquiry, and suggested that the selection criteria for the programme be made more flexible by extending the age range and accepting married women. A month later, Cochran had decided that in addition to being Lovelace's 'special consultant' and the programme's financier, she wished to become a Women in Space candidate. Her hopes, however, were short-lived, as a diagnosed heart problem disqualified her from participating in the tests.

In 1961 Lovelace invited twenty-four female pilots who had met the selection criteria to participate in his programme as potential astronauts. Of those invited, eighteen completed the Phase I tests. Twelve of them passed:

Myrtle 'K' Cagyle (38), when aged 12, was taught to fly by her brother. She also used a pillowcase as a 'parachute' when, as a young girl, she jumped off the roof of her house. At the time of the Phase I tests she had 4,300 hours of flying time, and was, according to *LIFE* magazine, a 'flight instructor'.

Jan Dietrich (36) As a teenager she gained her student's pilot licence, and in 1949 she graduated from the University of California at Berkeley. She then went on to become a corporate pilot. At the time of the Phase I tests she had obtained an airline transport licence, and had 8,000 hours of flying time.

Marion Dietrich (36) Jan Dietrich's identical twin sister, she too had both gained her student's pilot licence and graduated from Berkeley. However, unlike Jan she did not become a professional pilot, and instead became a professional writer and reporter, using her own time to fly charter aircraft. At the time of the Phase I tests she had 1,500 hours of flying time.

Mary Wallace 'Wally' Funk (24) Her first experience of flying was similar to that of Myrtle Cagyle when, at age 5, she wore a Superman cape and jumped off the roof of her father's barn. A graduate of Oklahoma State University, she went on to Fort Sill in Oklahoma, where she was a flight instructor. At the time of the Phase I tests she had had 3,000 hours of flying time. She was also the youngest member of the 'class'.

Sarah Gorelick (née Ratley) (29) A graduate with a degree in mathematics, physics and chemistry, she worked for AT&T as an electrical engineer. She also held a commercial pilot's licence, and competed in female flying races such as the 'Powder Puff Derby'. At the time of the Phase I tests she had 1,800 hours of flying time.

Jane Hart (41) was married to the then Senator of Michigan, Philip Hart, and was mother to eight children. She was the first female in Michigan to be licensed to fly

Jerrie Cobb stands in front of a model of the Atlas rocket. (Courtesy Al Hallonquist.)

helicopters, and, like Sarah Ratley, competed in female flying races. At the time of the Phase I tests she had 2,000 hours of flying time. She was also oldest member of the class.

Jean Hixon (39) As a WASP engineering test-pilot during the Second World War she had flown the B-25 bomber. After the war she became a flight instructor, and, in her own time, gained a degree in Elementary and Secondary Education. She then changed jobs, and became a teacher. At the time of the Phase I tests she had 4,500 hours of flying time.

Rhea Hurrle (née Woltman) (32) A college graduate, she also became a schoolteacher, but left her job to become a professional pilot. At the time of the Phase I tests she had 1,500 hours of flying time.

Irene Leverton (36) Having begun to fly at age 15, she tried, at age 17 (under age), to join the WASPS with a fake identity and log-book, but was unsuccessful. At the time of the Phase I tests she was a flying school supervisor and had 9,000 hours of flying time.

Bernice 'B' Steadman (37) Married to a lawyer, she both established and operated her own flying school and charter service. She also competed in female flying races. At the time of the Phase I tests she had 8,000 hours of flying time.

Gene Nora Stumbough (née Jessen) (26) A graduate with an English degree, she became a professional pilot and taught flying at the University of Oklahoma, where she was the only female flight instructor. At the time of the Phase I tests she had 1,450 hours of flying time.

Geraldine 'Jerri' Sloan (née Truhill) (33) Another early convert to flying, she was aged 4 when she had her first flying experience sitting in the cockpit of an aircraft that was taking her father on a business trip. At age 14 she lied about her age (she said she was 15) in order to qualify for a student pilot's licence. The deception worked – until her mother found out and duly packed her off to a Catholic girl's school for a year. Undeterred, she went on to become a professional pilot working under contract on a classified project to develop the Terrain Following Radar (TFR) and smart bombs. At the time of the Phase I tests she was married with two children, and had 1,200 hours of flying time.

Frances Bera, Virginia Holmes, Patricia Jetton, Georgina McConnell, Joan Meriam (née Smith) and Betty Miller each completed the Phase 1 tests, but did not pass; and of the other six, Marilyn Link declined to take the test, while it is not known whether Dorothy Anderson, Marjorie Dufton, Elaine Harrison, Sylvia Roth and Frances Miller participated or declined.

The Phase 1 tests – which were the same as those undertaken and passed by Jerrie Cobb – included an additional test to which the Mercury 7 astronauts were not subjected: the gynaecological examination. As Wally Funk, the youngest of the 'passed' group, recalled: 'They just tested everything from head to toe. They X-rayed everything, and found out more about my body than I knew. I was quite a shy person then, and I didn't know what things were all about. I found out in a real hurry that I was pretty naïve!'[20]

Furthermore, unlike the Mercury 7 astronauts, the women did not undergo the Phase I tests in a collective group. As a result of the staggered distribution of the letters of invitation by Lovelace, and because the tests had to be taken as and when they could be accommodated in the Lovelace Foundation's schedules, the women had to take the tests in pairs or on their own. But they were determined not to fail, as Jerri Truhill recalled: 'I knew one thing. If you yapped, you probably weren't going to pass, so I don't think that anyone who ever passed ever let out anything. I wasn't going to let them flunk me out for complaining, or jumping, or even acting like I was remotely sensitive to any of the tests. I remember Dr Kilgore said, not so long ago, that we didn't complain as much as the men. Well, we didn't complain at all!'[21] Dr Kilgore – a Lovelace physician who tested the women – also felt that they were extraordinarily intelligent and incredibly motivated.

With Jerrie Cobb having successfully passed the Phase III tests at Pensacola in May 1961, arrangements were made for the twelve women who had passed the Phase I tests to undergo the same tests in July. However, as this was the busiest time for some of the women, the tests were deferred to 18 September 1961. Realising that there was now an opportunity for some of the women to undergo the Phase II tests, Cobb ascertained from the VA that they were able to schedule the tests. Cobb then sent out letters of invitation, asking those women who had both the time and the money to cover costs to travel to Oklahoma City and take the Phase II tests. Rhea Hurrle and Wally Funk informed her that they were available. In her correspondence with the women, Cobb referred to them as 'Fellow Lady Astronaut Trainees' – thus coining the phrase 'The FLATS' to describe the group of twelve women who had passed the Phase I tests.

Rhea Hurrle took her Phase II tests between 31 July and 2 August 1961, while Wally Funk took her tests between 3 August and 5 August 1961. Both women were successfull, with Wally setting a new record of more than 10 hours in the isolation tank. As she recalled: 'This test [is used] to simulate weightlessness. Everything is sound-proof, light-proof, and so on, and all your senses are taken from you. The temperature of the room and the temperature of the water perfectly matched the temperature of your skin, so you didn't feel the water. You couldn't even feel it dripping off your fingers ... Finally, a voice came over the microphones that were right above your head and says, 'Wally, how are you doing?', and I said, 'Terrific.' They then said, 'What time do you think it is?', and I said, 'Well, I don't have any bodily needs here, I'm not hungry, I guess it's probably about 12 o'clock.' So I went out to the briefing room and they uncovered the clock, and said, 'Well congratulations, you've been in 10 hours and 35 minutes and broken the record."

With their attention now focused on the Phase III tests, the women fully expected to see each other for the first time at Pensacola on 17 September 1961. Sadly, however, it was not to be. With only days to go before the tests, Lovelace sent each of them a telegramme which simply read: 'Regret to advise arrangements at Pensacola cancelled. Probably will not be possible to carry out this part of the

Sarah Ratley tries on a space helmet at her farewell party at AT&T. (Courtesy Al Hallonquist.)

program.' The Navy had withdrawn its permission for the Phase III tests, and Lovelace's programme had come to an abrupt end.

Although little documentary evidence exists, Dr Weitekamp located a memorandum sent by Jackie Cochran to Vice Admiral Robert Pirie on 1 August 1961, regarding women astronauts. In the two-page memorandum, Cochran reiterates details of Lovelace's Women in Space programme and its future plans, which she had discussed with Pirie during a car journey. Since Pirie was Deputy Chief of Naval Operations (Air), the Navy headquarters at the Pentagon was aware of Lovelace's informally agreed plans for testing the women at Pensacola. Pirie, it seems, then wrote to NASA to ask if there was an official request or requirement for testing the women. Responding in a letter dated 2 October 1961, NASA Deputy Administrator Hugh Dryden stated bluntly: 'NASA does not at this time have requirement for such a program.'[22]

Consequently, both NASA and the Navy were able to use the missing 'requirement' to justify the termination of the Pensacola tests without either of them taking ultimate responsibility.

The House of Representatives

Jerrie Cobb was at Pensacola preparing for the imminent arrival of the women when she received the news that the tests had been cancelled: 'The first thing I did was to march into the commanding admiral's office at Pensacola. I said, 'I've just got this word, how come?' as just the day before he'd told me everything was all set for the women. And he said, 'Jerrie, I am so sorry. I didn't know about it yesterday, I just got the word myself from CNO [Chief of Naval Operations in Washington], and they say they've got to cancel it because NASA does not have a requirement for it.'

Jerrie then left Pensacola and went straight to Washington, where she knocked on every door possible to talk to congressmen and senators. She also spoke to people at NASA, on the basis that they had appointed her as a consultant.

As a result of her public exposure in the August 1960 edition of *LIFE*, and the subsequent media interest, Cobb became both a national figure and a representative for space and female astronauts. She was also viewed generally as potentially America's first woman astronaut. It was because of this high media profile that NASA Administrator James Webb had appointed her as a consultant to the space agency in May 1961. With NASA having just been tasked by President Kennedy to land a man on the Moon and bring him safely back to Earth before the end of the decade, Webb did not want anything that detracted from achieving this goal. He believed that by having Cobb report directly to NASA, any woman astronaut 'issues' could be dealt with internally rather than debated in public. Cobb, however, viewed the appointment somewhat differently: 'Since I am not a PhD with three different degrees, I assumed my appointment to NASA had something to do with women in space.'

Cobb's first consultative document to NASA, submitted in June 1961, had been a report in which she outlined a two-part proposal: Part 1 – a recommendation that research be continued officially by NASA so that women's potential contributions to space could be thoroughly investigated and measured; and Part 2 – a plea that the first woman in space be an American, not a Russian.

As far as Cobb was concerned, this was her raison d'être, and in March 1962 her campaign to have the Phase III tests reinstated was given a significant boost by the addition of Jane Hart who, in addition to being one of the twelve women denied the testing at Pensacola, was the wife of Senator Philip Hart of Michigan. Through Hart's endeavours, the two women had a meeting (also in March) with Vice President Lyndon B. Johnson, which Hart described to Dr Weitekamp as 'a miserable meeting. He kept saying there wasn't much that he could do about it.' Unbeknown to the two women, Johnson's assistant, Liz Carpenter, had composed a letter that she hoped the Vice President would sign and send to Webb as a means of giving them some encouragement. Although never sent, Dr Weitekamp found the unmailed letter, which asked whether NASA had disqualified anyone because they were a woman. Instead of signing it, Johnson had written in large letters, 'Let's stop this now!'[23] Those in the US government appeared to share the same views as NASA. However, as a result of Cobb and Hart's campaign, some senators began to question why women were not part of the astronaut programme. As a result, a special hearing in Congress was called.

The congressional hearing before the Special Subcommittee of the Committee on Science and Astronautics of the US House of Representatives on the Selection of Astronauts was held over two days in July 1962. Cobb and Hart, together with Jacqueline Cochran, testified on the first day, although it soon became evident that Cochran was actually testifying against them. In response to a letter from Cobb on 8 March 1962, concerning Cochran's personal views on a Women in Space programme, Cochran had written: 'Women will travel in space just as surely as men. It's only a question of when. I don't believe there is any distinction between men and women as to ability in the air, but it's the men who have received that long training as test-pilots and therefore it's natural that such men were selected for the present phase of the astronaut programme. Women for one reason or another have always come into each phase of aviation a little behind their brothers. They should, I believe, accept this delay and not get into the hair of the public authorities about it.'[24]

Then, a month later, in response to a letter from James Webb, dated 24 May 1962, Cochran summarised the views she had expressed in her letter to Cobb on 23 March (which she also attached): 'I think the space effort is so expensive and so great in national importance that it should not be changed to the slightest degree at this time to accommodate women because, as yet, there is no actual need; to put a woman into space first seems unlikely if, as alleged, Russian women have already been in training for a long time.'[25]

Continuing in the same tone during her testimony, Cochran stated that she did not think that suitable female astronauts should be sought by injecting women into the middle of an important and expensive astronaut programme. She also argued that a female astronaut programme would waste a great deal of money, as it would, by necessity, require a large group of women in order to offset attrition rates caused by marriage. Nevertheless, at the end of the first day, and in spite of what Cochran had said, Cobb thought that everything was going well for them, since the congressmen appeared to agree with their testimony. On the second day, however, NASA began its testimony, and it brought along two new heroes – flown astronauts!

Fifteen months earlier, on 12 April 1961, Russian cosmonaut Yuri Gagarin had become the first human to fly into space when he made one orbit of the Earth in Vostok 1. The US could only respond with Alan Shepard's sub-orbital flight, and it would be a further ten months before a US astronaut would orbit the Earth. The astronaut who made that flight was John Glenn, and he was now being asked to testify at the congressional hearing. Accompanying Glenn was astronaut Scott Carpenter, fresh from his own somewhat controversial mission.

As national heroes, Glenn and Carpenter's testimonies carried a great deal of weight with the congressmen, so when the astronauts stated that they had enough men in the programme, and that they did not need any more astronauts – especially women, who would slow down the programme – considerable damage was done to the women's cause. However, even after the astronaut's testimony, the congressmen still thought that having women in the programme was important: Cobb has said: 'At the end of the hearings – and it's a matter of record in the congressional hearings – the head of the committee told the NASA people to go back and come up with a programme to include women in space, but it was never done.'

The women's day in Congress had come to an end, and NASA would not be implementing (for the foreseeable future) a female astronaut programme. However, it might be of some consolation to Jerrie Cobb to learn that Dr Robert Seamans, Deputy Administrator of NASA from September 1960 to January 1968, believes that she was a gifted pilot who could probably have undergone training as an astronaut, had NASA not established a strict set of ground rules for Mercury. He also believed that it would have been possible, given sufficient time, to fly a woman on a Mercury mission, and that women's waste management would not have been a major factor – an argument levied against having female astronauts.[26]

Hypothesising on this subject, NASA's Joe Kosmo – who entered the Mercury programme in late 1961 after B.F. Goodrich had already fabricated the Mercury spacesuits – believed that had female astronauts been selected, then they could have introduced in-suit waste-management provisions, as they have for the Space Shuttle programme.[27] However, Dr Stanley White was glad that they did not have to deal with providing dual sex accommodation during the early missions, as they had all they could handle in providing the minimum essential environmental and living accommodation. Also, as the decision had been made to have the astronaut 'fly' the spacecraft, he felt that the utilisation of available weight and space to provide the flow of data, communications and controls to permit this to be achieved was a far more important issue than concerns about whether it was politically important to fly a woman.

I never said I was an astronaut!

With the Lovelace women consigned to being spectators of their country's space programme, the Soviet Union made a further propaganda coup on 16 June 1963, when Valentina Tereshkova, onboard Vostok 6 – an automated spacecraft – became the first woman to travel into space. However, in the BBC Radio 4 programme, *The Right Stuff, the Wrong Sex*, Christopher C. Kraft, the first Flight Director at NASA's Mission Control, expressed the opinion that Tereshkova was 'an absolute basket case when she was in orbit', and that the Russians were 'damn lucky to get her back ... She was nothing but hysterical while she flew.' He also said that NASA might well have been in a similar situation if they had sent a woman into space.[28] In an interview for the *Sunday Times Magazine*, Kraft is also reported as saying: 'Had we lost a woman in space back then, because we'd put a gal up there rather than a man, we would have been castrated. You've got to remember times were different, and women were thought of differently then.'[29] Dr Robert Seamans, expressing similar comments, had 'heard that Tereshkova was hysterical when she was in orbit,' but pointed out that she was not an aircraft pilot. He did not believe that properly trained women would become hysterical in space.

These are interesting comments given that NASA, as result of operative errors, nearly lost one of its male Mercury astronauts in space. Scott Carpenter – one of NASA's star witnesses at the congressional hearings – had failed to perform as expected while carrying out NASA's second manned orbital mission. Furthermore, he, like Glenn (another star witness), did not have a science or engineering degree, which was one of NASA's astronaut selection criteria. However, Kraft argued that Glenn was, at that time, a leading test-pilot, and that this met the selection requirement of having a degree or the equivalent of an engineering degree from an acceptable college. He was more critical of Carpenter: 'He did not have a test-pilot's training, and, to be perfectly blunt about it, he shouldn't have flown in space. But he got there, and fortunately he lived. Damned fortunate!'

Although he would never make another spaceflight, Carpenter had at least made it into space – something that Jerrie Cobb finally realised that she was never going to achieve. 'It was obvious they weren't going to do a thing, so that's when I decided to use my talent where it was needed. I went down to the Amazon jungles in South America and became a missionary pilot. Every single day is a joy down there, and I just love it!'

The other women – with the exception of Jane Hart, who was involved with the congressional hearings – returned to their pre-Lovelace lives after the Phase III tests at Pensacola were cancelled. They had sacrificed a lot for Lovelace's programme

Jerri Truhill. (Courtesy J. Truhill and Al Hallonquist.)

(some had lost jobs, and one was issued with divorce papers), and in the process they had pushed the envelope to the limit. With regard to Jackie Cochran's role in the termination of the programme, the views of the women are mixed. Jerri Truhill felt that Cochran was 'too old, and didn't want younger women to steal her thunder'[30] – a sentiment echoed by Lillian Kozloski and Maura Mackowski: '[Cochran] had a lifetime of achievements in a man's world, they say, and when age made it time to gracefully step aside, she just couldn't give a hand up to another woman.'[31]

Latterly, the women (including Jerrie Cobb) have become known as the 'Mercury 13'. However, Gene Nora Jessen – an amateur historian, and one of the thirteen – has some reservations: 'Mercury 13 was a light-hearted tag thought up by Hollywood producer James Cross. It implies that we had something to do with the Mercury astronaut programme, which we absolutely did not. I don't like FLATS either, since it says 'trainees', which we were not. Testees, maybe, not trainees. To put this in perspective, twenty-five women were invited to take a five-day astronaut-like physical exam. Thirteen women passed. I was never told we were astronaut candidates, though I enjoyed having even this kind of contact with the astronaut programme, and looked forward to the challenge of seeing if I were as physically fit as I thought I was. Then it was over. Many people are now inferring that we were astronauts who got dumped. That's not true at all, and I'm afraid that someone some day will say 'You are all frauds'. I want to yell, 'I never said I was an astronaut!''[32]

WOMEN AND THE GEMINI, APOLLO AND SKYLAB PROGRAMMES

Although NASA may have been less than receptive about having female astronauts working in space, this was not the case for women working on the ground, as computers, secretaries, nurses, and so on. The December 1961 issue of *American Girl* magazine carried a full-page message from President Kennedy, inviting women (and men) to participate in the country's space programme. Entitled 'Lots of Room in Space for Women', the invitation read: 'A message to you from the President. In our many endeavours for a lasting peace, America's space programme has a new and critical importance. The skills and imagination of our young men and women are not only welcome but urgently sought in this vital area. I know they will meet this challenge to them and to the nation with vigour and resourcefuless. John F. Kennedy.'

Not only did the Kennedy administration want women working at NASA, it also wanted them working in the aerospace industries that supported the country's space efforts. The following examples illustrate some of the contributions made by women to NASA's Gemini, Apollo and Skylab missions.

Bridging the gap between the end of Project Mercury and the start of Project Apollo was the Gemini programme. Launched atop a Titan II rocket, ten manned Gemini missions (GT-3 to GTA-12) were made between March 1965 and November 1966 (the first two flights having been unmanned). On 3 June 1965, during the Gemini 4 mission, Ed White became the first American astronaut to perform an

EVA (spacewalk). The Gemini spacesuits that White and his fellow astronauts wore were designed and manufactured by the David Clark Company, which had designed and manufactured the earlier partial pressure suits (PPS). Working on these suits were DCC seamstresses Helen Hoyen and Marie Porier, and in recognition of their contribution to NASA's manned space programme, both women were later presented with a Snoopy Award, given by astronauts to fellow workers.

In her article, 'The 13 Who Were Left Behind', Joan McCullough, wrote: 'There was already [*c.*1966] talk of the United States sending its first woman astronaut into space. Candidates Jan and Marion Dietrich – real-life twins – were being eyed as possibilities for a Gemini team.'[33]

Although the authors cannot confirm McCullough's statement, Robert Seamans has stated: 'On occasion I was asked in congressional testimony for my views on women astronauts, particularly for the Gemini programme. I explained that a flight in Gemini was equivalent to a flight in the front two seats of a Volkswagen for periods up to fourteen days. During that time the cockpit could get pretty grungy.' Voicing similar sentiments, Dr White felt that 'accommodations for both sexes onboard the two-person Gemini [capsule] would have been near impossible due to lack of space'.

The primary objective of Project Apollo was to land a man on the Moon and return him safely to Earth. To achieve this goal NASA needed a heavy launch vehicle, and developed the Saturn series of rockets. NASA's Marshall Space Flight Center, in Huntsville, Alabama, was responsible for all matters pertaining to the design, development, manufacture and delivery of the Saturn rockets, and two of the engineers who worked on the programme were Sara Cobbit and Doris Chandler. A series of unmanned launches would precede a manned mission designated Apollo 1, scheduled for 21 February 1967. But tragically, three weeks before the scheduled launch date a fire took hold in the Command Module during a test exercise on the launch pad, killing all three crew-members – Gus Grissom, Ed White and Roger Chaffee. As recalled by Mary Mahassel, a seamstress at David Clark Company: 'I was in Florida, at the Kennedy Space Center, at the time of the fire that took the lives of the three astronauts. I was there along with a pattern designer, to rework the gloves of Gus Grissom. Although [the astronauts] passed through the room, we were not introduced to them. It was an experience I shall never forget. To this day I can still recall that moment.'

After a full investigation into the cause of the fire, NASA inititated a number of alterations and modifications prior to resuming manned operations. In October 1968 the crew of Apollo 7 checked out the Apollo Command and Service Module in Earth orbit, to be followed two months later by the historic flight of Apollo 8 – the first manned flight to the Moon. During his eight-nation tour of Europe in February 1969, Apollo 8 Commander Frank Borman told reporters that it was a 22-year old female engineer who computed the trajectory that brought Apollo 8 back from the Moon.[34] Unfortunately, the name of the woman is unknown, although as Gene Kranz, former Apollo Flight Director at NASA Mission Control, recalls: 'There were a large number of women in the Maths-Physics Branch of the Mission Planning and Analysis Division. Their leader was Cathy Osgood. Two other women also come to mind: Shirley Hunt and Mary Shep Burton. During the early years, we armed these women with a manual calculator, and they punched away at the numbers all day long and plotted them on graph paper. They came with us when we deployed to the Cape and prepared the plot boards for launch. No single woman computed the trajectory. This was a team effort by a large and talented group.'[35]

During his European tour, a female reporter also questioned Borman about why NASA had no women astronauts. Borman responded that space 'was a hostile environment. We do not have lady tank drivers or aircraft carrier pilots. Why in space?' However, when it came to long space missions he thought that it would probably be the right time for lady astronauts to be included.

Following Apollo 8, in March 1969 the Apollo 9 mission tested the Lunar Module with a crew in space for the first time. In May 1969, Apollo 10 completed a full dress-rehearsal for a manned landing on the Moon, with its Lunar Module reaching a low point of 14,447 metres above the lunar surface. Two months later, Apollo 11 achieved the goal: the first manned lunar landing with only the third manned LM.

Built by Grumman Aircraft Engineering Corporation, the 'spider-like' LM was the first spacecraft to be built for the sole purpose of operating in the vacuum of space. Unlike the CSM, therefore, it did not have a smooth aerodynamic shape, and, more importantly, it did not have a re-entry heat shield. One of the few women working at Grumman on the LM's electronic equipment was Myrtle Holland, an electrician 'borrowed' from the Martin Marietta Company in 1966. At that time, local law prohibited women from working overtime, but as Myrtle was one of the few NASA-certified solderers, the law was changed, although Grumman were not particularly enthusiastic about employing women. Myrtle felt that women were better suited to the kind of work being undertaken, as their small hands were suited for accessing small places and for carrying out intricate wiring. The problem was further compounded, however, as all the work had to be done while wearing gloves – which was not an easy task, as very tiny wires were involved. Another female 'Grummanite' working on the LM, and who also had a position of responsibility, was Peggy Hewitt. As the director of Grumman's LM check-out at the Cape, Hewitt was responsible for testing the LM prior to its being enclosed in the Saturn V's conical shroud situated beneath the CSM.[36]

One of the critical phases during any Apollo Moon-landing mission was the rendezvous and docking of the LM with the orbiting CSM. To achieve this rendezvous above the lunar surface, a 'special radar' was built by RCA Aerospace Systems, and two female engineers – Amy Spears and Beverly Eckhardt – were responsible for key functions in its development.

Amy Spears studied electrical engineering at the University of Oklahoma, Purdue, and Cornell, where in 1948 she graduated with a Bachelor's degree. She later gained a Master's degree from Northeastern University. As the manager of the reliability assurance aspects for the LM rendezvous radar (which she had been since the start of the project) she was responsible for all quality control, and ensured that designers and sub-contractors did not violate any of the specifications set by RCA. Beverly Eckhardt, was a graduate of the University of Vermont, and gained her Master's degree from Brown University. In July 1969 she studied for her PhD at Tufts

University. A psychologist by profession, her first work on Apollo was with the people who made the flight simulators, to study problems in manual control of the lunar landing and rendezvous of the LM with the Command Module (CM). She was also involved in designing the ground equipment used to support and check out the complex Apollo communications hardware.[37]

Women were also involved in the manufacture of the spacesuits worn by the Apollo astronauts. Some of the International Latex Corporation seamstresses who worked on the Apollo spacesuits were Delema Austin, Delema Comegys, Doris Boisey, Michelle Trice, Julia Brown and Delores Zeroles. Another female ILC employee, Iona Allen, made the lunar overboots that Neil Armstrong wore when he made his historic 'one small step'.[38] To slow itself down after re-entering Earth's atmosphere, the CM deployed three large parachutes, and one of the master parachute riggers for Apollo was Norma Cretal. The Apollo programme ended with the sixth manned landing – Apollo 17 – in December 1972. We now await a manned return to the Moon, when the first woman might well leave her footprints in the lunar dust.

By the time that Project Apollo was underway, planning had begun for the inclusion of people other than test-pilots as astronauts, including women. In April 1964, Harry H. Hess, of the National Academy of Sciences, agreed to have his Space Science Board define the appropriate scientific qualifications for scientist-astronauts, while NASA took responsibility for the age and physical criteria. Thus, in October 1964 NASA announced that it was recruiting scientist-astronauts; and for the first time, an astronaut candidate did not have to meet the test-pilot/jet aircraft criteria,

ILC seamstresses Dolores Zeroles and Ceal Webb work on the Skylab parasol.

although selected candidates who were not qualified pilots would be taught to fly. Of the 400 applications that were screened, four were from women, but in June 1965, when NASA announced that it had chosen the six scientist-astronauts who would make up the Group 4 astronaut intake, all were male. The following year, on 26 September 1966, NASA and the NAS issued a further invitation for scientist-astronauts to participate in the Apollo Applications Program. This time there were 923 applications screened, including seventeen from women; but in August 1967, when NASA announced that it had chosen eleven of them for the Group 6 astronaut intake, once again they were all male.

The follow-on programme to the Apollo lunar missions was the Apollo Applications Program, which manifested itself in the form of Skylab – NASA's first Earth-orbiting space station. Launched unmanned on 14 May 1973, Skylab was subsequently home to three three-man crews. However, the aerodynamic forces acting on it during its ascent to Earth orbit were such that both its micrometeoroid/thermal protection shield and one of its two solar panel wings were torn off. Consequently, Skylab began orbiting the Earth without thermal protection or power (the second solar panel wing having failed to deploy due to its being stuck in the closed position). With no thermal protection to shield it from the full force of the Sun, the temperature inside Skylab began to soar to around 52° C. NASA therefore had to devise a method of drastically reducing Skylab's intolerably high internal temperature to an acceptable level for its future astronaut occupants, who would be living and working inside the space station for missions lasting a month or more. The solution that NASA arrived at required the fabrication of a deployable solar shield (a parasol), and it would take a woman's touch to make it a reality!

To assist in the manufacture of Skylab's emergency parasol, two of ILC's seamstresses – Delores Zeroles and Ceal Webb, were despatched, with their own sewing machines, to NASA's Marshall Space Flight Center, while the material for the parasol was flown from Houston to MSFC. Director Dr Rocco Petrone watched a seamstress work on the Marshall sail with growing agitation. As material passed through one of the seamstress' machines onto the floor, she used her foot to move it out of the way to make room for more. 'It just isn't right,' Petrone muttered, 'You're not supposed to kick flight hardware.'[39]

With the parasol completed, and a rescue mission devised, the first Skylab crew – Pete Conrad, Joe Kerwin and Paul Weitz – blasted off from the Cape on 25 May 1973. They successfully deployed the parasol, and the internal temperature began to fall. Within a couple of days it had stabilised at the more tolerable temperature of about 23° C, and the astronauts were able to spend 28 days onboard Skylab. They also managed to free the jammed solar panel wing, paving the way for the second Skylab crew – Al Bean, Owen Garriott and Jack Lousma – to spend 59 days onboard, and the third and final crew – Gerry Carr, Ed Gibson and Bill Pogue – to stay for 84 days. After the third crew left in February 1974, Skylab spent a further five years orbiting the Earth until it re-entered the Earth's atmosphere and broke up over Australia in July 1979.

In her article, 'The 13 Who Were Left Behind', Joan McCullough mentions the heat tolerance studies which were being carried out at Florida State University, and which Jerrie Cobb cited at the congressional hearings. As these studies showed that women could be comfortable at temperatures 16–26 degrees higher than those comfortable for men, McCullough asked the question: 'Need we have waited so long for Skylab to cool down?'

The same question was posed to Dr Joe Kerwin MD, a member of the first Skylab crew, who felt that there would have been nothing gained by sending a female crew to the stricken Skylab space station.[40] However, Dr Robert Seamans expected, in the early days of NASA, that the US would send women into space when stations were available. When asked about the possibility of female astronaut access to Skylab, Seamans responded by saying: 'There is no question in my mind that Skylab could have been configured for a mixed sex crew.'

In addition to carrying out scientific experiments on behalf of research scientists (both male and female) from within NASA and academia, the Skylab astronauts also carried out a number of high-school student experiments. Of the nineteen approved experiments, five were from female students, and the one which received the most press attention was the 'Web Formation in Zero Gravity' experiment by Judith S. Miles from Lexington High School, Massachusetts. This experiment, conducted by the second Skylab crew (SL-3), involved two spiders called Anita and Arabella.[41]

Project Apollo's final mission, in July 1975, was a symbolic linking in space of Apollo 18's CSM with Russia's Soyuz 19. The end of Apollo also saw the end of expensive disposable rockets. The mighty Saturn family of launch vehicles would be replaced by the Space Shuttle.

BEHIND THE SCENES

In addition to the noted women in the forefront of aeronautical research and scientific efforts, there were thousands involved in the space programme from its earliest days. In his autobiography *Carrying the Fire*, published in 1974, Michael Collins acknowledged the unsung work of female workers who assembled the crews spacesuits that kept the astronauts alive in period of unpressurised flight activities, including spacewalking. 'My favourite ladies with their gluepots', he called them. Throughout the history of human spaceflight, women have fulfilled supporting roles on the ground – both in preparing the crews and hardware for flight, in controlling the missions, in analysing the data, and in assisting in the development of procedures and equipment.

Small steps for women

Despite the fact that women were not being selected for astronaut training for the pioneering US manned space programmes (Mercury, Gemini and Apollo), there were scores of women in unsung support roles, working in the background in various fields and research work that significantly contributed to the US space programme. This reflected the continuing participation of women in science, engineering and technology demonstrated over the previous century.

Some of the wide-ranging roles filled by these pioneering support women early in the space programme included the following:[42]

Evelyn Anderson joined NASA Ames in California in 1962, to work on human hormones, which control involuntary body activities. By studying the human reactions to extended weightlessness – such as apprehension in an artificial atmosphere – she studied the release of insulin into the blood at times of stress.

Dr Jimmie Flume overcame physical disability to work in the space programme, studying the processes in the body that create our immunity from disease, and comparing the processes on Earth and during spaceflight, leading to a broader understanding of the human immunity system. She used healthy tissue cells from test subjects, and utilised ground-based and space-borne experiments to investigate how potentially lethal microbes might multiply in large numbers in conditions of weightlessness, radiation exposure or periods of high g loads.

Lee Curry Rock assisted in the development of protective coverings for pressure suits at Wright–Patterson AFB, Ohio.

Julie Beasely was an experimental physiologist who worked at NASA Ames Research Center studying the response in animal behaviour against the forces of high g loads.

Margaret Jackson was an environmental physiologist at the Manned Spacecraft Center, Houston, during the Gemini programme. She was assigned to the

Margaret Jackson becomes the first woman to experience weightlessness in a C131-B aircraft. (Courtesy USAF.)

development of life support systems in the Crew System Division, and assisted in the development of different life support systems by varying the temperatures, pressure and types of gauges in the system, and monitoring and analysing the reading from test subjects wearing prototype pressure garments or working in mock-up space cabins. In 1941 she had joined the Aero Medical Laboratory at Wright Field, Ohio, as an assistant in altitude studies, and was involved in the early simulations of 'free floating' onboard the Air Force C131-B aircraft flying parabolic curves. She was fascinated by reports from the pilots who conducted these tests, and volunteered to be a test-subject herself. She is credited as being the first woman to experience extended weightlessness, floating around in an aircraft cabin. She later reported that it felt 'fantastic ... One feels completely buoyant ... like floating in seawater, only completely free.'[43]

Nan Glennon enrolled in a mechanics course whilst at the University of Southern California, with the hope of understanding the workings of her car. However, it was a physics course for engineers; but she enjoyed the lectures so much that she stayed on, and became the first woman graduate of the department of engineering at USC. After working in the oil industry she joined Space Technologies Laboratories (later TRW Inc) to analyse proposals, sales and contractors for computers, having first learned the then relatively new skill of programming a computer. At TRW she coordinated the engineering data involved in the development of spacecraft from the initial drafting designs, the models, the production line and the finished products.

Bea Finkelstein investigated the provision of suitable food for crews in space. A professional nutritionist, she was convinced that providing the astronauts with three

Dorothy B. John checks information put into a digital differential analyser. (Courtesy USAF.)

good meals a day would help them in dealing with the challenges of spaceflight, and prevent their becoming irritable and possibly making mistakes. She had been interested in nutrition since she attended college, and her older sister, taking a Doctor of Medicine degree, inspired her to pursue a career in that field. She joined the USAF as a civilian, evaluating the food programme for the armed services, and working with aero-medical staff on the importance of nutrition for pilots. In 1956 this work led to thoughts of providing food for future space explorers. This included not only the evaluation of types of food, but also the methods of packaging and serving in a weightless environment. Finkelstein became the Mercury astronauts' dietician, and provided the food for them to evaluate as part of the learning curve of these pioneering space missions. Called 'Bea's Diner' by the astronauts, they helped establish the experience and knowledge required to produce the comprehensive menus used today.

Lieut Colonel Elizabeth Guild, USAF, was assigned to the development of protection again the extreme noise levels produced by rocket engines. Her father was Dr Stacey Guild – one of America's leading specialists on the human ear. His work always fascinated her, and she pursued an academic career to follow his research on the structure and diseases of the ear. After working in his laboratory she earned her degree and joined Johns Hopkins University, where she became one of a team of doctors that were developing methods and procedures aimed at preventing hearing loss in young children. During the Second World War she joined the Women's Army Auxiliary Corps and then the Women's Army Air Corps, and decided to combine her experience in medical skills and experience in the Air Corps and apply it to problems in aviation medicine dealing with hearing. Working while directly exposed to excessive noise and vibration levels in and near aircraft in flight and on the ground, she worked on noise levels and vibrations generated by engines, the slip-stream, radio sound waves, shock waves, and the effects of sonic booms on the human ear. She later worked at Wright Patterson AFB Biological Acoustics Branch, evaluating the noise and damage limitation effects from jet engines, rockets and missiles, space launch vehicles, turbo-pumps, and supersonic flight.

Pat Rydstrom was a Northrop Corporation scientist who researched into the effects of spaceflight, specialising in the effects on living cells exposed to sudden impact, sudden increases in speed, extended acceleration, weightlessness, vibration, heat and cold. She was, like many others, not fond of school except the biology classes, and in particular veterinarians, who allowed her to assist in animal hospital work and helped her decide on her future career.

Barbara Short investigated aerodynamic stability of various designs of spacecraft, firing minute modules shot out of gun-like devices in wind tunnels at Ames Research Center, Moffett Field. She worked on designs for Earth orbital re-entry shapes – including the testing of pellet-sized models of the Mercury capsule – and on early designs of lunar re-entry shapes for the Apollo programme.

Edith Olson was a civilian chemist working for the US Army in Washington, where she researched the miniaturisation of electronic components that were installed in

Bea Finkelstein – food technician and nutritionist.

early missiles and later larger rockets and a host of commercial and domestic products. As a young girl she was encouraged by her family to use a microscope and to look under stones and in water to find 'bugs'. In the 1950s she completed her research and academic studies to become an inorganic chemist.

Margaret Townsend was, like many of the women supporting the space programme, both a mother and a scientist. She worked as a senior electrical engineer at NASA Goddard Space Flight Center, supervising the development of the computer that flew onboard the Tiros and Nimbus meteorological satellites. She was the only daughter of an engineer, and became the first woman engineering graduate from George Washington University. She then went on to work for nine years at the Naval Research Laboratory in Washington, where she carried out research in anti-submarine warfare. She joined NASA in 1959.

Merna Dawson was an analytical chemist involved in the development of the rocket research aircraft programme at Edwards AFB. Her work involved the laboratory testing, analysis and evaluation of a variety of propellants for various research programmes, and in determining compatibility and safety.

Annette Chambers worked at at the Aerospace Corporation, California, and during early manned spaceflights was located at the Cape as a Guidance Project Engineer. She was responsible for writing the mathematical instructions used in the computers onboard manned spacecraft.

Helen Mann was, in the late 1950s, one of the first women to work at Cape Canaveral. She worked in the impact predictor section, determining the impact point of engines of early missiles which malfunctioned, and informing the range safety office to destruct the projectile. She also monitored and recorded missile trajectories, elevation and range, updating the recovery vessels on predicted impact points.

Mary Hedgepeth was in charge of a group of mathematical data analysts at Edwards AFB, analysing the data from a previous flight of a rocket research aircraft, extracting data on its actual performance over predicted performances in velocity, acceleration, direction and orientation, and developing mathematical computer programmes to interpret this data in order to update and predict future flights.

Eleanor Pressly was a mathematics teacher who worked as a computer in a radio research laboratory during the Second World War. She later joined the Naval Research Laboratory, and became the liaison officer between the designers of several sounding rockets, the aerospace manufacturers and the experimenters, in determining the criteria for placing experiments on the sounding rockets, and for which rocket was most suitable for which experiment.

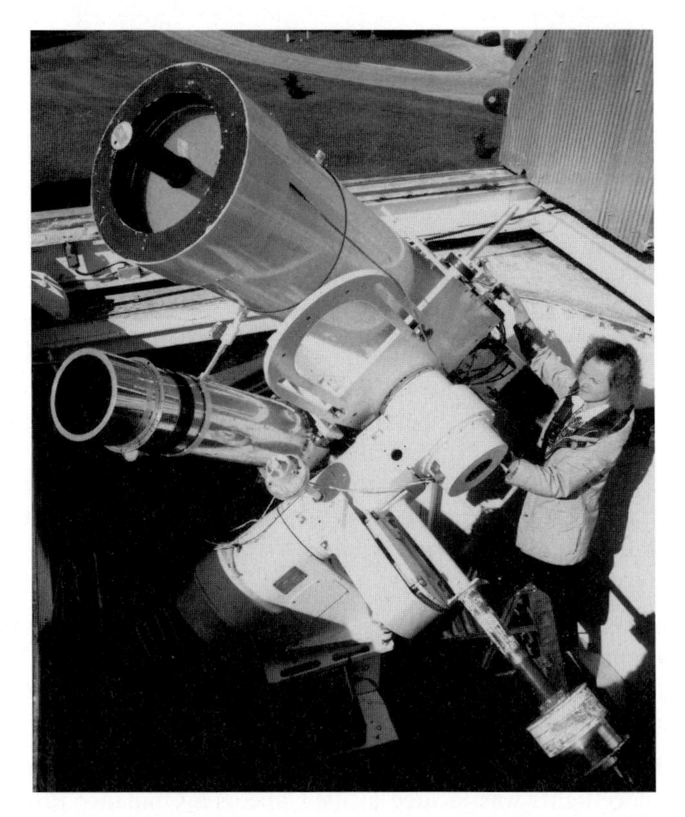

Dr Mona Hagyard, of the MSFC, who expressed interest in flying as a PS on the Shuttle and participated in Spacelab ground simulations in the 1970s.

Virginia Norwood worked at Hughes Aircraft Company as head of the Microwaves Group, which developed antennae for use in the Syncom communication satellite series. Like several of these pioneering women of the space age, she was inspired by her parents – notably her father (a former Army engineer), who introduced her to machine shops and mathematics. During her school years this was built upon by the influence of several teachers, who inspired her to develop her love of engineering and mathematics. She graduated from the Massachusetts Institute of Technology with a BS degree, married her Professor, and joined Hughes in 1954.

Nancy Roman was perhaps one of the more prominent female personalities in the space programme during the 1960s. As a child she always wanted to be an astronomer, and pursued academic studies to achieve that goal, becoming an astronomy teacher and researcher in radio astronomy, including conducting measurements of the Moon by radar. In the early 1960s she became the head of NASA's Observational Astronomy Program in the Office of Flight Development, working on the Orbiting Solar Observatory (OSO), Orbiting Geophysical Observatory (OGO), Polar Orbiting Geophysical Observatory (POGO), Elliptical Orbiting Geophysical Observatory (EGO), and Orbital Astronomical Observatories (OAO) that obtained images and data on deep-space phenomena, gathered important information on the make-up of space and the effects of 'space weather' (radiation, dust, particles, solar wind debris) on other deep-space probes, and investigated the relationship between the Sun, Earth, Moon and planets. Roman was also involved in the development of several astronomical experiments and instruments flown on the X-15 rocket research aircraft, on satellites, and on manned spacecraft, including the Apollo Telescope Mount flown successfully on Skylab during 1973–74.

Marcelline Chartz supervises the writing of computer programs.

Ann Eckels Bailie was a mathematician working on interpreting the data from Vanguard 1. She was a member of the team that noted that the perigee of the satellite was different when it flew over the Earth's northern hemisphere than it was over the southern hemisphere. This was originally thought to be human error, but analysis of the data revealed that the Earth is asymmetrical.

Dr Mildred Mitchell was one of the growing number of women in the new field of bionics, helping develop an artificial muscle experiment termed the 'nail bender'.

Geneve B. Barnes worked in public affairs, including during the Apollo 11 mission.[44] She had a background in the federal service, working in the Office of Naval Material and the Judge Advocate General offices before becoming a secretary in the Office of Programs at NASA HQ in Washington in 1962. In 1963 she moved to the Office of Public Affairs, where she worked as a secretary for eight years. She was primarily involved in organising special events for NASA at the White House, astronaut award ceremonies, and appearances and tours. She served as a 'protocol officer' at KSC during the early Apollo missions, and accompanied the crew of Apollo 11 and their wives on the Presidential World Tour of Goodwill. This international event – codenamed Project Giant Step – used the Vice President's aircraft to visit twenty-two countries in thirty-eight days between 29 September and 5 November 1969. Barnes then fulfilled a number of administrative roles over the next twenty-four years. When Neil Armstrong became the Deputy Associate Administrator for Astronautics in the Office of Aeronautics and Space Technology, she served as his public affairs assistant. She also served as administrative secretary to the Associate Administrator for the Office of Aeronautics and Space Technology; administrative secretary to the Associate Administrator for Equal Opportunity Programs; administrative officer in the Office of Aeronautics and Space Technology; management analyst in the Office of Management Operations; astronaut appearances coordinator in the Office of Public Affairs, working and touring America with the first five Shuttle crews; and astronaut international appearances coordinator. After thirty-two years with NASA and a total of forty-one years in federal service, she retired in 1994. After leaving NASA she worked as a volunteer in the e-mail section of the Presidential Mail Office in the White House.

Lola Morrow began work at NASA as a travel clerk in 1962, and then, upon the request of the astronauts, became their secretary at the Cape in 1964, backing up the office secretaries in Houston. She also arranged transport and accommodation for the astronauts during their trips to Florida, and was responsible for the administration of their activities at the Cape.[45] Immediately after the Apollo 11 mission, she left NASA for a position in public relations.

In addition to the hundreds of women in medicine, engineering and science, a support team of secretaries, librarians, archivists and public affairs specialists help record and distribute the facts and figures on the space programme to the general public. Some of the early pioneers in this field were Ida Young, a talented oil painter, who worked for NASA as an editor of technical reports, including the account of John Glenn, the first American to orbit the Earth; Dorothy Morris, who worked at the

Lewis Research Center, Cleveland, as a research librarian, converting thousands of pages of technical reference material into more accessible summaries; Grace Kennedy Winn, who worked at MSC in Houston, arranging media interviews, and press and community group tours of the facility; Dorothy Whidden, an artist and illustrator who worked at Lewis Research Center interpreting the latest ideas for space science experiments into artwork for pamphlets and leaflets; Rita Rapp a food technologist working on the development of food for the Apollo, Skylab and early Shuttle programmes; and Iva 'Scotty' Scott, who worked as secretary to the first Director of the Manned Spacecraft Center (Robert Gilruth), and later in Public Affairs at JSC during the transition from the Apollo programme to the Shuttle programme.

THE PRIMARY BACK-UP CREW: THE ASTRONAUT FAMILY

Crew assignments to a spaceflight include assignments of a prime crew, and normally one or more back-up crews who can take the place of one or all the flight crew-members in the event of accident and illness. Ask most space explorers who were the

Shannon Lucid with her husband and family following the conclusion of her record-breaking stay onboard Mir in 1996.

primary back-up team in preparation of their mission, and most would reply: 'The family who let them fly their missions'. Although several astronauts have presented accounts of their marriages in their published biographies, and have naturally called upon their wives for contributions, relatively few wives have personally written of their time 'in the space programme'. Some, however, have recorded the loss of a husband in a spaceflight tragedy: Gus Grissom on Apollo 1 (Betty Grissom, *Starfell*, 1974), and Rick Husband on *Columbia* (Evelyn Husband, *High Calling*, 2003); or loss of a daughter, Christa McAuliffe (Grace Corrigan, *A Journal for Christa*, 1993); and the daughter of Mercury astronaut Scott Carpenter helped her father with his autobiography (Kris Stoever, *For Spacious Skies*, 2002). The inside story of the roles and exploits of space explorers' families has yet to be written – and deserves to be written. A complete account of the 'astronaut wives' and 'cosmonaut wives' in the space programme is beyond the scope of this book, but the role they play in the history of space exploration is important and indispensable.

The 'astronauts' wives club'

Shortly after the selection of the second group of NASA astronauts in 1962, Marge Slayton – the wife of Mercury astronaut and coordinator of astronaut activities, Deke Slayton – suggested that it might be a good idea if each of the astronauts' families were to become better acquainted with each other. She therefore organised

The 'astronauts' wives club': (standing, left–right), unknown, Susan Cooper, unknown, Jo Schirra, Jane Conrad, Jan Evans, Joan Roosa, Pat White, Betty Grissom, Marilyn Lovell, Betty-Anne (Chris) Kraft, Kathy England; (seated, left–right), Sig Sjoberg's wife, Marge Slayton, Louise Shepard, Barbara Gordon, Dottie Duke. Beth Williams was also present. (Undated; AIS archives.)

the 'astronauts' wives club' – initially with the help of Sue Borman, the wife of newly selected astronaut Frank Borman.[46] This unofficial 'club' meet occasionally for many years during the 1960s and 1970s, but drifted apart during the late 1970s as the original astronauts selected in the 1960s moved to new goals outside of NASA, although dinners and socials gatherings still took place for several years. With the newer astronauts arriving for the Shuttle programme, the astronaut community changed; but it remained close, with the families of crew getting together during the preparation for flight, and attending the launch and landing events and ceremonies. Of course, the 'family of NASA' closed inwards during a time of tragedy – such as with the loss of the Apollo 1 (204), *Challenger* and *Columbia* crews.

Any account of the role of 'women in space' should include the astronauts' wives, daughters, sisters, mothers and aunts. When any family member witnesses a launch, the person they are watching being rocketed into space is not just an astronaut or cosmonaut, but is a husband, son, brother – or simply 'dad'. Since the 1980s, of course, this has also included watching a wife, daughter, sister, aunt or 'mom' leaving Earth and loved ones – sometimes for many months.

'Extremely pleased, honoured and proud'

Several astronauts have written autobiographies that praise the support of their wives and families in their quest for space, and several accounts, in print and in motion pictures, have tried to portray the rollercoaster ride of excitement, pride, anguish, fear and relief that any family goes through during the course of every spaceflight. Since no mission is routine, the threat of danger and loss is present from before launch until after landing. Coupled with the long period of training, the intensity of this preparation and the extended period of absence while travelling to contractors, fulfilling public appearances, and, more recently, an increase in international travel, places stress and strain on all relationships – sometimes to breaking point. Though naturally feeling all the human emotions, during the early missions the wives were required to mask their fears and doubts, and were required to deliver a standard reply to reporters' questions on their thoughts on the progress of the mission. They were extremely pleased, thoroughly happy and immensely proud. However this public face of 'a perfect wife' began to crack after the loss of the Apollo 1 crew in a pad fire in January 1967, after the Apollo programme had achieved its goal of landing a man on the Moon in 1969, following the almost fatal Apollo 13 mission of 1970, and with the end of the Apollo era in the mid-1970s.

In the early phase of the programme, an astronaut's image was part of the role. Talk of marital problems or divorce was not something that NASA wanted to be headline news – especially during a mission – and it was not until the late 1960s that the first astronauts' divorces were unavoidably announced. It has been suggested that this was an influencing factor in some careers, affecting their dedication and commitment to the programme. A difficult home life could not have been easy, and like any cross-section of a population, some handled it well, while others did not. In addition, the loss of astronauts in accidents seemed to place the families in the difficult position of either remaining around the space centre and the programme, or moving on.

John Young with his second wife Suzy, *c.* 1982.

With admiration and love

Of the many astronaut biographies written over the past four decades, many have indicated the numerous difficulties that the families have endured whilst 'dad' was away training and preparing for the mission, or the stresses of watching loved ones being blasted off into space. In most of these books, the wives and families of the astronauts are mentioned. Many of these books were written by the former military pilot astronauts of the Apollo era – those whose life was suddenly adjusted from the almost unknown life before NASA to a high-profile life after selection. In the dedications, the astronauts have acknowledged the efforts and sacrifices of the wives and families during the time they were actively involved in the space programme. Walter Cunningham (*All American Boys*, 1977) praised his wife and children 'for coping so beautifully during the period I was having the time of my life, and for their patience during the four years it took to write about it.'

Hiding the anxiety

Spaceflight is, and will always be, hazardous. Whilst astronauts and cosmonauts are aware of the risk, and accept it, it is not always so easy for the family, though many have been associated with a military family life. Many have lived on bases where the testing of high-performance aircraft was a daily job, and the loss of friends and colleagues in accidents was unnervingly routine. In Frank Borman's autobiography (*Countdown*, 1988) he devotes a chapter to his wife, who accepted his drive and his strong beliefs in duty, honour and achievement to his country, to the USAF, and eventually to NASA. Early in their marriage in the early 1950s, she read a copy of

Pat White and Pat McDivitt talk to their husbands onboard Gemini 4 in June 1965.

The Army Wife, and took every word to heart: instructions about 'wearing her party manners at all times', joining in social events with other officers' wives, keeping the children well-behaved, taking all the emotional punches from enforced setbacks, disappointments, separations, and still wearing a smile and deferring to the rules of the military life. However, although one of a thousand wives at Edwards AFB some years later, she was suddenly propelled from almost obscurity to the elevated position of a small group of women in America: the astronauts' wives. As Borman recalled: 'She simply continued playing her Air Force role – the perfect military wife who never complained, kept any fears to herself, raised two fine boys of whom I was inordinally proud, and subordinated everything to my career advancement ... My priorities were hers.'

Although the early days were exciting for any new astronaut and his family at Houston, the realities of being 'an astronaut' involved being out of town for days at a time, hours upon hours on meetings, training, simulations and more meetings, interspersed with the stress of a space mission or two, soon became evident. Then there were the tragedies of losing not only colleagues of your husband, but also husbands of your closest friends. During the 1960s, eight astronauts were killed in accidents; but it was the loss of the three Apollo 1 astronauts in January 1967 that to many wives brought home the reality of disaster. The image of a perfect wife, in a perfect marriage with a perfect husband, could not be sustained, and from the late 1960s the effort began to flag. The options to deal with such stress were limited, and the inability to cope often resulted in increased medication, alcohol, illness, divorce, or, in the case of Apollo 1 astronaut Ed White's wife, Pat White, suicide in 1984. The 'divorce bubble', as it was called, burst in the astronaut community in 1969, with the parting of several astronauts and their wives.

In 1990, Apollo 16 astronaut Charlie Duke published his autobiography,

Moonwalker. It was co-written by his wife Dorothy, who provided a chapter on her experiences as an astronaut's wife. Dottie Duke wrote that stress in the marriage began almost as soon as they exchanged vows, in that her husband's career came first – a career that saw him always rushing to study, to a new assignment, to a meeting . . . After graduate studies for an advanced degree, then test-pilot school and his love of aircraft, came along NASA and a whole new passion: spaceflight. In 1966 he was selected with the fifth group of astronauts at the height of the Apollo programme. The next six years were an even harder struggle, during which one assignment rolled into another, and then another. Dottie Duke reasoned that things would change *after* he flew in space; but it did not – and not for some time, until he left the space programme in 1976. She summarised her role as the wife of a crew-member as 'to make sure that my husband was taken care of in such a way that he could do the best job possible. I tried not to bother him with mundane burdens at home. Most [astronaut] wives cut the grass, took out the garbage, and kept the house and kids in order. That was our contribution to the US effort in space.'

This supporting role continues in the current programme, but now also includes the encouragement and support of husbands of female members of the astronaut corps. It also, of course, has included a number of international female space explorers who are married to husbands who are not astronauts or cosmonauts.

REFERENCES

1 Correspondence from Dr Stanley White, 1994.
2 Loyd S. Swenson, Jr, James M. Grimwood and Charles C. Alexander, *This New Ocean: A History of Project Mercury*, NASA SP-4201, 1966.
3 Stanley C. White, MD, telephone interview with Lillian D. Kozloski, November 1988.
4 William K Douglas, MD, telephone interview with Lillian D. Kozloski, November 1988.
5 See Ian A. Moule, 'Astronauts Interviewed: In Remembrance of Things Past', *Spaceflight*, **36**, September 1994.
6 http://quest.arc.nasa.gov/women/bios/ohara.html.
7 Colin Burgess, 'America's first astronauts' nurse: An interview with Dee O'Hara, *Spaceflight*, **43**, 7, July 2001.
8 Linda M. Plush, 'Origins, Founding and Activities of the Space Nursing Society', *Journal of Parmacy Practice*, **16**, 2, July 2003.
9 Margaret A. Weitekamp, 'The Right Stuff, The Wrong Sex: The Science, Culture, and Politics of the Lovelace Woman in Space Program, 1959–1963', PhD dissertation, Cornell University, UMI Number 3011230, May 2001.
10 Jerrie Cobb and Jane Rieker, *Women Into Space: The Jerrie Cobb Story*, Prentice Hall International, 1963.
11 Jerrie Cobb Fact Sheets, Jerrie Cobb Foundation.
12 Lillian D. Kozloski, 'The Wrong Stuff', paper presented at the Society for History of Technology, Sacramento, California, October 1989.

13 Letter from Donald Flickinger to Jerrie Cobb, 7 December 1959, discovered by Dr M. Weitekamp in the Jerrie Cobb Papers held by the Ninety-Nines International Organisation of Women Pilots, National Headquarters, Will Rogers Airport, Oklahoma City.

14 'Reminiscences of Ruth Nichols' (1960), in the Columbia University Oral History Collection.

15 'Betty Skelton Frankham', NASA Oral History interview by Carol L. Butler, Cocoa Beach, Florida, 19 July 1999.

16 Ben Kocivar (production) and Bob Sandbury (photography), 'The Lady Wants To Orbit', *LOOK*, **24**, 3, 2 February 1960.

17 Memorandum from Donald Flickinger to W.R.'Randy' Lovelace II, 'Action Memorandum', 20 December 1959, discovered by Dr M. Weitekamp in the Jerrie Cobb Papers held by the Ninety-Nines International Organisation of Women Pilots, National Headquarters, Will Rogers Airport, Oklahoma City.

18 Jerrie Cobb, interviewed by Nicola Humphries, WIA Conference, Orlando, Florida, 18 March 1999,

19 Reporting of Dr Lovelace's announcement in 'From Aviatrix to Astronautix', *Time*, 29 August 1960.

20 Wally Funk, interviewed by Nicola Humphries, WIA Conference, Orlando, Florida, 20 March 1999.

21 Jerri Truhill, interviewed by Nicola Humphries, Richmond, Texas, 12 March 1999.

22 Letter from Hugh L. Dryden to Vice Admiral R.B. Pirie, 2 October 1961, discovered by Dr M. Weitekamp in the Dryden chronological files, NASA.

23 Correspondence from Dr Robert C. Seamans Jr to Teresa Kingsnorth, 10 February 2003.

24 Letter from Jacqueline Cochran to Jerrie Cobb, 23 March 1962.

25 Letter from Jacqueline Cochran to James Webb, 14 June 1962.

26 Correspondence from Dr Robert C. Seamans Jr to Teresa Kingsnorth, 10 February 2003.

27 Correspondence from Joseph Kosmo, 2004.

28 *The Right Stuff, The Wrong Sex*, presented by Sue Nelson, BBC Radio 4.

29 Bob Graham, 'I'd Prefer to Send a Monkey into Space than a Bunch of Women', *Sunday Times Magazine*, 15 November 1998.

30 Correspondence from Jerri Truhill, summer 1998.

31 Lillian Kozloski and Maura J. Mackowski, 'The Wrong Stuff', *Final Frontier*, May/June 1995.

32 Correspondence from Gene Nora Jessen, 9 June 1998.

33 Joan McCullough, 'The 13 Who Were Left Behind', manuscript, December 1973.

34 'Borman Doesn't See Why Women Would Want 'Hostile Environment' of Astronaut', *Denver Post*, 13 February 1969.

35 Correspondence from Eugene F. Kranz to Teresa Kingsnorth, 10 March 2003.

36 Charles R. Pellegrino and Joshua Stoff, *Chariots for Apollo: The Untold Story behind the Race to the Moon*, TAB Books, 1985.

37 RCA News/Defence and Space 'Special Apollo Edition', July 1969.

38 Lillian D. Kozloski, *US Space Gear: Outfitting The Astronaut*, Smithsonian Institution Press, 1994.
39 'Skylab Rescue: Improvisation vs Race With Time', in 'Skylab Missions: 15th Anniversary', Skylab stories by Bob Lessels, Public Affairs Office, *Marshall Star*, 11 May 1988.
40 Correspondence from Dr Joseph P. Kerwin to Teresa Kingsnorth, 24 February 2003.
41 Leland F. Blew and Ernst Stuhlinger, *Skylab: A Guide Book*, NASA Marshall Space Flight Center, 1973-0-500-721.
42 Mary Finch Hoyt, *America's Women of the Space Age*, Atheneum Press, 1966.
43 *Ibid.*, p. 32.
44 'Before This Decade is Out . . . ', edited by Glen E. Swanson, NASA SP-4223, 1999 pp. 223–243.
45 Gene Farmer and Dora Hamblin (with the Apollo 11 astronauts and families), *First on the Moon*, Michael Joseph, 1970, pp. 46–49.
46 Deke Slayton with Michael Cassutt, *Deke*, Forge Books, 1994, p. 123.

The rocket-plane and the Space Shuttle

It was by no means preordained that mankind's first forays into space would be as occupants of 'tin-can' capsules launched atop expendable rockets. It was simply a quirk of fate – a quick fix for the problem posed by the Soviet/American space race. It had been generally assumed that mankind's journey into space would be achieved by the logical evolution of the aeroplane, integrating rocket technology with aeroplane development to create a reusable, rocket-powered winged aircraft for space transportation (a spaceplane). This is a view held by Jerri Truhill, one of the women from the Lovelace Women in Space programme: 'I felt that the US would go with the X-15 rocket-plane. I still think that's the way they should have gone. That's what was going to take us into space, not rockets!'[1]

A NEW TYPE OF AEROPLANE

The first documented flights of rocket-propelled manned aircraft, involving the attachment of solid-powder rockets to unpowered gliders, took place in Germany during the late 1920s. On 8 December 1927, Max Valier, a leading Austrian authority on rocketry and space travel and a founder member of the German Verein für Raumschiffahrt (Society for Space Travel), together with Fritz von Opel, a German 'Henry Ford' who manufactured cheap cars, and Friederich Wilhelm Sander, a manufacturer of 'practical' powder rockets, agreed an ambitious research programme involving ground-based vehicles, aeroplane models, and, ultimately, manned flights of aircraft, all propelled by rockets. The work would initially be carried out using 'solid' rockets, but they also envisaged the development of liquid-fuelled rockets and their subsequent application to manned flights.[2]

Having successfully completed the first part of their programme with manned tests of a rocket-propelled car, Opel Rak II, on 23 May 1928, attention switched to experimenting with rocket-propelled model aeroplanes. Directing this experimental work would be the brilliant young aerodynamicist, Alexander Martin Lippisch, who would later design the world's first rocket-fighter aircraft. The model chosen by Lippisch was a scaled-down version of his tailless glider, the Storch (Stork).[3] Tests were carried out in early June, and, on the basis of experience and knowledge gained from the Storch rocket models, the decision was made to attempt a manned flight with a full-size version.

On 11 June 1928, Alexander Lippisch's glider the Ente (Duck) took off from the Wasserkuppe (one of the Rhone Mountains in Western Germany), propelled by a pair of Sander solid rockets and an initial catapult launch using a rubber rope. Piloting the Ente was Friedrich Stamer, an experienced glider pilot who had assisted Lippisch with the earlier Storch model rocket tests. The Ente flew for approximately 60 seconds and covered more than three-quarters of a mile (1.2 km). Unfortunately, its solid rockets did not provide enough thrust, and Stamer's flight is now regarded as being a normal glider flight – albeit with the added curiosity of two burning rockets!

Inspired by Stamer's rocket flight attempt in the Ente, a young Julius Hatry contacted Lippisch with the intent of building and flying his own rocket-propelled glider. Lippisch duly agreed to help, but the work came to the attention of von Opel, who wanted to make the first 'official' rocket-propelled flight and pressurised Hatry into selling him the glider. On 30 September 1929, Fritz von Opel took to the skies in the Opel Sander Rak-1 Hatry Flugzeug (Hatry Aeroplane) rocket-propelled glider, taking off from Rebstock Airport, near Frankfurt. He was airborne for nearly 75 seconds, and travelled some 5,000 feet (1,525 m). Watching from the ground was his fiancé, Frau Sellnik, who was a pilot in her own right, and one of only six German aviatrixes. She had acted as an aviation consultant to von Opel in the months preceding his flight. However, earlier that month, on 17 September 1929, Julius Hatry had test-flown the rocket-propelled Rak-1, and in doing so had become the world's first pilot of a rocket-plane!

Although there were other recorded rocket-propelled glider flights, they were essentially imitations of the rocket-glider research flights of Stamer, Hatry and von Opel. The next practical steps would be born in the storm clouds of the Second World War. At the same time as the rocket-glider experiments were being carried out, a theoretical study for a reusable space transporter was being formulated by another Austrian, Dr Eugen Sänger. To many people, Sänger is regarded as being the father of the reusable space transporter.

Dr Eugen Sänger and his Silbervogel

For Dr Sänger – an aeronautical engineer conducting rocket-engine tests at the Technische Hochschule in Vienna – the manned exploration of space would be achieved by a logical progression of steps: stratospheric (sub-orbital) rocket-plane; spaceplane (orbital); space station; interplanetary spaceship; interstellar spaceship.

To this end, in February 1933 Sänger publicly released details of his concept for achieving the first of these steps – the stratospheric rocket-plane – when the *Deutschösterreichische Tageszeitung* (*German–Austrian Daily Paper*) published his essay 'On the construction and performance of rocket-planes.' This first public release was quickly followed by Sänger's publication of his own book, *Raketenflugtechnik* (Rocket Flight Technique), which provided detailed technical information on his first theoretical high-altitude stratospheric rocket-plane, or, as he would later call it, Silbervogel (Silver Bird).[4]

Sänger's rocket-plane would have a petrol and liquid-oxygen rocket-propulsion system to attain speeds of about 10,000 km/hr (about Mach 10) and altitudes of 60–70 km. Its body had a projectile-like shape, pointed in front and blunt at the back where the exhaust (rocket) nozzle was located. The profile of the wings had to be as thin as possible, with sharp leading edges, so that the aspect ratio could be kept low because of the negligible resistance of the wing-tips. However, Sänger had also realised, very early on, that the most difficult point of ascending into space was not in the aerodynamics of the flying body but in its rocket propulsion system. Hence he spent the majority of his working life addressing the issue of propulsion.

As a result of his published works, in 1936 Sänger was invited by the Hermann Goering Institute – the research laboratories of the Luftwaffe – to undertake a ten-year experimental programme in Germany, to further develop his concept for a stratospheric rocket-plane. To aid him in his work, specialised research facilities and laboratories – known as the Research Institute for Rocket Techniques – were constructed at Trauen, on Luneburg Heath in Lower Saxony. (The area at Trauen had been chosen by Eugen Sänger, and he had also been responsible for the construction plans.) For security reasons, the site at Trauen was called the Flugzeugprufstelle (Aircraft Test Site), and as a further security precaution, Sänger was given an assistant's job at the University of Braunschweig, where he was also given an office that he would never use.

In 1937 Sänger's small team was boosted by the arrival of two new members, one of whom was a brilliant young female mathematician: Dr Irene Bredt. Her arrival heralded the beginning of a life-long collaboration (both professionally and privately) with Dr Sänger – a collaboration which would eventually lead to marriage and the Sänger–Bredt Racketenbomber.

Dr Irene Bredt and the Racketenbomber

Irene Reinhild Agnes Elisabeth Bredt was born in Bonn on 24 April 1911. As a schoolgirl growing up in Cologne, she read the book *Peterchens Mondfahrt* (Little Peter's Travel to the Moon), and saw Fritz Lang's ground-breaking cinema film *Frau im Mond* (*The Girl in the Moon*), which had Hermann Oberth as its technical consultant. (Irene Bredt and Hermann Oberth would later become good friends, and Oberth would regard her as being the most scientifically educated woman he ever met.) Then, at age 11 Irene had her first contact with rockets when she saw one of Fritz von Opel's rocket cars (between tests) at an exhibition in Cologne. This led her to began thinking about the possibilities that rockets could offer.

In 1937, having obtained a PhD in physics from the Friedrich-Wilhelm University in Bonn, the newly qualified Dr Bredt went for an interview at the Aircraft Test Site in Trauen. Before the interview, someone had told her that the work at Trauen would involve rockets, but this only added to her determination to secure the job. She was, moreover, so determined that during her interview she revealed that she knew they were working on rockets (although when she made this bold statement she thought that they were working on powder rockets like those manufactured by Sander). With approval from Dr Sänger, who was always looking for possible new assistants, Irene was duly offered the position of scientific researcher.

Dr Irene Sänger-Bredt. (Courtesy Hartmut Sänger.)

Bredt arrived at Trauen on 1 October 1937, and was given responsibility for working on thermodynamic and gas-kinetic problems of liquid-propelled rockets, with particular emphasis on non-balanced reactions. The work itself was focused on understanding two important areas of interest to Sänger and his Silver Bird project: the aerothermodynamics of hypersonic flight and the reaction within liquid-propelled rocket engines.

Bredt also had a precise method of working that suited Sänger. He would indicate the direction the work needed to take, and she would carry out the calculations. To help with these she used a mechanical Brunsviga calculator – about 15 kg of 'cogs and gears'. Initially she only assisted Sänger, but with the publication of his Technical Report, *Gaskinetik sehr grosser Fluggeschwindigkeiten* (Gas Kinetics of Very High Flight Velocities) in May 1938, their professional working relationship evolved into one of collaboration. The report was also notable for presenting, for the very first time, formulae and numerical values for atmospheric forces affecting vehicles at altitudes where the atmosphere can no longer be regarded as a continuous medium.

In October 1938, work began on the construction of a 1:100 scale model of the Silver Bird. The non-corrosive steel model, which was tested in a wind tunnel, had a half-ogival (domed) shaped body profile with a flat bottom. It also had swept-back wings with a thin, wedge-shaped, cross-section and horizontal tail fins that had small vertical surfaces protruding from their tips. However, because of its domed-shaped body profile and flat bottom, the Silver Bird wind tunnel model was given the nickname Flat Iron, and this was used by all the members of Sänger's team.

To reduce the fuel weight of his Silver Bird, Sänger proposed that his stratospheric rocket-plane be launched from atop a rocket-propelled sledge accelerated along a straight horizontal steel rail track. However, at that time very little was known about how the sliding metal surfaces would perform at the high launching speeds required. One of Sänger's overriding fears was that the heat generated by friction would be impossible to control – which put in doubt the whole idea of saving weight by using a sledge-assisted take-off. The burden of responsibility for resolving this potential show-stopper was given to Irene Bredt.

It was evident to the team at Trauen that the project needed to acquire information on the amount of friction between the sledge and the rail, considering the very high sliding velocities and the subsequent high braking forces. Bredt therefore visited the Instituts für Technische Stromungsforschung (Institute for Technical Flow Research) in Berlin, with the intent of obtaining reliable values for sliding friction and lubricating procedures. Unfortunately, the highest sliding velocity values obtainable from this specialist research institute were only a fraction of those required by Sänger and Bredt. Undeterred, Sänger, with assistance from Bredt, proposed his own solution to the problem by conducting sliding-friction experiments at Trauen. These experiments would involve firing a stainless-steel bullet at about 800 m/s, into a new type of spiral sliding test-rig. The first of the sliding experiments began on 2 June 1939, and the results indicated that there were no major obstacles to the concept of using a rocket-propelled sledge on a metal rail track.

It was also during 1939 that the relationship between Bredt and Sänger took on a more personal nature, as they began spending their free time together. But the peaceful life of those living and working at Trauen was to be rudely interrupted with the outbreak of the Second World War. Although he had never intended his Silver Bird to be used for military purposes, it became evident to Sänger that if they were to continue their work at Trauen, it would have to be coupled with Germany's war objectives. The Silver Bird would now evolve into the Rakentenbomber (Rocket Bomber), or Rabo.

In 1941 Bredt became head of the Physics Department at Trauen. At the same time, the manuscript for the now famous Sänger–Bredt report, *Ubereien Raketenantieb für Fernbomber* (A Rocket Engine for Long-Range Bombers), describing the Rabo, was completed. It was submitted to the Reichsluftfahrtministerium (State Ministry for Aviation) on 3 December 1941, for approval; but it was not well received, and in the autumn of 1942 the order was given to stop all work on the project.

Ainring and the Sänger–Bredt report

With work on the Raketenbomber project at an end, Bredt, Sänger, and colleagues from Trauen, were transferred to the Deutschen Forschungsanstalt für Segelfug (German Research Institute for Gliding Flight) at Ainring in Upper Bavaria. Here they continued to work on ramjet propulsion and in preparing designs to demonstrate how the new technologies could be used to meet Germany's war objectives, such as the development of a ramjet fighter.

While at Ainring, Bredt assisted Sänger with the development of an outline for the complete thermodynamic theory of the ramjet engine. She was also responsible for evaluating the flight tests of Dornier Do-17Z and Do-217 aircraft fitted with Sänger's prototype ramjet engine and piloted by Paul Spremberg. Observing Spremberg's flight in the Do-217 (as a representative of the Ainring Institute's Director, Professor Walter Georgii) was Friedrich Stammer, who in 1928 had piloted the rocket-powered glider the Ente.[5]

Although the Rabo project had officially ended, Sänger and Bredt managed to persuade Georgii to publish an abridged version of their Sänger–Bredt Report, and official approval to publish was granted on 22 September 1944. The report – which covered fundamental areas such as the aircraft, launching, climb, gliding flight and landing – also included the prerequisite military requirements with areas covering projection of bombs and types of attack.[6]

The Sänger–Bredt antipodal rocket bomber

In their report, Sänger and Bredt proposed a fleet of a hundred reusable rocket-bombers that would be able to level large cities at arbitrary places on the surface of the Earth in a few days. The rocket-bomber would take off using a ground-based, rocket-propelled, sledge that would accelerate the rocket-bomber to 500 m/s in 11 seconds along a 3-km long, perfectly straight, horizontal steel rail track. Towards the end of the track, having achieved the desired take-off speed, the rocket-bomber would separate from the sledge and continue to climb to a height of 50–150 km, using its own internal rocket motor.

The duration of the climb would have been 4–8 minutes, during which time all of the onboard fuel would usually have been consumed. At the end of the climb, the pilot would have switched off the rocket motor and prepared for an unpowered return to Earth. However, instead of re-entering like a ballistic missile or capsule, the rocket-bomber would have been able to 'bounce' back into space, as a consequence of its wings descending into the denser layers of the atmosphere. The same re-entry manoeuvre would then have been repeated further along the flight path when the rocket-bomber once again struck the denser layers of the atmosphere and bounced off. Like a flat stone skipping across the surface of a still pond, this ricocheting trajectory would have allowed the rocket-bomber to extend its range to become an intercontinental bomber; but, just as the stone eventually stops bouncing and sinks to the bottom of the pond, so the laws of physics dictate that each time the rocket-bomber struck the denser layer of the atmosphere, a fraction of its kinetic energy would be consumed. Each successive bounce achieved by the rocket-bomber would have become smaller, until eventually it would have gone into a steady gliding flight back to Earth.

From a military perspective, a major (if not fundamental) flaw of this extended-range technique was that prospective targets had to be located at low points in the flight path, and all the important allied cities (for flights from German soil) were located under a peak. However, there was one exception: New York – which could have been bombed from a low point with the rocket bomber 'flying' to an antipodal point in Japan or part of the Pacific then controlled by the Japanese. Sänger and Bredt had also calculated that if they increased the maximum velocity of the rocket-bomber to 7,000 m/s, then instead of landing at an antipodal point, it would have been possible to travel all around the world, landing back at its take-off point some 3 hrs 40 min later.

Although in special cases a landing speed more than 200 km/hr would have been possible, the preferred landing speed for the rocket-bomber would have been 150 km/hr. Sänger and Bredt considered it to be a glide landing, and reasoned that they could not count on there being an experienced pilot at the controls.

Paperclip, marriage and attempted kidnap

With the collapse of the German Reich and the end of the Second World War, in 1946 Bredt and Sänger were offered work by the French Ministry of Aviation (under Operation Paperclip) as consulting engineers for the Arsénal de l'Aéronautique at Paris-Chatillion. Bredt also worked as a voluntary consulting engineer at the French rocket manufacturing company MATRA until 1954. In 1951 she married Sänger, and their only child, Hartmut, was born on 15 March 1952. However, their post-war lives may have had a very different outcome, as their Sänger–Bredt Report had fallen into the hands of the Russians.

The German authorities, having published a hundred copies of the Sänger–Bredt Report, had duly declared them as a state secret before dispatching them to a drawn-up list of recipients. At the end of the war, these reports were either found by or given to Western intelligence as investigations were carried out. Similarly the same thing happened in the East, and in May 1945 the Russians discovered a copy at Peenemünde.[7] News of the Sänger–Bredt Report eventually found its way to Stalin, who, intrigued by its possibilities, convened a meeting, in April 1947, that resulted in Lieut-Colonel Grigory Tokayev and Colonel Servov being ordered to find Bredt and Sänger and bring them to Russia in a 'voluntary–compulsory' manner.

Heading first to Berlin, the Russians spent several months searching in vain throughout Germany and Austria before finally returning to Russia, their mission a total failure. A year or so later, in September 1948, Tokayev took an opportunity to defect to the West. Settling in England, he began writing his memoirs, which were published in his book *Stalin Means War*. In 1951, Bredt and Sänger were given a copy of this book – and so learned of Stalin's plan to 'kidnap' them.

Wings, rockets and wallpaper girls

While Bredt and Sänger were conducting their rocket research for the German Luftwaffe at Trauen, the Ballistics Branch of the German Army's Waffenprufamt (Army Weapons Department) was conducting its own rocket research at Peenemünde. Overseeing the Army's rocket development work was the commander of the Peenemünde complex, General Walter Dornberger, and his civilian Technical Director, Wernher von Braun. On 3 October 1942, having had two previous failures, von Braun and his team achieved the first successful demonstration flight of their A-4 rocket. Reaching a maximum altitude of 85 km (53 miles) on its ballistic trajectory, the 14-m long liquid-fuelled rocket had become the first man-made object to pass through the sound barrier and enter the rarefied layers of the upper atmosphere. On the ground a delighted Dornberger congratulated von Braun with the words: 'Do you realise what we accomplished today? Today the spaceship was born!'[8]

Unfortunately for the inhabitants of London, space exploration was not on the agenda of the German High Command when they renamed the A-4 the Vergeltungswaffee Zwei (Vengeance Weapon 2). The V2 carried a 975-kg warhead in its nose cone, and it was launched at England.

Work on the rockets at Peenemünde was not entirely a male preserve, for like the Allies, Germany had put its female population to work. While some of the women sent to Peenemünde carried out the more traditional roles of secretaries and administrative clerks, those with specialist or technical skills were given work as draughtsman, illustrators and computers. These Messfrauen (measurement girls) worked in a large room under the direction of a male supervisor, Dr Paul Schroder. Here, with the aid of slide rules and mechanical calculators (like that used by Irene Bredt), they derived data from the rocket test flights and reduced the vast amount of data that streamed continuously from the static rocket tests. A further fifty women, involved in calculating the trajectory of the rocket flights, were known as the Tapetenfrauen (wallpaper girls) because of the large number of paper rolls they used while calculating the trajectories![9]

Not content with developing unmanned versions of their A-4 rocket, von Braun and his team also investigated the possibilities of creating a manned 'A' version by replacing the warhead with a pressurised cockpit and adding a tricycle landing gear and flaps for a conventional runway landing. However, like the Sänger–Bredt antipodal rocket-bomber, von Braun's piloted winged rocket would be only a paper dream. But this was not the end of the story for Peenemünde and its association with winged rockets. Its skies would reverberate to the sound of a piloted production rocket-plane known as Me 163 Komet.

The bat

In the late 1930s the German Aviation Ministry commissioned Alexander Lippisch, by then a Doctor, to design a high-speed research aircraft that would be used as a test-bed for the small Walther R-1-203 liquid-fuelled rocket motor. Deriving inspiration from his Storch rocket models, Lippisch produced a tailless, swept-back-wing (Delta-wing) aircraft. Designated the DFS 194 in honour of Lippisch's association with the Deutschen Forschungsanstalt für Segelfug (German Research Institute for Gliding Flight), it was 6.4 m long and had a wingspan of 10.7 m. Flight testing of the rocket-propelled DFS 194 was carried out at Peenemünde by test-pilot Heini Dittmar during the summer of 1940.

Having successfully demonstrated that the DFS 194's rocket system and airframe design were sound, work began at the Messerschmitt factory, in Augsburg, on the design and manufacture of twelve pre-production rocket-fighter variants. Although slightly smaller than their predecessor, these bat-like Me 163A rocket-planes would be propelled by the more powerful 750-kg thrust Walther R-11-203 'cold' rocket motor.

In October 1941, Generaloberst Ernst Udet, impressed by what he had seen of the Me 163A flight trials, approved the plan by Messerschmitt to develop and construct seventy operational interceptor variants of the Me 163A. Designated the Me 163B and called the Komet, these rocket fighters would be propelled by the even more powerful 1,500-kg variable thrust Walther R-11-211 'hot' rocket motor. It was also proposed that the first operational Me 163B fighter group would be in place by the spring of 1943, and that the prototype Me 163As would act as trainers.

To assist with the test-flying programme, Dittmar was joined by test-pilot Rudolph Opitz; but in November 1941 Dittmar sustained a severe spinal injury while landing an Me 163A, and Opitz became the prime test-pilot. Dittmar was the chief civilian (Messerschmitt) factory experimental test-pilot on the Me 163 programme, while Opitz was the chief military experimental test-pilot for the programme. With Dittmar injured, Opitz temporarily assumed both roles. Aircraft that were in regular series production received their initial pre-delivery acceptance flight checks by factory production acceptance 'test-pilots' – which presented an opportunity for Hanna Reitsch to fly the Me 163B (glider only). The chief of the Regensburg production plant requested that she be a production acceptance 'test-pilot' for his sub-division.

Hanna Reitsch: German test-pilot

Hanna Reitsch was born in Hirschberg, Silesia, on 29 March 1912. Although raised with the belief that it was her duty to become a good German wife and mother, the young Hanna became interested in medicine through observing her father's work as an eye specialist. Then, in her early teens, she announced to her parents that she intended to be missionary doctor. Her father approved, but she wanted to become a *flying* missionary doctor – and he certainly did not approve. But they came to an agreement. In return for not mentioning anything to do with flying until she had passed her school leaving certificate, her father would permit her to learn to fly a glider.

Having dutifully adhered to her part of the agreement. Hanna was duly sent to the nearby Grunau Gliding School in 1931. As the only girl on the course, coupled with her small stature of just over 5 feet and her weight of only 6 stone 6 lbs (41 kg), she found that her desire to fly was not taken seriously. Undeterred, however, she persevered and successfully passed her 'A', 'B' and 'C' tests to gain her glider pilot's licence. In 1932 she set (unintentionally) a five-hour endurance record for women, and followed this with an altitude record (again unintentionally) in 1933. Wearing only a light summer dress and sandals (it was an unplanned flight), she had allowed herself to be drawn up into a violent storm cloud, where, as a result of the icing-up of her instruments and windows, she had for a short time lost control of the glider.[10]

It was also during this time that Hanna Reitsch began her medical studies in Berlin; but her thoughts were preoccupied with flying. Having again persuaded her parents that her career as a flying doctor in Africa would greatly benefit from her being able to fly a powered aircraft, she enrolled at the amateur flying school at Staaken, in Berlin. Once again she was the only female in the class, and once again her determination resulted in her gaining her 'powered' pilot's licence. She had also shown that she was not afraid of hard work and dirt when, as a direct result of wanting to know how aeroplane engines work, she spent a Sunday stripping down and rebuilding a discarded aircraft engine.

All of this was achieved to the detriment of her medical studies, and despite unexpectedly passing her examinations to go to the Medical Faculty at Kiel University she would eventually abandon her studies to pursue her overriding passion for flying. In June 1934, at the invitation of Professor Georgii, she joined the German Research Institute for Gliding Flight at Darmstadt-Griesheim. Georgii had first met Reitsch when she competed in her first Rhone Soaring Contest at the Wasserkuppe in the summer of 1933. Although beset with bad luck, and not attaining a place in the results of the competition, her determination and persistance endeared her to Georgii, and he asked her to join his expedition (which included Heini Dittmar) to South America to study thermal conditions. She accepted, and while in Argentina she performed a long-distance gliding flight that resulted in her becoming the first woman to be awarded the Silver Soaring Medal.

When she arrived at the German Research Institute Reitsch had no specific tasks, and therefore conducted meteorological flights, long-distance flights and altitude flights with Heini Dittmar. Within a few weeks she had also set another new long-distance world record for women with a glider flight of more than 100 miles. Later that year, having had a successful trip to Finland to train glider pilots, she was given permission to attend the semi-military German Civil Airways School at Stettin, near the Baltic Sea; and as previously, she was the sole female at the school, as it had hitherto accepted only male candidates. Undeterred, however, she successfully learned to fly the school's large powered aircraft.

In 1935, having returned from Stettin, Hanna became a member of the Institute's Gliding Department, run by Chief Designer Hans Jacobs. The following year she began testing dive-brakes, which had been designed by Jacobs to increase a glider's

Hanna Reitsch and Rudy Opitz. (Courtesy Rudy Opitz.)

stability in the air while limiting its top speed, even in a vertical dive. Having successfully completed the testing, she then demonstrated their effectiveness as a safety device in front of an invited group of Luftwaffe generals, including Generaloberst Ernst Udet, Director of Luftwaffe Equipment. Duly impressed by what he saw, Udet requested that they be fitted to certain German military aircraft, and Reitsch presented a further demonstration to Germany's leading aircraft manufacturers. As a result she was awarded the honorary title of Flugkapitän. Again, this was the first time that such an award had been given to a woman; and it was the first time that it had been awarded to a pilot conducting aeronautical research.

Under orders from Udet, Reitsch was seconded to the Luftwaffe Testing Station at Rechlin. as a test-pilot, in September 1937. Here she was able to fly every type of military aircraft that was stationed at the base, although as usual it was made clear to her that she was encroaching on an all-male preserve. While at Rechlin she had the good fortune of being asked by a fellow test-pilot, Karl Franke, to fly him to Bremen to continue his flight-testing of Professor Focke's prototype helicopter. She then had the further good fortune to be mistaken as Franke's co-pilot by Professor Focke, which Franke generously accepted. In her usual manner, she set about learning all she could about this novel way to fly, and on her very first test-flight she carried out an untethered hover some 300 feet above the ground. She had achieved another female first, and was awarded the German Military Flying cross – which itself was another female first. But this was not the end. In recognition of her flying achievements – which included experiments with a cutting device to sever the restraining cables on barrage balloons over London, which were a danger to German aircraft – she was awarded the Iron Cross (Second Class) by Hitler on 28 March 1941. However, her most life-threatening challenge would be during a production acceptance flight in the Me 163B Komet.

In the summer of 1942, the German High Command in Berlin gave Reitsch permission to fly the pre-production Me 163B Komet. Duly arriving at the Messerschmitt factory site at Augsburg, she carried out four unpowered transition flights in the pre-production Me 163B, under instruction from the Komet's prime test-pilot, Rudy Opitz.[11] Her fifth Me 163B flight took place at the Messerschmitt factory site at Oberstraubling, near Regensburg – although her version of what transpired on that day is not an entirely true account of what happened.

Rudy Opitz was the only qualified Me 163 pilot present when Reitsch had her accident, and he believes that she acquired her fifth flight through a recommendation from the head of the Regensburg plant, who suggested that she fly the first production (Me 163B) aircraft that came off that line. Consequently, in October 1942 she went to Oberstraubling, near Regensburg, to fly the first production Me 163B Komet rocket-fighter that had come off the Messerschmitt factory production line. This 'acceptance' flight would be unpowered; but the Messerschmitt factory had not prepared for flight-testing the (Me 163B) aircraft because they had never had one of their pilots 'transitioned', and consequently did not have the tow plane with the tow hitch. Opitz was therefore ordered to fly from Augsburg to Regensburg, so that he could personally tow Reitsch in the Me 163B.

Opitz recalled that the Me 163B had no adjustable seat and no adjustable pedals, and as Reitsch was rather small in stature, she had a problem both reaching and having full control of the pedals. To overcome this problem, heavy cushions were put at the back of her seat. Unfortunately this meant the shoulder harness then did not fit, but she continued with the flight despite only being restrained by the lap belt. The Me 163B that Reitsch flew did not have any guns or rocket engine installed, although it had a gun sight. As Opitz remarked: 'It was just a glider!'

With Opitz at the controls of the Me 110 tow plane, Reitsch was made ready for her flight in the Me 163B. When airborne behind the Me 110, she tried to release the undercarriage trolley (landing being achieved by means of a landing skid) at a height of 30 feet, but due to a technical failure it had become jammed. Her dilemma had been noted by Willi Elias, who was onboard the Me 110, and he informed Opitz that she had not dropped the trolley. Opitz signalled the situation to Reitsch by repeatedly lowering and retracting the Me 110's undercarriage. As he later recalled: 'There was no radio installed in her Me 163B [navigation radios were not fitted until much later], and no electrical instruments, only pneumatic ones [altimeter and airspeed indicator] and a compass.'

At about 10,500 feet, Reitsch cast off the tow-rope, and then tried, unsuccessfully, to rid the Me 163B of its unwanted trolley. As it was unthinkable to abandon such a valuable aircraft by bailing out, she decided to attempt a landing, and after reassuring herself that the aircraft was controllable she began her descent.[10, 12] Opitz had been in a similar situation on his first flight in the Me 163A, and he had landed with the trolley still attached to the aircraft. Apart from a long landing, the flight had been otherwise uneventful and, fully expecting a similar outcome, Opitz made a fast descent in the Me 110 and then rolled along so that he could both observe Reitsch's whole approach and landing, and be there at the end of the runway when she rolled-out long. From his vantage point, Opitz watched as Reitsch attempted to make a 'trolley-attached' landing. As he recalled: 'She didn't have enough altitude to complete her final turn. She set herself up extremely tight so that it was a very sharp 180-degree turn to land, and she never reached final approach. Out of that turn she had to level off, because the ground came up and made the first ground contact with the Me 163B level. Basically, she was too low! After the first ground contact, the aircraft briefly became airborne again, only to strike the ground a final time. Final contact with the ground was with the aircraft level, but the deceleration in the freshly ploughed field was very high, and the aircraft started to tip up on its nose. It decelerated such that when it stopped it was in a near vertical position. The aircraft then fell back into its normal wings level position. It did not fall onto its back, and Hanna never somersaulted as she claimed. The aircraft was standing maybe 100 feet from the edge of the runway, not really lined up but standing level. Apparently during the sharp deceleration and without having the shoulder harness, Hanna had slammed her face into the gun sight. It was – and still is – my judgement that the landing which I observed should have caused no injury whatsoever to a pilot who had the proper full harnesses on. After the landing I was one of the first to arrive. Hanna was already out of the aircraft, having raised the canopy and made her own way out. She had her hanky in front of her face, and very quickly people and cars came. Basically, the Me 163B was undamaged and standing upright.'

Reitsch's injuries were fairly severe. She had a crushed nose, six fractures of the skull involving compression of the brain, and displacement of the jaw-bones. For five months she lay in the Hospital for the Sisters of Mercy, in Regensburg, fighting for her life. Then, in March 1943, having been given an artificial (reconstructed) nose, she was discharged from hospital – her only wish being to fly again. Rejecting offers to convalesce at a sanatorium, she made her own way to a friend's isolated summerhouse in Saalberg, wher she successfully overcame her constant headaches and giddiness by means of roof climbing, pine-tree climbing and mountain walking. To regain her powers of concentration and focused thought, Reitsch sent for her secretary, to whom she could dictate something each day. Confident that she was sufficiently recovered, Reitsch began to secretly fly again, her main concern being the effect that rapid changes in pressure, brought about by sudden changes in height, would have on her head and brain. But having performed dives, spins, turns and aerobatics, she found that her fears were unfounded, and declared herself fit to resume test-flying once more.

During the night of 17/18 August 1943, RAF bombers severely damaged the rocket test site at Peenemünde. As a consequence, the Eropbungskommando 16 (Test Commando 16) – which had been set up at Peenemünde to both prepare the Me 163B for operational service and train the pilots to fly it – was forced to move to Bad Zwischenahn. It was here, in the autumn of that year, that Reitsch arrived to carry out her rocket flight.

All would-be Me 163B rocket-plane pilots had to undergo a flight training programme consisting of three phases, and including a pressure-chamber test for an altitude of up to 30,000 feet. The first phase of the training was designed to give the trainee pilots a grounding in flying a glider, and was achieved using a Habicht (Hawk) glider with progressively smaller wing spans. Beginning with a wing span of 30 feet, this was then reduced to 24 feet and finally to 18 feet. With the smaller wing span the glider had a much higher landing speed, akin to that of the Komet.

Having gained experience of flying the glider, the trainee pilots then learned to fly the unpowered Me 163 as a glider behind the Me 110 towing aircraft. This was then followed with water-ballasted Me 163A flights, when special 'fuel' tanks were progressively filled with water to increase its landing speed. Upon completion of Phase 2, the trainee pilots then made three rocket-powered flights in the Me 163A, with progressively larger amounts of fuel onboard. At the end of the Phase 3 flights, they were qualified to undertake powered flights in the somewhat heavier Me 163B. However, Opitz does not believe that Reitsch completed the entire flight training programme.

As part of her preparation for making a rocket powered flight in the Me 163A, Reitsch decided that she needed to become accustomed to the noise that the little rocket-fighter produced when its rocket engine was operating. She therefore sat in the Me 163A while the ground crew carried out their tests on the rocket engine, and continued doing so until the noise no longer frightened her. As a result, her mind was clear, and she was able to focus on any emergency that might occur during a powered flight.

Reistch was by then ready to become the first female to fly the reusable Me 163A rocket-plane – a feat that would not be bettered until some 50 years later, when Eileen Collins became the first female pilot to fly the Space Shuttle. As flatulence could be a problem, Reitsch, like all Komet pilots, had to adhere to a strict diet that included food such as toasted white bread with scrambled eggs. Then, having suited up in her fireproof overall, protective boots and gloves, she climbed into the cockpit of the Me 163A and put on her oxygen mask, as the Me 163A (unlike the Space Shuttle) did not have a pressurised cockpit.

Sitting inside the Me 163A, it was all that she could do to hold on to the machine as it rocked under a succession of explosions. Although powered (like the Space Shuttle) by extremely volatile liquid fuels, the Me 163A's highly concentrated hydrogen peroxide propellant had a far more macabre property. It could burn away any organic material that came into contact with it for more than a few seconds. On more than one occasion, sufficient hydrogen peroxide leaked into the Komet's cockpit to ensure that the ground crew were greeted by the pilot's skeletal face!

Undeterred by the inherent dangers, Reitsch roared into the air and began climbing at an angle of 40–50º. Within about 2 minutes she had reached an altitude of 30,000 feet, and within 3–3½ minutes all of the Me 163A's onboard fuel had been exhausted. (The Space Shuttle's three main engines burn for 8½ minutes in order for it to reach orbit). With its fuel spent, Reitsch glided her little rocket-plane back to Earth, and landed at a speed of 170 km/hr. Unlike the other trainee pilots, she would not go on to make a powered flight in the Me 163B, but, stepping out of the Me 163A, she had become the 'fastest woman in the world'.

In recognition of both her Komet flights and the production acceptance accident, Reitsch was awarded the Iron Cross (First Class) by Hitler in February 1944. However, with the war coming to an end, it would be the victorious Allies who would both seize and exploit Germany's advancements in rocket propulsion and aeronautical engineering. As a result, the technological leap from the reusable, subsonic, rocket-powered Komet to the world's first reusable, supersonic, rocket-powered Space Shuttle orbiter would be made in America, not Europe.

MACH 1 AND BEYOND

The speed of sound is not constant. It varies with temperature and hence with the altitude at which an aircraft is flying. (The speed of sound at sea level is 341 m/s, while at an altitude of 10 km it is 301 m/s). Therefore, to link an aircraft's speed to that of the speed of sound at the altitude at which it is flying (the local speed of sound), a mathematical ratio is used: the Mach number, in honour of the nineteenth-century Austrian physicist, Ernst Mach, who conducted research into the aerodynamic behaviour of artillery shells. It is the ratio of the speed of the aircraft to the local speed of sound.

During the aerial battles of the Second World War, many pilots experienced violent buffeting of their aircraft while performing high-speed dives, together with the aircraft's control surfaces either locking or behaving in the opposite sense. The pilots were encountering 'compressibility effects' brought about by their aircraft approaching the speed of sound. When an aircraft flies at speeds below Mach 0.7, the air molecules disturbed by the presence of the aircraft radiate the disturbance information at the speed of sound and, in essence, forewarn those lying ahead of the aircraft of its presence, allowing the molecules to move out of the way of the aircraft as it passes among them. This flow regime is referred to as 'subsonic'. However, as the aircraft flies closer to the speed of sound, the air molecules cannot move out of the way fast enough, as the radiated disturbance information is not reaching them fast enough. Consequently, they begin 'piling up' in front of the aircraft (the air is being compressed), increasing aircraft drag. When the aircraft's speed equals the speed of sound, the air molecules in front of the aircraft do not have any prior warning of the presence of the aircraft, and therefore do not have any time to move out of the way. As a result, the compression of air in front of the aircraft becomes a shock wave, which, like the bow wave in front of a boat, impedes the forward movement of the aircraft. It was both the formation and strength of the shock wave, together with the power required by the aircraft to push through it, that led to the idea of a 'sound barrier.'

The Germans, however, had found a way to delay the onset of compressibility effects by sweeping back an aircraft's wings, making them more dart-like. By this device, with an aircraft travelling at high speed, flow over the wings is reduced to the component perpendicular to the wing's leading edge. Using this, the swept-winged, rocket-powered Komet – the fastest aircraft of the Second World War – was able to reach a speed of Mach 0.84 before it began to be buffeted by compressibility effects. But, in the aftermath of the war, it would be the USA and not Europe that would be pre-eminent in pushing back the boundaries of both high-speed and high-altitude manned winged flight.

Looking similar to Sänger's 1933 stratospheric rocket-plane and von Braun's manned 'A' winged variant of his V2 rocket, the bullet-shaped, straight-winged Bell XS-1 (Experimental Supersonic) was the US Army Air Force's contender to penetrate the enigmatic sound barrier. The XS-1 – or, as it would later be called, the X-1 – was a small rocket-powered research aircraft that was air-launched from under the bomb-bay of a modified B-29 bomber. All the fuel carried onboard the X-1 could therefore be saved for speed and altitude tests. On 14 October 1947, with Air Force Captain Charles 'Chuck' E. Yeager as pilot, the X-1 made its assault on the sound barrier. Taking off from Muroc Air Force Base in southern California, the X-1 was dropped from the B-29 at 20,000 feet. Yeager then flew the X-1 under its own rocket power to an altitude of 42,000 feet, where it reached a top speed of Mach 1.06. The X-1 – which would be followed by second-generation variants – had shown that manned, winged flight at supersonic speeds was possible. This later resulted in a series of rocket research aircraft, including the Douglas D-558-II Skyrocket and the Bell X-2, which achieved speeds beyond Mach 3. Between 1959 and 1968, North American Aviation's X-15 was used for the investigation of high-speed and high-altitude flight, beyond Mach 5 and above 50 miles.

Men were engineers; women were computers

Just as the Messfrauen had supported the rocket research work at Peenemünde, the rocket-plane research flights at Muroc were also supported by female computers. In December 1946 the first two women computers – Roxanah B. Yancey and Isabell K. Martin – arrived at the NACA Muroc site where flight testing of the X-1 had already begun. Although both women had degrees in mathematics, their role as computers would involve the laborious task of transcribing flight data (measuring the deflections on oscillograph film) recorded by the aircraft's onboard instrumentation (that recorded the aircraft's acceleration, air speed, altitude, control positions, pitch, yaw and roll rates, wing loads and so on), and then reducing the data into standard engineering units that could be used by the engineers to analyse the aircraft's flight characteristics. The preference for using women rather than men as computers appears, in part, to have been a result of the work being laborious – which is given further credence by the NASA Dryden Flight Research Center website, which states that 'at least part of the rationale for using women seems to have been that the work was long and tedious, and men were not thought to have the patience to do it.' However, Isabell Martin did not stay long. She had left Muroc by April 1947,[13] and so Roxanah Yancey, together with two new computers – Phyllis Rogers Actis and Dorothy Clift Hughes – plus Walt Williams' (head of NACA Muroc) secretary Naomi Wimmer, were the only females at the NACA Muroc Flight Test Unit when Yeager made his historic Mach 1 flight in the X-1 on 14 October 1947. Yancey – whom Yeager has acknowledged working with – would in 1949 co-author a NACA report on 'The Static-Pressure Error of Wing and Fuselage Airspeed Installations on the X-1 Airplanes in Transonic Flight.' (The transonic region lies between Mach 0.7 and Mach 1.2.)

By the early 1950s the number of computers at the NACA High-Speed Flight Research Station (as it was then called) had grown considerably. The computers – who by then were under the direction of Roxanah Yancey – were recruited during visits to NACA sites at Langley, Lewis and Ames, and from the colleges in California. Beverly June Swanson (Cothren), who graduated in 1947 with a degree in mathematics, was working in the Loads Divison at Langley when she was recruited by Yancey. Having spoken to Yancey in the main office, Cothren decided that she was ready for a little adventure, and arrived at Muroc in autumn 1949.[14] Another computer from Langley was Mary 'Tut' Hedgepeth, who had a mathematics degree and was working as computer in the Flight Research Division. Hedgepeth had married in 1947, and her husband was offered the job of setting up a photographic laboratory at Muroc, which he accepted. Consequently, she was granted a transfer to Muroc, where she and her husband arrived in November 1948.

Another means of recruiting computers was by word of mouth, and one of the women recruited in this manner was Betty Scott Love. She had originally wanted to be an airline hostess, but at that time one of the requirements of the job was qualification as a registered nurse. She had therefore undertaken extensive studies in science and associated fields at high school and junior college, and as a result had graduated from Antelope Valley Junior College with a good science background. She intended to go to nursing school at the Los Angeles County Hospital, but in the

Some of the female computers at NACA Muroc during November 1949.

event she married and raised two children. However, in January 1952, a friend's husband who worked at Muroc informed Love that they were looking for computers, and encouraged her to apply. She did not have a mathematics degree (the prime requirement), but Yancey told her: 'Well, you've had some maths, and you've got quite a bit of science. We'll try it.'[15]

The computers though, had their own views on what level of mathematics was required. According to Cothren, they simply used equations in the work, and there was nothing theoretical or advanced. Hedgepeth agreed: 'It was the sort of job that NACA would like to have people with degrees, because if something more critical in the way of calculations came up, we could do it. But a lot of the measuring stuff didn't require a college degree.'[16] This is a view which, to some extent, is corroborated by Love: 'Anyone with any kind of maths – I think it could have been even high school maths – could have done what NACA really wanted done, except that we were doing loads of work. All of the ladies that did the loads were mathematicians.'

Muroc (now Edwards AFB), in the Mojave Desert, was selected for research flying on the basis of its remoteness (for safety and security), its good flying conditions, and its large dry lakes that could act as runways. Due to its desert location, however, the computers – especially those coming from Langley on the east coast, with its humidity and vegetation – were exposed to a very alien environment: a remote, desolate terrain with sand, wind, tumbleweed and Joshua trees. Before she left Langley to start work at the NACA High-Speed Flight Research Station, Hedgepeth had been told: 'When you go out there, you have to pound a stick in the ground at night and hang your shoes over them so the rattlesnakes and things won't get in' However, her fears were alleviated by her husband, who had been at Muroc for training missions during the Second World War.

As a single woman arriving at Muroc, Cothren was housed in a women's dormitory at NACA's South Base, although she had little time to use it: 'When I first got there, they were working on the X-1, and we were working everyday, even Saturdays and Sundays. So for the first six weeks I didn't do anything except work.' There was no compulsion for women like Cothren to live in the dormitory, but at about $12 per month for a single room with shared kitchen and bathroom, it made economic sense to stay there. As a married couple, Hedgepeth and her husband were initially put into old Navy housing that had been built during the Second World War. Having, on their arrival, been informed that it had been cleaned, she found that everything was coated in thick dust. She would later discover that no matter how frequently everything was cleaned, it was soon recoated in dust. The wind even blew sand through the sealed and non-opening windows.

The computers fared little better with their working environment, which was tightly controlled. They were expected to start work promptly at 7.30 am and finish at 4.00 pm, although in the summer they mostly worked from 5.00 am until 1.00 pm, as this allowed them to take advantage of the cooler early morning. During the day the computers were expected to work at their desks, their only breaks being when they went to the rest room or fetched coffee from the urn, which involved walking through the hangar. In fact, some of the computers who normally did not drink coffee took to drinking it just because it gave them the opportunity to stretch their legs.

Engineers would come to the computers single-roomed office, where they would hand their requests for data to Roxanah Yancey. She would fill out the data sheets,

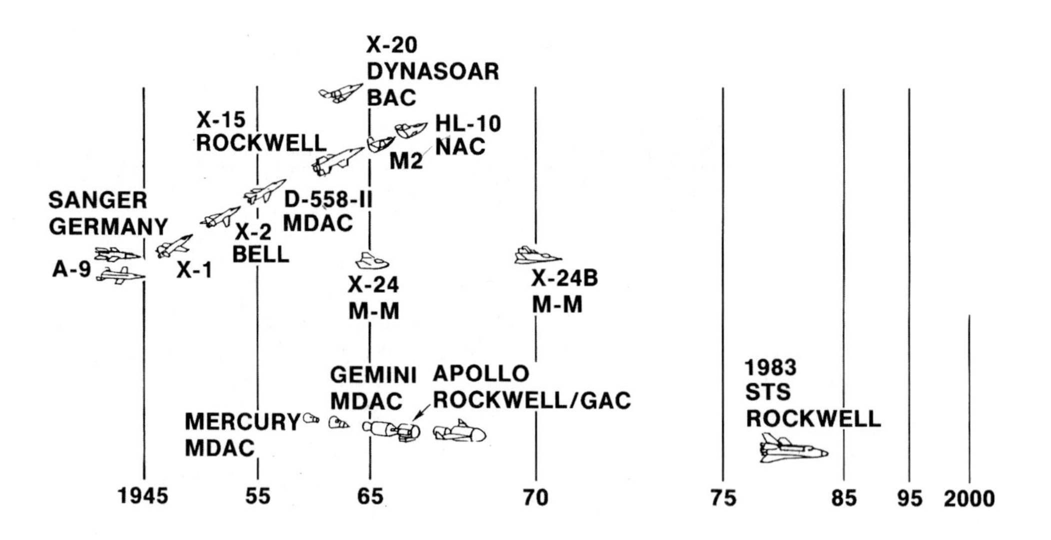

Rocket-plane development compared with NASA's ballistic capsule programmes.

and tell the computer what was supposed to go in each column, although this was not always the case. Cothren recalled working alone on whatever she was doing, with Yancey assigning her to an engineer who would bring the material to Yancey to hand on, or who would just work individually with the computer. Having been given a data sheet by Yancey, the computer would begin working at her desk, using a large mechanical Frieden calculator that was covered at night to prevent its being coated with dust. The first column in the data sheet would be for recording data from the oscillograph film, which would involve the computer using a light-box and measuring device. With the data duly recorded, the computer would then work her way across the columns on the data sheet, following each column's instructions until she had finally calculated the required engineering unit. Yancey, however, was particular about how the computers worked. Everything had to be done by one person and checked by another, and probably double-checked by Yancey. But the computers considered Yancey to be fair, and did not think that she ever tried to overwork or overload anyone.

Hedgepeth left NACA in 1953, and went to work for the Air Force unit at Muroc. During her time as a NACA computer she worked on several X-planes, including the X-2. Cothren left in 1953 and went back east with her husband. Not only had she worked on the X-1 and D-558-II Skyrocket, but she had achieved what many of the computers wished for: having her name included in a report (John P. Mayer, George M. Valentine and Beverly J. Swanson, Flight Measurements with the Douglas D-558-II (BUAERO No 37974) Research Airplane: Measurements of Wing loads at Mach Numbers up to 0.87, NACA-RM-L50H16, December 1950). Love, however, remained at Muroc, and worked on the rocket-powered X-15.

Brassières, capstans and rocket-plane pilots

In addition to solving the problems associated with designing and building high-speed, high-altitude research aircraft, the engineers and scientists also had to resolve the issues of keeping the pilot alive and functioning. As one ascends from the surface of the Earth – which is, in effect, at the bottom of a 'sea of air' – the atmospheric pressure reduces and the air becomes thinner (with decreasing oxygen content). The problem of decreasing air pressure with increasing altitude was overcome by pressurising the pilot's cockpit to that of a lower altitude, while the issues and solutions relating to the supply of oxygen to the pilot can be summarised thus:

- Up to an altitude of 10,000 feet a pilot can breathe atmospheric air.
- Up to an altitude of 40,000 feet a pilot can breathe pure oxygen from a mask.
- Up to an altitude of 50,000 feet a pilot can 'pressure'-breathe pure oxygen from a mask.
- At 63,000 feet the air pressure is equal to the vapour pressure of the dissolved gases in a human body. Consequently, at 63,000 feet the body heat of an unrestrained pilot is sufficient to make his blood boil (explosive decompression).[17]

To overcome the life-threatening consequences of a high-altitude cockpit

A group photograph, taken in October 1947, of the NACA Muroc staff posing in front of the X-1 and the B-29.

pressurisation failure, it was reasoned that pilots would have to wear some sort of pressure suit (or space-suit).

In 1943, the University of Southern California built its own human centrifuge as part of the continuing research into the development of g-suits, which were needed to protect pilots from the strong centrifugal forces generated by high-speed turns and other aircraft manoeuvres. These g-suits were tight-fitting garments fitted with rubber bladders that, when inflated with gas, applied an external pressure to the pilot's abdomen and legs, preventing the migration of blood from the brain to the lower regions of the body, which, if not corrected, would lead to the pilot losing consciousness. One of the staff at USC, Dr James (Jim) Paget Henry, saw great potential in the inflatable bladder principle of the g-suits as a means of creating a pressure suit, and therefore contacted David Clark, of the David Clark Company, in Worcester, Massachusetts, to ask whether he would like to collaborate on developing the DCC g-suit into an altitude (pressure) suit. Clark, however, declined, as he was committed to working on the g-suits, but he provided Henry with a number of US Navy Z-1 coverall g-suits, a quantity of material and thread, and an industrial sewing machine. He also offered him the services of his former principal experimental seamstress, Julia Greene.[18]

Julia Greene was, in the words of David Clark, 'a genial Irish lady' and 'the best producer of the (DCC) jungle hammock,' and he had chosen her to assist him with the development of his experimental g-suit. However, just as the DCC had gone into full production mode for the Navy g-suits, Julia's husband accepted a job at Lockheed, and they had moved to Los Angeles. David Clark therefore contacted Julia, who agreed to work with Henry on his altitude suit.

While Julia Greene was working on the embryonic altitude suit, Helen Lester – a highly skilled clothing designer with General Electric's blanket division in Bridgeport, which had produced the prototype 'Lamport g-suit' – visited David Clark. In developing his g-suit, Dr Harold Lamport, of the Yale University Laboratories, had devised a scheme that employed bladders in the form of tubes that ran along the outside of the suit in a longitudinal direction. Around these external tubes, known as capstans, were interdigitating tapes that were attached to the fabric of the suit. Consequently when the pneumatic capstans inflated, the tapes tightened the suit both evenly and (for the pilot wearing it) comfortably. Upon its completion, the prototype Lamport g-suit had been the subject of centrifuge testing at both Wright Field and the Mayo clinic, but on both occasions had failed to provide any measurable protection against g forces. However, Dr Baldes, at the Mayo Clinic, had been impressed with the capstan design, and had asked the Lamport–General Electric team to show it to David Clark. It was in response to this request that Helen Lester had visited the DCC.

Having shown David Clark a sample section of the Lamport g-suit, Lester told him of the failed centrifuge tests, which she thought had been unfair. Clark then informed her that there was another centrifuge capable of testing the suit – the one at USC. Lester and the rest of the Lamport–General Electric team went immediately to USC where, once again, the suit failed to provide any measurable protection. However, as Henry was there, it is conjectured that he saw the Lamport g-suit and picked up its capstan idea as being suitable for applying a counter-pressure on the arms and legs of his altitude suit.

In 1946, Henry and Greene completed work on the altitude suit – which, in fact, was a partial pressure suit (PPS), as it covered only the pilot's body as far as the neck, wrists and ankles. Designated the S-1, Henry's PPS consisted of an internal, inflatable, double-walled bladder that covered the torso and abdominal areas, with external capstans running down the arms and legs. Using himself as the guinea pig, Henry successfully demonstrated the potential of the S-1 PPS by reaching simulated altitudes of up to 90,000 feet in an altitude chamber at Wright Field. Witnessing the demonstration was David Clark, who had accepted an offer by the USAF to adapt and modify Henry's prototype S-1 PPS for ease of manufacture. The following day he telephoned Greene, who agreed to return to Worcester for three to four weeks and help with the production of a commercial version of the S-1 PPS.

Greene's experience of working on Henry's prototype suit was invaluable, and her initial advice to Clark was to start afresh. Heeding her advice, work began with Clark putting on boiler suit-styled underwear upon which the optimum spots on the elbows, shoulders, hips and knees for attaching the capstans were marked through trial and error. Work then focused on creating the patterns for the shell of a suit that had to be both loose and comfortable for a pilot in the seated position, but still left the arms free while flying. This they accomplished using 'cut and try' methods – first with paper and then with fabric – and it was during this phase of the work that Greene's skills and patience were invaluable.

Betty Love stands in front of the X-1E at Edwards AFB – more than forty years after the aircraft was flown.

With the patterns created and the material cut to take advantage of the fabric's bias, or give, work began on stitching it all together. As there would be no cross-seams, the decision was made to use a French seam – and once again Greene's sewing skills came to the fore (and she even had to teach David Clark how to make such a seam). Attached to the protruding French seams would be the capstans, and this required Julia staying on for a further three weeks to assist with the capstan assemblies. The helmet for the PPS would be supplied to the DCC by Wright Field, which had tasked one Alice King with its construction.

Alice King Chatham: the enigma

According to Ethlie Ann Vare and Greg Ptacek,[19] Alice King Chatham – a well-known sculptor in Dayton, Ohio – was called upon by the USAF during the Second World War to design and fabricate a leak-proof rubber mask for the new fighter aircraft pilots who were reaching ever-higher altitudes. She was then assigned to the X-1 project, where she hand-made the helmet worn by Chuck Yeager when he broke the sound barrier (a helmet that covered the whole head and was worn with a PPS). Following this she worked for NASA on the design of space helmets.

This chronology of events is repeated in an obituary of Chatham, who died, aged 81, on 8 July 1989. In this it is stated that several of her animal sculptures are in the Dayton Art Institute, and that she hand-made the helmet worn by Captain Yeager when he first broke the sound barrier. It also states that she later designed space helmets for astronauts, and that her inventions for NASA include a space bed, stretch-knit garments, and various restraints and tethering devices. Mention is also made to her forming Alice King Chatham Medical Arts, which designs and manufactures physiological equipment for both humans and animals.[20]

However, according to information gleaned from the USAF Flight Test Center, throughout the X-1 programme Yeager wore a helmet that he made himself. He apparently cut the dome off a Second World War tank helmet, and fastened it to a leather flying helmet with an arrangement of snaps and straps (in the improvatory mode typical of those days). Furthermore, the X-1 was not a 100,000-feet aircraft; it was primarily built to research high-speed flight, not altitude. (The X-1 once reached 69,000 feet once, but most of its work was at much lower altitudes). Following on from the X-1 were the A, B, C, D, and E variants, and Yeager *did* wear a pressure suit and helmet during the X-1A and subsequent high-altitude programmes. Yeager himself has stated that he made his own helmet, which he wore on the first supersonic flight in the X-1 on 14 October 1947, and that he wore the T-1 PPS in the X-1A on his Mach 2.44 flight at 80,000 feet on 12 December 1953.[21]

A search of the telephone directories at the former Wright Field has shown that in 1948 a Miss A. King was working at the Engineering Division of the Aeromedical Laboratory. In 1951 the section changed its name to the Physiology Branch, and in 1954 to the Engineering and Development Branch. At that point, Miss A. King disappeared from the telephone book, and there is no listing of a Mrs A. Chatham. However, the fact that a Miss A. King is listed gives credence to David Clark's recollections of an Alice King at Wright Field, as he mentions that it was Alice King who designed and constructed a hand-stitched fabric helmet that covered the whole

USAF test pilots Arthur 'Kit' Murray (standing) and Chuck Yeager wearing the DCC T-1 PPS pressure suit in 1954.

head and incorporated a sealed faceplate that also covered the ears. A further document, 'On Partial Pressure Suits' – which appears to be an early 1960s transcript of an oral history by David Clark – states that a Joshua Chatham married an Alice King, that Joshua Chatham's health failed, and that Alice was reported to be working in the Space Department at Douglas, thus producing a credible link between Alice King and Alice King Chatham. In the words of Bruce Hess, Staff Historian at Wright-Patterson Air Force Base: 'What Alice did exactly is currently debatable, but she undoubtedly was involved in those pioneering steps involving manned space exploration and flight.'

Further improvements and refinements of the commercial S-1 PPS by the DCC resulted in the T-1 PPS, which also featured the new, two-piece, clam-like, K-1 hard-hat which snapped over the tight-fitting fabric helmet. The first operational use of the T-1 PPS occurred in 1949 with the advent of the high-altitude flights of the X-1, and it was on one such test flight that the untried PPS saved both the life of the pilot and, ultimately, the X-1. On 25 August 1949, with Air Force Major Frank 'Pete' Everest as pilot, the Plexiglas canopy of the X-1, which had earlier developed a small crack, failed at 69,000 feet. With the loss of cockpit pressurisation, Everest's T-1 PPS immediately inflated, allowing him to pilot the X-1 down to a 'safe' altitude of 20,000 feet. Everest had become the first person to use a PPS in an emergency situation. The T-1 PPS subsequently became the principle emergency altitude suit for all the early experimental rocket-plane pilots, and would also save the life of Chuck Yeager on 12 December 1953, when he lost control of the X-1A at 70,000 feet. As the X-1A fell out of the sky, Yeager was slammed around the cockpit with such force that his white K-1 hard-hat fractured the canopy and his T-1 PPS immediately inflated. Using all of his piloting skills, Yeager managed to regain control of the X-1A and bring it down for a safe landing.

The T-1 PPS, and its subsequent variants, were only ever designed as an emergency 'get-down-quick' suit, and therefore did not meet the requirements of the pilots assigned to the X-15, which would fly to the edge of space. What these pilots required was a full pressure suit – and the USAF contract to develop the full pressure suit for the X-15 programme was awarded to DCC!

Julia Greene's contribution to the development of the PPS is best summed up by David Clark, who has stated that he probably would not have tackled the project or succeeded without her.

David Clark Company and the two Jacquelines

While the rocket-planes at Edwards were pushing back the barriers of both speed and altitude, two women from either side of the Atlantic Ocean were vying for the title of 'fastest woman'. The women in question were Jacqueline Cochran – who, having been trained by her friend Chuck Yeager, became the first woman to break the sound barrier, in an F-86 Canadian Sabre, at Edwards AFB in May 1953 – and Jacqueline Auriol – who became the second woman to break the sound barrier, in a Mystere IV, in August 1953. Both women had links with the DCC.

As Joseph Ruseckas, former Vice President in charge of Research and Development for the DCC, recalled: 'I had the privilege of measuring Jacqueline

Cochran for her g-suit. Chuck Yeager – her mentor for her record attempts – insisted that she wore a g-suit, and sent her to the David Clark Co. The suit was made and delivered ... By some unusual circumstances, due to high military classification, Madame Auriol also came to the David Clark Co. She was measured and fitted with a partial pressure suit. I have no knowledge as to whether she ever used it in flight.'[22]

Margaret McGrew and the Mach 2.8 missile

During the lull between the rocket-powered flights of the Bell X-2 and the North American X-15, one woman, with the collusion of her work colleagues, broke through the sound barrier in a rocket-propelled winged projectile. She was Mrs Margaret Wood McGrew, and her flight in the Bomarc (Boeing and Michigan Aerospace Research Center) surface-to-air missile (SAM) was the fulfilment of a final wish.

McGrew graduated from Mills College, Okland, California, in 1930, and went on to take advanced studies in mathematics, engineering and aerodynamics at the Universities of California, Alabama and Washington. A dedicated physicist, she was able, by means of her qualifications and drive, to advance in an area that was strictly dominated by men. In 1954, having had a variety of jobs ranging from propeller research at the California Research of Technology to teaching mechanics and stress analysis at the University of California (including spells at the Boeing Aircraft Company in Seattle and Parsons Aerojet in Florida), McGrew joined the Recording Company of America (RCA) at the Air Force Missile Test Center in Florida. By 1955 she had become an operational planning manger, and had her own staff.[23]

The Air Force Missile Test Center encompassed both Cape Canaveral and Patrick Air Force Base, with missile preparation and launching being performed at the Cape Canaveral facility, while the vast majority of the paperwork was carried out at Patrick Air Force Base about 15 miles south of the Cape. At that time, RCA was contracted by the Air Force to run the technical aspects of the Air Force Missile Test Center's range (electronics equipment), and it is the belief of Al Hartmann – a research associate working with the University of California on its Florida Space Coast History – that McGrew's job most probably involved her scheduling when the various missiles could use the range, based on priority and the availablility of equipment. One of the missiles being launched at that time was the Bomarc, but as it was still in its developmental phase the launches would have been made primarily by Boeing, the missile's manufacturer.

It was also in 1955 that McGrew was diagnosed as having an incurable cancer. Undeterred, however, she continued with her work, and it is reported in the *New York Times* that male associates said she insisted in coming in to work at times when they doubted they would have been able to stand on their feet. On 10 January 1956, Margaret McGrew died at the age of 46, and after a small funeral service held in the town of Melborne – where she had lived with her husband and two children, Margo and Robert – her body was cremated.

Margaret – or Maggie, as her colleagues at the Air Force Missile Test Center knew her – had been interested in aircraft development since her teenage years. She had also become a 'missile buff', and before her death had asked her friends at work if they would carry out a final wish for her. Although it would not be easy, they

The Bomarc A, launched on 2 February 1956, carried the ashes of Margaret McGrew – arguably the first female to exceed the speed of sound. (Courtesy Al Hartmann.)

agreed to make it happen. On 2 February 1956, a routine test launch of an 'A' variant of the Bomarc missile (CIM-10A) was carried out. Known by the Air Force as SN# 54-3053, and by Boeing as 623-13, the Bomarc 'A' was launched from its Cape Canaveral silo, and headed out eastwards over the sea. The Bomarc ultimately exploded, and Margaret's ashes – which had been sealed inside by her friends – were scattered majestically over the Atlantic Ocean. Thus her final wish was granted.

EVOLUTION OF THE SPACE SHUTTLE

In December 1968, in Bonn, John V. Becker graciously accepted the Eugen Sänger Medal for outstanding contributions to the field of reusable spacecraft, on behalf of NASA's X-15 team. Addressing the audience – which included Irene Bredt and her son Hartmut – Becker acknowledged the contribution of the Sänger–Bredt antipodal rocket-bomber in the genesis of the X-15: 'Professor Sänger's pioneering studies of long-range rocket-propelled aircraft had a strong influence on the thinking which led to initiation of the X-15 programme. Until the Sänger–Bredt paper became available to the US after the war, we had thought of hypersonic flight only as a domain for missiles. From this stimulus there appeared shortly in the United States a number of studies of rocket aircraft investigating various extensions and modifications of the Sänger–Bredt concept. These studies provided the background from which the X-15 proposal emerged.'[24]

Flights to the edge of space

The next practical step in the development of a reusable orbital spaceplane was the construction and flight testing of three experimental X-15 rocket-planes. Between June 1959 and October 1968 these aircraft completed a total of 199 free flights up to an altitude of 66.75 miles and speeds up to Mach 6.7. This research was directed not only to the development of hypersonic aircraft, but also carried experiments related to the Apollo–Saturn programme. The X-15's legacy to the Space Shuttle included:[25]

- The development and operation of the wedge tail (hypersonic control surfaces) on a manned aircraft.
- The development and operation of the first large restartable, man-rated, throttleable rocket engine (XLR-99).
- The development and operation of dual control systems: aerodynamic (wings) and reaction controls (small thrust rockets for pitch, yaw and roll control outside the viable atmosphere).
- High-altitude 'dead-stick' landings when more than 200 miles from the landing site.
- Correlation between experimental wind-tunnel test data and flight-test data.
- High-quality flight simulations for training 'astronaut' pilots.
- The development of the first full pressure (space) suit and the taking. of physiological measurements on the pilot during operational flights.

Rose Lunn and the X-15

Since she was about six years old, Rose Elizabeth Lunn had always wanted to study at the Massachusetts Institute of Technology. Having graduated from Broadway High School, she enrolled in the then Aeronautical Engineering Department (now Aeronautics and Astronautics Department) in the College of Engineering at the University of Washington in 1933. In 1937 she graduated as the University's first female aeronautical engineering student, at the top of her class. She was then awarded a scholarship, which allowed her to fulfil her childhood wish of studying at MIT.[26]

Upon completion of her three-year-long MS degree in aeronautical engineering at MIT, Lunn became the first recipient of Zonta International's Amelia Earhart scholarship award – a $4,000 scholarship that allowed her to continue studying at MIT for her doctorate. Having gained her ScD, she left MIT and began working at Curtiss-Wright in Buffalo, New York, where she had the job of setting up a Flutter and Vibration Group. However, she was not enamoured of the New York winters, vowed not to stay for a second one, and moved to Los Angeles, California, where she began what was to be twenty-two years of dedicated work with North American Aviation.

Once again, Lunn was tasked with setting up a Flutter and Vibration Group at NAA, to which she then added a Vibration Laboratory for ground and flight testing, an acoustical group, and an analogue computer facility. Her work included the X-15. Because of the decision to use the B-52 as the replacement mothership, the X-15,

Rose Lunn, an aeronautical engineer assigned to the X-15 programme.

unlike the earlier rocket-planes, could not be mounted under the bomb bay, as the B-52's landing gear prevented this. This led to the then controversial decision to mount the X-15 onto a pylon on the starboard wing, between the B-52's fuselage and inboard jet engine. Lunn – who had a reputation in NAA of 'being right' – expressed concerns over the proposed wing-mounting of the X-15 by indicating that the vibration from the B-52's inboard jet engine had the potential to severely damage it. Consequently, a detailed study was undertaken to ascertain the full extent of engine vibration effects on the X-15.[27]

Women at NASA FRC and the X-15

By the time of the X-15 flights, the NACA High Speed Flight Research Station at Muroc had been renamed the NASA Flight Research Center (FRC), and four of the early female computers – Mary Little, Roxanah Yancey, Katherine Armstead and Betty Love – had moved on to engineering posts. Betty Love was working in the Aerostructures Branch where her title changed from computer to aeronautical research engineering technician. For the first time, Love was able to ask questions relating to the 'what and why' of her work; and not only was she given answers, but she was also taken to see the aircraft related to her work.[28]

Love also assisted Jim McKay, who had moved from Langley to NASA's FRC and was responsible for determining all of the landing loads on the X-15. During the course of this work, the engineers wanted to ascertain data such as the landing loads, and so Love was tasked with finding the film that contained information they were seeking and then both reduce it and plot it. These plots, together with the accompanying text, were then sent to editorial, where the final text and figures for the reports were generated. However, as Love had been responsible for producing the initial plots that had been used to generate the final figures for the reports, her name was added to authors of the report.

During her time working on the X-15, Love kept a logbook that was an extension of a reference book she was already keeping. While working on the X-planes, the engineers were always wanting technical drawings, round-the-clock sets of photographs (taken from different angles), tables of dimensions, and similar data. However, having provided this for them a couple of times, Love began setting up a file which contained all of this sort of information. For her X-15 logbook she recorded such items as the flight number, the date, the pilot, the maximum Mach number, the maximum altitude, and, in some cases, the maximum angle of attack. Her X-15 logbook (covering all three X-15s) has proven to be a valuable source of information, and has been cited in several books.

One curious job assigned to Love was that of assisting FRC's illustrator, Mr Fiskan, with the making of an X-15 movie entitled *Pathway to the Stars*, which featured a day in the life of X-15 pilot Joe Walker. Having put together the film's storyboard, Love was then tasked with finding the necessary film footage in the photography laboratory. These short clips of film were then spliced together to produce the final cut of the film. However, as much as Love enjoyed working on the film, she wanted to remain anonymous, and her name does not appear in it; and neither, rather peculiarly, does Mr Fiskan's.

Love later spoke of her career with the X-15: 'I was part of that team and enjoyed doing my part in obtaining the large amount of knowledge from the X-15 flights that was used in other projects to further the history of flight.'[29]

In the summer of 1952, college student Harriet J. DeVries (Smith) began working as a summer aide with the female computers at the NACA High-Speed Flight Research Center (HSFRC). Following her graduation from Bakersfield College, Smith spent two summers working as an aide at NACA's HSFRC while continuing her education at the University of California at Berkeley. After graduating from Berkeley in 1954, at age 20, with a degree in physics and mathematics, she returned to the HSFRC to begin full-time work as its only female engineer. However, she found herself doing very much the same kind of work that she had done as an aide, although this time her own projects made the work rather more interesting. She was, however, still regarded by some of her male colleagues as a female computer, and found herself being used by one of the male engineers as his own private computer – which she considered would not have happened had she been a man.[30]

Smith also felt that at the time she was probably better than the average engineer in mathematics, and much of her early work therefore involved theoretical calculations. For example, although she was not part of the X-15 project team,

Harriet Smith at work at NACA FRC, with fellow research engineers Clinton Johnson and Richard Banner, in April 1958.

she was called upon to calculate the flow fields around the aircraft. She has also cited an incident with the X-15 as indicative of the discrimination against women at that time: 'A big X-15 conference was held at NASA's FRC, and they were going to serve lunch out in the hangar. But then the Director, Paul Bikle, decided to have the female employees, including the engineers, wait the tables. However, another female engineer, Bertha M. Ryan, who knew Bikle personally, went and told him that she was clumsy, and that she was 'liable to spill soup over a guy and then later have to work with him in a professional capacity, and he wouldn't take her seriously.' Bikle, having thought about it, saw that Ryan was right, and both she and I were exempted from being waitresses.' Both Smith and Ryan would be involved in the Lifting Body programme, which would investigate the concept of using the body of an air-vehicle, as opposed to its wings, to generate lift.

At the end of May 1963, college student Sheryll Goecke (Powers) arrived at NASA's FRC as one of two female aerospace engineering co-ops from Iowa State University. Under the co-operative programme, Powers would alternate between work periods at NASA and periods of study at the university. However, as a co-op student she would take a year longer to graduate than would an ordinary student; but this was tempered by the fact that she was both gaining work experience and being paid at the same time. One of the first projects that Powers worked on was the X-15, where, having worked on the base drag of the aircraft, she wrote a NASA technical report.[31]

Although there were still female computers at NASA's FRC when Powers started as a co-op, she felt that there was no confusion between them, although she noted that the co-ops always complained about receiving all of the tedious work, even though that was all they could be given and was what they expected. However, just as Harriet Smith was remembered for being a computer when she returned as a full-time engineer, so Powers felt that people still remembered her as the co-op when she returned. After working on the X-15 project, Powers, who was in the Vehicle Aerodynamics Group, moved onto other projects, one of which would involve the first Space Shuttle, *Enterprise*.

Following the end of the Second World War, the US Air Force – which believed that any future global conflict would be dictated by the country that controlled the 'high ground' of space – decided to expand its theatre of operations into this new frontier. As a result, it initiated a number of studies into a manned, winged, space bomber–reconnaissance aircraft, which would evolve the antipodal rocket-bomber concept proposed by Sänger and Bredt. On 10 October 1957 these studies were consolidated into a single programme called Dyna-Soar, which would involve the development of a manned spaceplane capable of being boosted into orbit (DYNAmic ascent) and then gliding (SOARing flight) back to Earth. On 20 May 1958 the Air Force and NACA signed an agreement for NACA's participation in the Dyna-Soar programme, which was seen as a successor to the X-15. However, events and politics (not least the US commitment to landing a man on the Moon within the decade), coupled with ever-growing complexities and costs, conspired to result in the cancellation of Dyna-Soar on 10 December 1963.

With the cancellation of the X-20 Dyna-Soar, the Air Force switched its attention to NASA's lifting bodies. In the early 1950s work was carried out into the heating effects on a body re-entering the Earth's atmosphere from either Earth orbit or the return leg of a mission to the Moon. As a result of this work, H. Jullian 'Harvey' Allen, of NACA Ames Aeronautical Laboratory, introduced the 'blunt-body' concept. That is, for a blunt object re-entering the atmosphere (at hypersonic and supersonic speeds), a strong, curved (detached) bow shock wave is formed ahead of the blunted surface which effectively slows down the airflow passing directly over it, and in turn reduces the surface heating by (skin) friction. Although Allen's concept significantly reduced the effects of surface heating by skin friction, the peak surface temperatures were still high enough to melt the heat-resistant metal alloys that would be used to construct the space capsules. This problem was overcome by the application of a sacrificial ablative heat shield, which was thick enough to not be totally vapourised by the high surface temperatures. However, the scientists at Ames wondered whether it would be possible to take a vehicle designed for the rigours of launch and re-entry and then shape it in such a way that instead of a ballistic splashdown in the sea, it could be aerodynamically flown from orbital (hypersonic) speeds down to subsonic speeds for a safe (horizontal) landing on a conventional runway. Hence the 'lifting body' concept was born.

In 1957 Dr Alfred J. Eggers, at NACA Ames, derived the first lifting body shape: a modified half cone with a flat top and a blunted nose. The shape was then further refined to a 13-degree half-cone with a blunt, rounded nose. However, as result of wind-tunnel tests – which had shown that at subsonic speeds the shape (designated

M2) had a tendency to tumble end over end – the top and bottom of the half-cone were 'boat-tailed' to produce an aerofoil tear-drop shape. To this final M2 configuration was added a protruding 'bubble' canopy and two short, stubby tail fins. The series of lifting bodies, of various configuration, were flown between 1963 and 1975, and included the M2-F1 (Manned, Modification 2, Fuselage No. 1), the M2-F2, the HL-10 (Horizontal Landing, tenth concept), and the X24-A. These vehicles were used to investigate the projected descent corridor of a lifting body returning from space, in both powered and unpowered descent, approach and landing.

The lifting body programme had proven that it was possible to land an unpowered returning spacecraft on a conventional runway. This would be put to the test by the first manufactured Space Shuttle orbiter *Enterprise*, during the 1977 Approach and Landing Tests programme.

Women at NASA FRC and wingless flight

Bertha M. Ryan grew up in Newton, on the outskirts of Boston, Massachusetts. She had become interested in aeroplanes at an early age, wanted to be involved in aviation, and began flying lessons while still at high school. She took her first flight in 1945 – a flight arranged by her mother, who was opposed to her daughter's desire to fly, and hoped that it would frighten and discourage her. This ruse, however, proved ineffective. Bertha learned to fly at Framingham Airport, Massachusetts, made her first solo flight, in a Taylorcraft aeroplane, in October 1945, and went on to earn her private pilot's licence flying a Piper J-3 Cub at Norwood Memorial Airport, Massachusetts.[32]

Having learned that a knowledge of mathematics is essential in aviation, Ryan enrolled at Emmanuel College, Boston. From here she went to MIT, and although her mother discouraged her from going into engineering, she switched from mathematics to aeronautical engineering. However, although her mother did not want her to pursue these choices, neither did she stop her. She also passed on to her daughter two pieces of motherly advice which are as relevant today as they were then (and perhaps even more so): 'You can do anything you want if you work hard enough', and 'Don't be afraid to be different.'

Following her graduation from MIT, Ryan went to work for the Douglas Aircraft Company in Santa Monica. However, this was not her first choice, as she had heard about the aircraft and flight testing that was being carried out at the NACA High-Speed Flight Research Station, and she wanted to work there. Unfortunately, the NACA recruitment officer at MIT was from Langley, and he advised her that she would not want to work at the HSFRC, as it was in the desert, and that instead she should work at Langley. But she did not want to work in Virginia either, and therefore went to work in the aerodynamics research group at Douglas.

Ryan worked at Douglas for about four years, and then left to join her first-choice aviation research organisation, the renamed NASA Flight Research Center. Starting work on 31 December 1959, she spent her first year working on sonic boom studies before moving on to the Lifting Body programme. Initially she was only the second person assigned to the project, and spent her time gathering and analysing data. Then, as flight-test data came in from the lightweight M2-F1 lifting body, she was able to compare it against experimental wind-tunnel test data, simulations, and calculations. This she found very satisfying, as it allowed her to ascertain whether her calculations were accurate. Also, as information needed to be disseminated quickly, most of her technical reports were informal, and were designed to disseminate the information. But it is her belief that she had at least one formal NASA technical report published.

Ryan's connection with the M2-F1 even extended to her recreational activities. During her time at MIT she had taken up gliding, so before she left to work for Douglas she had a glider kit sent ahead. She then built the Schweizer SGS 1-26A glider herself, and flew it during her first summer at Douglas (1956). Initially she had begun building the glider in a friend's garage in Santa Monica, but had moved it to Gus Briegleb's glider operations outfit in El Mirage for its final construction and first flight. She then flew from El Mirage for several years, and came to know Gus very well. Another glider pilot who flew from El Mirage was NASA FRC Director Paul Bikle, so when it came to the construction of the plywood airframe for the M2-F1, Gus, with his experience of designing and building gliders, was given the contract. Ryan's gilding activities also brought her into contact with glider pilot and Komet rocket-plane pilot Hanna Reitsch.

Although not part of the M2-F1 flight test team, Ryan (who was involved in analysis) was able to go onto the lake bed to observe what was happening. She was also able to take her own photographs of the flight tests; and it happened that her photographs were the only ones available for showing others what had been achieved during the first few M2-F1 test flights. Consequently her photographs were used, although soon afterwards a NASA photographer was sent out.

Ryan continued to work full-time on the Lifting Body programme, but left to work for the Naval Weapons Center at China Lake before the X-24A and M2-F3 lifting bodies flew. During her time at NASA she met a number of test-pilots who would later become famous – one of whom was Neil Armstrong. Together, they took a heat transfer course from the University of Southern California, and Armstrong 'put it to practical use, of course. I'd be stuck on some problems or something, and he'd tell me how to do them. He was a smart guy, a good choice for the job he had.' While Ryan was at NASA, her mother finally accepted her daughter's choice of career.

Another female engineer involved in the Lifting Body programme was Harriet Smith, who, like Ryan before her, had worked on the first lifting body, the M2-F1. However, Smith was hampered in her work by a directive from director Paul Bikle which prevented women from going into the hangar. Citing this as yet another example of the prejudice against women at that time, Smith recalled that Bikle – whom she described as a 'real male chauvinist' – did not want women in the hangar because it bothered the men. As a result, she had to liase with the technicians and mechanics via a male engineer, which was a very unsatisfactory way of working. Like Ryan, Smith left the programme before the X-24A or M2-F3 flew.

Enter *Enterprise*

On 5 January 1972, President Richard Nixon gave NASA official approval to proceed with the next phase of US manned spaceflight: the development of the Space

Transportation System (STS). Two months later, on 15 March, NASA selected the configuration for the STS launch vehicle. Known as the Space Shuttle, it would be a three-component vehicle consisting of a reusable manned spaceplane (Orbiter) that would house three reusable Space Shuttle Main Engines (SSME), a large expendable External Tank (ET) that would be attached to the underside of the Space Shuttle Orbiter and house the liquid oxygen and liquid hydrogen propellants needed to power the SSMEs, and two large partly reusable Solid Rocket Boosters (SRB) that would be attached to the side of the ET and provide additional thrust during the Space Shuttle's boost phase of flight.

The first Space Shuttle Orbiter (OV-101) was rolled out of Rockwell International's Palmdale factory (US Air Force Plant 42), California, on 17 September 1976. NASA had intended to name 101 *Constitution*, but a petition by fans of the science fiction TV series *Star Trek* led to President Ford naming 101 *Enterprise*.

While NASA was endeavouring to catch up with the Soviet Union's lead in space achievements, *Star Trek* was portraying a future in which space travel was an everyday occurrence, and humanity – which had resolved its Earth-bound differences and disputes – had formed an alliance with other intelligent life-forms to explore the 'final frontier.' Since both the *Star Trek* concept, and thus the pilot, involved a futuristic starship (*Enterprise*), creator Gene Roddenberry went to great lengths to make the vehicle believable, and NASA, North American (manufacturers of the X-15 rocket-plane and the Apollo Command Module), and other leading US aerospace companies assisted in its design. Having created his starship, Roddenberry needed to work on her crew – the human element of his show and the one to which his audience could relate. However, unlike the prevailing attitude at NASA, Roddenberry chose to explore the possibility of having a female in a position of responsibility in the operation of the starship. Known as Number One, this female would be the Executive Officer onboard the *Enterprise*, and one of her roles would be to command the vessel when the Captain was not on board. Unfortunately, stereotypical roles for women were also included in the show – the most brazen being the Captain's voluptuous yeoman.

On 12 December 1964, shooting began on the one-hour pilot. Two months later, in February 1965, the final edited film was delivered to NBC. It was rejected. The main criticism was that the show was 'too cerebral', and that the viewing public would not understand what was happening. NBC then made the unprecedented decision of asking Roddenberry to shoot a second pilot, but with changes – one of the casualties being the female Number One. Having screened the pilot to selected audiences, NBC found that they did not like a 'tough, strong-willed woman', and so Roddenberry was told to remove her character, which he duly did. Number One's ice-cold, emotionless characteristics and logical reasoning were transferred to Mr Spock – another character that NBC did not like.

The second pilot, 'Where No Man Has Gone Before', was delivered to NBC in January 1966, and one month later the show was given the green light. *Star Trek* would debut as a series in the autumn of 1966. Further changes were made to the *Enterprise*'s multi-racial crew, which saw the introduction of two new regular female characters, the most radical of which (at that time) was the casting of the ship's Communication Officer, Lieut Uhura, as an African female. As a mirror of NASA's Mercury days (Dee O' Hara) the *Enterprise*'s male medical doctor was assigned a female chief medical assistant.

Star Trek would run for three seasons before being cancelled in 1968, a year before man landed on the Moon. After its cancellation, however, the show's popularity began to increase through TV syndication, and this would become self-evident at the roll-out of NASA's first Space Shuttle Orbiter in the mid-1970s. In the American psyche, space fact and fiction would be forever merged when the first Space Shuttle Orbiter was named *Enterprise* by President Ford. It was a fitting tribute both to a series which tried to show women in a positive role, and a real vehicle – the Space Shuttle – which would see the fulfillment of all the Mercury 13's dreams (with the exception of a walk on the Moon), including the first US female astronaut, the first US black female astronaut, the first US female spacewalk, the first female astronaut-pilot, and the first female astronaut commander.[33]

Before NASA could certify the Space Shuttle system as spaceworthy, it needed to ascertain the gliding qualities of the Orbiter. Therefore *Enterprise* –which would be used to carry out a series of atmospheric flights as part of the Approach and Landing Test (ALT) programme – had not been configured for spaceflight, and had dummy SSMEs, RCS and OMS pods designed to replicate both the geometry and mass distribution of the Orbiter. Following a programme of unmanned tests during February–March 1977, three manned captive flights, each with a two-man crew, were followed by five free flights to demonstrate that a non-powered gliding orbiter could return from space and land safely.

US clearance for females on Space Shuttle flights

The Space Shuttle, with its possible complement of up to seven astronauts, would permit astronauts with no pilot training to venture into space for the first time in the US manned space programme. This new type of NASA career astronaut, the Mission Specialist (MS), would be selected from the engineering and science fraternity. In addition to assisting the Space Shuttle Commander and Pilot (both NASA career Pilot astronauts), the MS would be responsible for the coordination of onboard Space Shuttle operations in crew activity planning, consumable usage (fuel, water, food, and so on) and experiment and payload operations. NASA would also permit another new category of astronaut to venture into space onboard the Space Shuttle. Known as the Payload Specialist (PS), this type of non-career astronaut would be a professional scientist working in either the physical or life sciences area. Selected for a particular mission by either the payload sponsor or customer, the PS would carry out specific Shuttle-based experimental work. Unlike the career Pilot and MS astronauts, the PS did not have to be a US citizen.[34]

Another change in NASA's entry requirements for Space Shuttle astronauts was the physical fitness of candidates. The earlier 'ballistic' rocket flights of the Mercury, Gemini, and Apollo programmes had subjected the astronauts onboard to very high physical forces, but this would not be the case for astronauts onboard the Space Shuttle, who would be subjected to a more benign 3 g. NASA therefore eased the

The twelve USAF nurses selected for fitness for spaceflight studies in 1973.

physical entry requirements. Prospective Space Shuttle astronauts would only have to be in good health, and age limits for MS and PS candidates were also lifted.

These changes, coupled with a change in social attitudes, would finally give US women (in the first instance) access to space. However, one issue was still unresolved: female adaptation to weightlessness. Despite the many studies that had been conducted into the physiological responses of men to both simulated and real weightlessness, there was little or nothing relating to women. Acknowledging this omission were Dr Harold Sandler, Chief of the Biomedical Research Division at NASA Ames Research Center, and Dr David Winter, Director of Life Sciences at NASA HQ, who in the early 1970s began to accrue much-needed information on the physiological responses of women to weightlessness as simulated by bed rest. Data generated from male astronauts in space had shown that bed rest was a valid simulation of the effects of weightlessness on the human body.[35]

In 1973, Sandler and Winter recruited twelve US Air Force nurses for the first US investigation into female fitness for space travel. (Air Force nurses were deemed the ideal test subjects on the basis of their medical and flight training). The idea of using nurses in the study came from Dr William F. Winter, Director of Medical Research at NASA Dryden Research Center, who would act as the liaison between NASA and the US Air Force's Brigadier-General Claire Marie Garrecht. In 1972 Garrecht had become Command Nurse for Tactical Air Command at Langley Air Force Base, and was so enthusiastic about the joint programme of NASA–USAF nurse cooperation that she provided two active nurses from her own staff. She also instructed Major

Dixie Lee Childs at Hamilton Air Force Base to recruit ten nurse Air Force Reserve Officers from the San Francisco Bay area.

The twelve volunteer nurses, aged between 23 and 34, were assessed as being well adjusted psychologically and in prime physical condition. They were also subjected to medical and gynaecological examinations, both before and after the study. Three months prior to the study they all stopped taking oral contraceptives, to prevent any contamination of the study's biomedical measurements. During the study, the nurses would wear a 'biobelt', designed to transmit both body temperature and heart-rate information to a central receiving–recording device.

For five weeks the nurses were confined to the Human Research Facility at NASA Ames, where their environment (temperature, and exposure to simulated daylight) was tightly controlled. The first fourteen days, known as the Control Period, was spent undergoing a series of tests involving acceleration tolerance (centrifuge testing), Lower Body Negative Pressure (LBNP) exposure, and a Collins bicycle ergonometer, to establish baseline measurements.

The control period was followed by seventeen days of bed rest involving eight of the twelve nurses with the highest tolerance to the centrifuge and LBNP testing. During this phase of the study the eight nurses – referred to as subjects A to H, as their names were withheld – were given one pillow and were not allowed to exhibit excessive muscular movement. They were allowed to raise themselves up on one elbow at meal times, but all bodily functions and washing had to be carried out in a horizontal position. They were even given prismatic spectacles so that they could read while lying down.

The other four non-bed rest nurses, referred to as subjects I to L, were used as Ambulatory Controls and underwent the same tests as the bed rest nurses so that direct comparisons could be made. On the last day of the best rest phase, all the nurses carried out centrifuge, LBNP and bicycle ergonometer tests.

The final part of the study was a five-day Recovery Period. On the third day the nurses carried out their last centrifuge tests, followed on the fifth day by their last bicycle ergonometer test. LBNP testing was carried out five and 90 days after the end of the bed rest phase.

When compared against the bed rest results for men, the nurses were found to show a similar deconditioning, but with some differences. The study also indicated that the four non-bed rest nurses showed signs of deconditioning, which was attributed to the stress of confinement. Overall, the results of the study indicated that women were capable of coping with exposure to weightlessness, and might be more sensitive subjects for evaluating countermeasures to weightlessness and developing criteria for assessing applicants for Shuttle missions.

Winter and Sandler followed the Air Force nurse study with one involving three different age and gender groups: 35–45-year-olds, 45–55-year-olds, and 55–65-year-olds. Each group would comprise 50% male and 50% female subjects, and the study would operate on a single-sex rotational basis (a male age group followed by a female age group). However, even though the study was entitled Shuttle Re-entry Acceleration Tolerance in Male and Female Subjects Before and After Bed Rest, the main focus of attention would be on the women. Consequently, the study was named 'Housewives in Space.'[36]

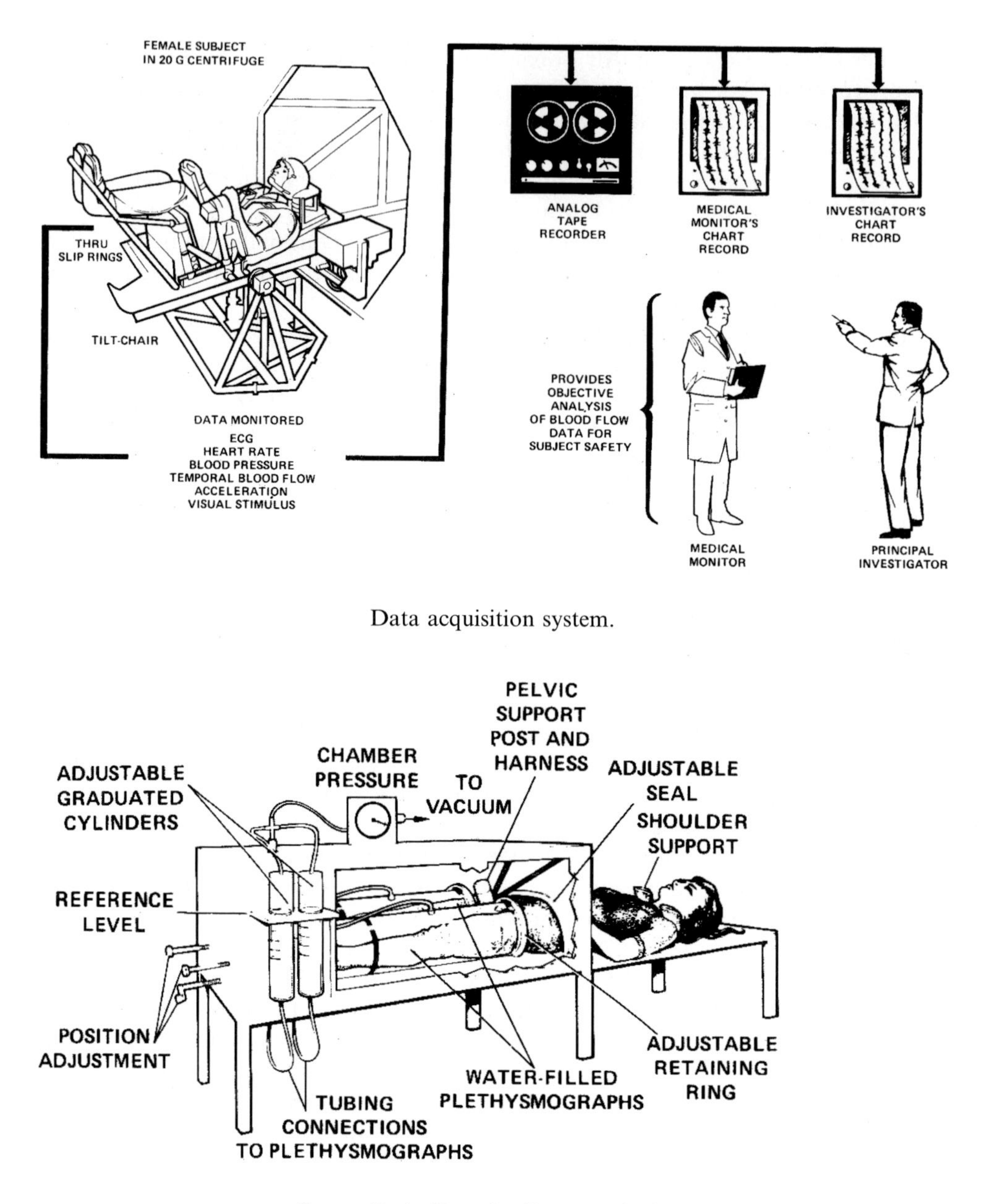

Data acquisition system.

Lower Body Negative Pressure device.

In 1976 NASA sent out an invitation for potential candidates, and for those that responded an extensive screening process followed. First they were interviewed to ensure that, like the nurses, they were well adjusted psychologically and had a real purpose in life. Those who successfully passed the interview were subjected to two weeks of intense physical testing involving the centrifuge, the LBNP, and running on

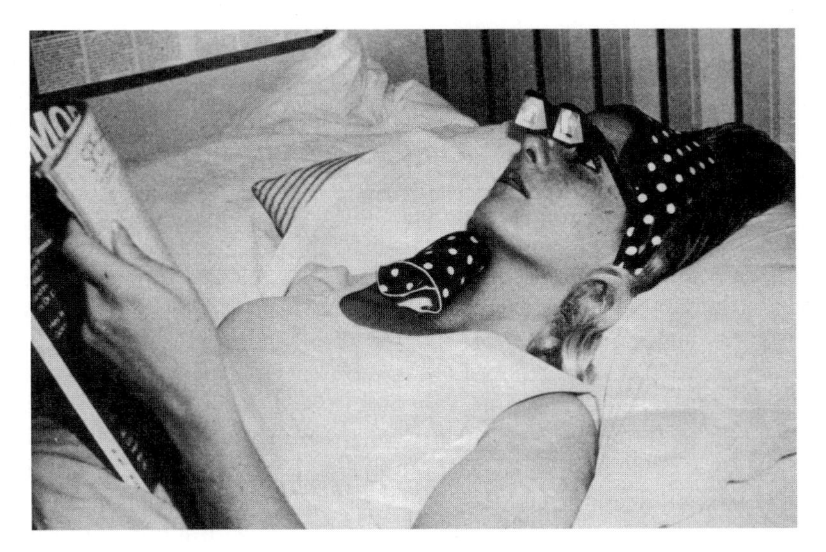

Bed rest subjects had to remain in a horizontal position, and were provided with prismatic spectacles for reading.

a treadmill. This was followed by a further four to five weeks of similar testing, but this time the candidates were forbidden from taking any alcohol, cigarettes, caffeine drinks, or medication. The eight to ten final subjects per age group were then selected for the bed rest study, which would run from 13 April 1977 to 28 April 1981.

In all, twenty-seven women were selected for the bed rest study, including Dale Graves, a licensed pilot, Captain in the US Naval Reserves, and scientist at Stanford Research Institute; Arline Fitzgerald, a 59-year-old widow; Lynn Luthi, a 45-year-old housewife; and Donna Howell, a 65-year-old retired administrative secretary.

Each group spent 28 days confined in the Human Research Facility at NASA Ames where, like the nurses, they wore 'biobelts' and underwent a similar battery of physical tests. The bed rest phase began on the tenth day of their incarceration, and lasted for ten days.

The results of the study reinforced those of the earlier nurse study. Furthermore, the women bonded together into a group that was more cohesive and much more supportive than anyone at NASA had anticipated. NASA therefore found no reason why women could not venture into space, and the only additional medical test for female astronaut qualification would be a pelvic examination. All that NASA had to do was select suitable candidates for its Astronaut Corp.

Nichelle Nichols and NASA's minority astronauts

In 1975, for the first time in its history, NASA granted permission for a member of staff to speak in an official capacity at a *Star Trek* convention, and Dr Jesco von Puttkamer (NASA's Director of Sciences) gave a presentation at a convention in Chicago. Sitting in the audience was one of the stars of the TV show, actress Nichelle Nichols, who, as a result of what she was hearing, began to question the phrase 'United States space programme.' To Nichelle – a black African-American woman –

the 'United States space programme' appeared to be the preserve of white, male-only astronauts, whereas she believed that it should epitomise the American ideal that anyone in the US can achieve that to which they aspire. Consequently, Nichelle felt that if it was to truly be the 'United States space programme' then it should also encompass women and minority astronauts.[37]

Nichelle Nichols was born in Robbins, Illinois, in 1932. A talented singer and dancer, she would rise to prominence through her portrayal of the African Communications Officer, Lieut Uhura, in the 1960s science fiction TV series *Star Trek*. At the time of Nichelle's casting, the only roles offered to black actresses were stereotypical maids or housekeepers, but the role of Uhura (whose name is based upon the Swahili word for 'freedom') was different. Here was a non-stereotypical black female character that was both seen and treated as an equal by the multiracial crew of the twenty-third-century starship *Enterprise*. However, in twentieth-century America, racial prejudice was still rife and, having endured its 'veiled insidiousness' off set, Nichols tended her resignation to the show's producer (and her friend), Gene Roddenberry upon completing the first series of *Star Trek*. But fate intervened in the guise of a chance meeting between Nichelle and the civil rights leader Dr Martin Luther King. King was a fan of the show, and told Nichols that she could not leave the cast, as Uhura was both a role model for black children and a positive image for the perception of black people by other races. Nichols duly reconsidered her position, and promptly withdrew her resignation. Uhura continued with her 'five-year mission' onboard the *Enterprise*, and inspired, among others, a young black girl called Mae Jemison – who would one day become the first black female astronaut.

Following von Puttkamer's presentation at the *Star Trek* convention in Chicago, Nichols – who had become interested in NASA and the 'United States space programme' – was invited to visit NASA's HQ in Washington, where she and two of her co-stars, James Doohan (Mr Scott) and George Takei (Sulu), were introduced by Von Puttkamer to NASA Administrator Dr James Fletcher. Nichols then went on to visit some of the other NASA centres, and in 1976 Dr Kerry Joels, then working at the NASA Ames Research Center, arranged for her to fly on one of the Kuiper Airborne Observatory missions. 'On that mission, water vapour was discovered in the atmosphere of Saturn. As we watched the telescope-tracking monitor showing the star-field, a meteor went through the field of view. One of the astronomers reported it as a sighting of a Klingon ship. Without pause, Nichelle pushed the microphone button on her oxygen mask and replied, 'Hailing frequencies are open, Captain."[38]

Later that year, on 17 September 1976, Nichols and other members of the *Star Trek* cast were present at the roll-out of the first Space Shuttle orbiter, *Enterprise*. Nichols was also invited, in 1976, to the Jet Propulsion Laboratory in Pasadena, where she witnessed the Viking lander touch down on the surface of Mars.

In January 1977, at the annual joint board/council meeting of the National Space Institute (NSI), Nichols gave a speech that was to play an important part in both her life and the recruitment of minorities into NASA's astronaut programme. In this speech – entitled 'New Opportunities for the Humanisation of Space' – Nichols, who was on the NSI's Board of Directors, voiced the concerns and criticisms that had been levelled against the 'United States space programme' by the women and

Gene Roddenberry and members of the *Star Trek* cast attend the roll-out of the Shuttle (OV-101) *Enterprise* in September 1976.

minorities whom she had met during her travels. Shortly afterwards she was invited by NASA's Associate Administrator for Space Flight, John Yardley, to discuss some of the issues raised and solutions proposed in her speech.

A year earlier, on 8 July 1976, NASA had announced that it was recruiting both Pilot astronauts and the new type of Mission Specialist (MS) astronaut. As the MS role offered the chance for non-pilots to become NASA career astronauts, John Yardley and the rest of NASA's higher management had become rather mystified by the small number of applications from women and minorities. However, it was of little surprise to Nichols. During the course of her meeting with John Yardley and other NASA management personnel, including African-American Dr Harriet Jenkins, NASA's Assistant Administrator for Equal Opportunity Programs, she talked herself into becoming a recruitment contractor of minorities for NASA's Astronaut Corp. However, she informed those present that in accepting the assignment (contract) her credibility was at stake, and if she found suitably qualified women and minorities for the astronaut programme who would subsequently not be selected, then she would 'personally file a class-action suit against NASA.' She was not going to be used to attract publicity and then have NASA say later that despite all its efforts it could find no qualified women or minorities. NASA concurred.

In February 1977, Nichols' company, Women in Motion Inc, signed the recruitment contract with NASA, which would run until the end of June. Having already immersed herself in NASA's programmes, visited the various centres, interviewed NASA astronauts and employees, undergone a modified form of astronaut training, and absorbed the culture, she began her mission to find suitable minority astronaut candidates. As Dr Joels recalled: 'NASA put Nichelle front and centre at a series of conferences, professional meetings and recruitment venues.'

For Nichols this was more than just 'star power'. She visited high schools and colleges across America, and spoke to many young minority professionals and scientists. She also made a *Space is for Everyone* recruitment film with Apollo 12 astronaut and Moon-walker Al Bean, and was given her own authentic blue NASA astronaut suit. At the end of June, with assistance from Dr Joels, Nichols put together the final report for her NASA contract. Among the qualified women or minorities who responded to her outreach recruitment drive were Guion Bluford, who, as an MS on the Space Shuttle STS-8 mission (30 August to 5 September 1983), would become the first African-American in space; Frederick Gregory, who would later be appointed Deputy Administrator of NASA by President George W. Bush; and Judith Resnik, Ronald McNair and Ellison Onizuka, who would all perish in the *Challenger* accident (28 January 1986).

In 1978, Joels left NASA Ames to establish the new Education Division at the Smithsonian's National Air and Space Museum. At that time there was very little participation by minorities in museum attendance and school visits, and to address this problem Joels asked Nichols to create and write a twenty-minute educational film. Entitled *What's in it for Me?*, the film featured Nichols (as Lieut Uhura) and a diverse group of middle-school children. A year later, on 6 December 1979, the first *Star Trek* feature film, *Star Trek: The Motion Picture* (which had Jesco von Puttkamer as its technical consultant) premiered in Washington. At the film's

Nichelle Nichols examines Apollo-era spacesuits at NASA MSFC with members of the CVT-4 Spacelab simulation crew.

Nichelle Nichols tries an Apollo-style spacesuit helmet, assisted by Carolyn Griner.

Nichelle Nichols and Judy Resnik in 1984. (Courtesy NASA/Nichelle Nichols.)

reception, which was held at the NASM, Nichols received the American Society of Aerospace Education's Friend of the Year Award for her educational film.

Five years later, in October 1984, Nichols was presented with NASA's Public Service Award by astronaut Judith Resnik in the auditorium of the Wilbur Cohen Building in Washington. Also present was her close friend, Dr Joels. What impressed him most was the number of rank-and-file NASA employees who attended the award ceremony. Rather than being a VIP-only occasion, Nichols was embraced as one of their own. Several recounted first meeting her and being attracted to NASA by that experience. They also mentioned her sincerity and honesty, as it would have been easy to see her effort as a publicity stunt.

In evaluating Nichols's contribution and her place in space history, Joels offered the following thoughts: 'It might be said that Nichelle was a role model for both women and minorities. Her pioneering role on *Star Trek* included the first interracial kiss on network television, and the first time a minority woman had an integral role in a science fiction crew on network television. Concurrent with the intensity of the Apollo programme, this became melded in the public mind as an acceptable human future, and that is why I believe that *Star Trek* achieved the status it did in popular culture. It humanised space travel at a time that it was becoming a reality for the general population. Thus Nichelle became a powerful image. Also going on in the popular culture of the late 1960s was the woman's movement in the US. Women were pursuing social equality, and there was a heightened awareness of feminine potential. In those days many women, even those in higher education, were still somewhat limited in career choices, and many became nurses or teachers. A decade later, doctors, lawyers, business careers and more were all equally common. Many women saw Nichelle as a vision of the feminine future, integrated into high technology, participating on an equal footing, yet still retaining those qualities that make women unique.'

Spacelab

The main premise for the Space Shuttle was its promise of routine access to space for both humans and hardware – the latter being carried aloft inside the orbiter's cavernous payload bay. The payload bay would also accommodate a purpose-built laboratory known as Spacelab, which would allow non-astronaut scientists to carry out experiments in microgravity. Spacelab would be ESA's contribution to the Space Shuttle programme, and NASA signed an agreement to that effect in 1973.

Spacelab consisted of two principal elements: a pressurised 'crew' module and U-shaped pallets, which could be configured for any Space Shuttle mission carrying out research into astronomy, Earth observation and remote sensing, life sciences, materials sciences, space sciences and other fields. The pressurised crew module, which allowed the astronaut-scientists to work in a shirt-sleeve environment, was a cylindrical structure that could be arranged either as one core segment (Short Module) containing both the laboratory area/fixings and the subsystems, or as two segments (Long Module) comprising the core segment and the experimental segment with extra laboratory space. In either version, access to the pressurised crew module was by means of the Spacelab Transfer Tunnel (STT) attached to the orbiter's cabin. The U-shaped pallets were platforms for mounting experiments requiring exposure to the vacuum of space. Using the pallets and the pressurised modules, Spacelab could be configured in a variety of combinations.

To support ESA's work on Spacelab, and hence reduce the costs associated with conducting experiments in space, a cylindrical Spacelab-like structure, the General Purpose Laboratory (GPL), was added to the Concept Verification Test (CVT) programme at NASA Marshall Space Flight Center, in Hunstville, Alabama. The CVT programme was set up to allow the involvement of scientists (Principal Investigators) at the beginning of development planning for their Spacelab experiments, and the GPL allowed them to test their Spacelab experiments in an environment that closely simulated the environment of space.

On 16 December 1974, an all-female crew began a five-day simulated space mission in the GPL. Designated 'CVT Test No. 4', the GPL's Materials Science Payload consisted of eleven experiments that the women themselves had created. The four women scientists – all employees of NASA MSFC – were Ann F. Whitaker, Carolyn S. Griner, Dr Mary Helen Johnston and Doris Chandler.

Ann F. Whitaker (*b.*1939) graduated from Berry College, Georgia, in 1961 with a BS degree in physics and mathematics. In 1968 she received an MS degree in physics from the University of Alabama. A physicist by profession, in June 1963 Whitaker joined NASA MSFC, where she specialised in lubrication and surface physics, organic semi-conductors, and solar cells. Her first project involved the Crawler – a large tracked vehicle used to transport the mighty Saturn V rockets from the Vehicle Assembly Building to the launch pad. Whitaker was a member of the team responsible for the Crawler's lubrication system and the redesign of the bearing system. She also worked on the Skylab thermal shield problem, and had science demonstrations performed by the last Skylab crew (Skylab 4, 16 November 1973 to 8 February 1974). She was involved in the testing and evaluation of materials in vacuum, and under electron, proton and ultraviolet radiation in order to select the thermal shield materials that could survive the space environment, and also in the development of the bicycle ergometer that was used by all three Skylab crews. According to Whitaker: 'Spacelab provides an excellent opportunity for the scientific community – both men and women – to conduct their own experiments in space. They can change their procedures in real-time if necessary, and adjust to changing conditions. They can adjust or repair their equipment on the spot. I want to fly in Spacelab because I have some ideas I would like to test.'[39]

Carolyn S. Griner (*b.* 1945) graduated from Florida State University in 1967 with a BS degree in astronautical Engineering. As a 15-year-old schoolgirl she had decided to apply her love of science and mathematics to the US space programme after witnessing the launch of John Glenn's Mercury/Atlas rocket. She joined NASA MSFC in 1964 as a materials and structures engineer, and later gained progressively higher positions of responsibility within the Science and Engineering Directorate. She was also Principal Investigator Interface for the Apollo–Soyuz Test Project. 'Recent sounding rocket flights have confirmed our belief that certain materials processing can be done only under weightless conditions, or at least under very low gravity influence. The Spacelab offers the opportunity for us to develop this technology for the benefit of everyone on Earth. This, to me, is too important to be ignored. We must pursue this avenue of research in space.'[39]

Carolyn Griner, Ann Whitaker and Mary Helen Johnston wear Apollo-era suits for a publicity photograph.

Dr Mary Helen Johnston (*b.* 1945) graduated from Florida State University in 1966 with a BS degree. In 1969 she gained an MS degree, and in 1973 was awarded a PhD in metallurgical engineering by the University of Florida. Johnston had worked at MSFC as an engineering cooperative student from 1963 to 1968, and joined NASA MSFC as a metallurgist in 1968. 'There is unlimited opportunity to produce unique materials in space – metals, crystals, medicine – by taking advantage of the vacuum, weightlessness and solar energy we have out there in Earth orbit. Spacelab is a laboratory no scientist would want to miss using. Any scientist would want to go. The laboratory in space is like a golden door that must be opened because it leads to a bright future – to unlimited benefits for everyone on Earth.'[39]

Doris Chandler, an engineer, worked (with Sara Cobbit) on the Apollo/Saturn programmes at NASA MSFC.

The four women spent about eight hours a day inside the GPL, where, for the most part, they worked on their individual materials science experiments. Keeping the place in good order was a group effort.[40] Following the completion of the test, Whitaker – who had enjoyed working in the GPL – was quite surprised that it received so much publicity. Three of the women – Griner, Johnston and Whitaker – also reported their work in a NASA Technical Memorandum (X-73320) entitled 'The Concept Verification Testing of Materials Science Payloads', published in June 1976. In this report, the women highlighted the value of having experienced personnel conducting the experiments: 'The value of the well trained scientist crew was emphasised during Test IV. Several minor equipment malfunctions occurred during test week that were repaired onboard by the respective PI. At least two experiments would have been lost by Day 2 of the mission if it had not been for the fact that the crew was extremely knowledgeable concerning their experiment hardware as well as the science to be obtained from them.'[41]

To enhance their prospects of being selected as astronauts, Whitaker, Griner, and Johnston undertook additional training. All three women completed scuba-diving training, which allowed them to use MSFC's Neutral Buoyancy Simulator – a very large water tank used to simulate the dynamics of weightlessness. This is achieved by the attachment of ballast weights so that the diver is prevented from either rising to the surface or sinking to the bottom, and therefore remains in a state of neutral buoyancy. Having been suitably ballasted, the women were able to gain first-hand experience of the problems associated with carrying out experiments in space. Whitaker – who believes that the NBC work was after the CVT test – loved training in the NBC, and found it great fun.[42]

The women also flew in NASA's KC-135 'zero gravity' aircraft in which, for short periods, they were able to experience weightlessness while flight-testing prospective space hardware. To gain yet further insight into conducting experiments in space, they also completed pressure suit (spacesuit) training, and were checked out in a high-altitude chamber at a USAF base.

In addition to this training, Whitaker, Griner and Johnston also began weekend flying lessons under the instruction of fellow MSFC employee Mel McIlwain; but in spite of their dedication and endeavours, none of them were selected by NASA to become Mission Specialists (career astronauts), although Griner was a finalist. There was, however, one other route that the women could pursue: that of a (non-career) Space Shuttle Payload Specialist.

NASA's first Space Shuttle/Spacelab flight was the STS-9/Spacelab 1 mission (1983), which, in addition to its two Mission Specialists, would require two Payload Specialists – one from NASA and one from ESA. Whitaker – who had a desire to run her own experiments in space – was the sole female candidate from a pool of six for the selection of NASA's Spacelab 1 Payload Specialist (PS). She was not selected, but as she later recalled: 'I enjoyed the selection process, which involved applying and being interviewed by the experimenters on Spacelab 1 or their reps, and undergoing the physical and psychological tests at JSC. I chose to pursue what I considered mainstream engineering rather than wait for an opportunity to fly, which may or may not be available.

Although she never made it into space, Whitaker had a tri-biology experiment on Spacelab 1, materials exposure experiments on various Shuttle flights, and the Noah's Arc experiment (a very large number of materials) on the Long Duration Exposure Facility (LDEF). All of these experiments were related to materials behaviour in space. In 1989 she was awarded a PhD in materials engineering by Auburn University, Alabama. Like Griner, Whitaker gained progressively higher positions of responsibility within MSFC's Science and Engineering Directorate, and has been awarded NASA's Exceptional Service medal and Exceptional Engineering Achievement medal. In 1992 she was nominated by NASA for the Women in Science and Engineering Lifetime Award. She also remembers meeting Nichelle Nichols when she visited MSFC, and she and the other members of

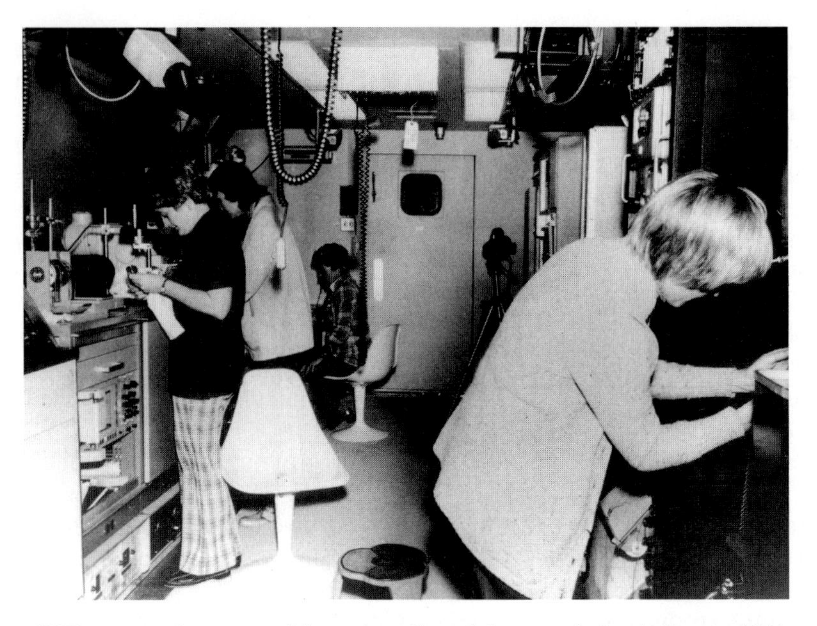

CVT test no. 4 crew participate in a Spacelab ground simulation, *c.* 1974.

the CVT test no. 4 also accompanied Nichols on a visit to the Stennis Space Center in Mississippi. Whitaker found Nichols to be very enthusiastic, and noted that she was interested in the possibilities of telemedicine.

NASA's second Space Shuttle/Spacelab flight was the STS 51-B/Spacelab 3 mission (1985), and as a result of her work in the Materials and Processes Laboratory, in which she was responsible for research into crystal growth and metal alloys for the Materials Processing in Space Program and the Space Shuttle programme, Dr Mary Helen Johnston was selected as one of the four Spacelab 3 Payload Specialists.

Carolyn Griner, who was the Chief Engineer and Payload Operations Director for Spacelab 3 Integration, also never made it into space. In 1987 she was appointed as Manager of the Operations Office, Space Station Projects Office, and in November 1988 was appointed to the Space Station Freedom programme at NASA Headquarters, where in April 1990 she was appointed Director of Space Station Operations and Utilisation. In this role Griner was responsible for systems and payload operations, utilisation oversight, and strategic planning for the International Space Station. Griner has written many technical papers, including 'Space Station Operations: The Integrated Concept', 'Space Station Overview Studies, 1988–1990', and 'Space Station: An International Space Laboratory', and has been awarded NASA's Exceptional Leadership Medal, Exceptional Service Medal, and Distinguished Service Medal. Her last appointment was as Deputy Director of NASA's MSFC.

SUPPORTING THE SHUTTLE

From the mid-1970s, more women began to feature in prominent positions in the space programme, both at NASA and in supporting industries working on the development of the Shuttle, its test and operational flights, and the creation of the International Space Station.

In March 1979 NASA news reports featured the first group of women to occupy consoles at NASA JSC for the forthcoming Shuttle programme. At the age of 18, Jackie Parker worked in support of the Data Processing System console during ascent and the initial orbits, monitoring the operation and response to the five onboard computers.[43] Another early member of the Flight Control team on Shuttle flights was Second Lieut Diana Langmade, USAF – one the ninety-eight women who formed the first female graduating class of the USAF Academy in May 1980, and assigned to the Air Force Manned Spaceflight Support Group at JSC. She was also the first female Air Force officer to be assigned to JSC. The assignment at JSC was for 3–5 years, and included working at the thermal systems console in Mission Control, Houston, during the early Shuttle flights.[44]

Suiting up for the Shuttle

Although the David Clark Company had continued to supply protective suits to the US experimental rocket-plane pilots (A/P 22S-2 full-pressure suits were worn by the X-15 pilots and the Lifting Body pilots), its only contribution to NASA's manned space programme had been the Gemini suits and the Apollo Block I suits (pre-1967 Apollo 1 fire). However, with the demise of NASA's manned expendable rockets and the introduction of the reusable Space Shuttle, DCC's services were again required. In addition to full-pressure suits for the rocket-plane pilots, DCC was also supplying them to the USAF pilots flying the high-altitude SR-71 Blackbird and U-2 spyplanes, and in 1978 it released its S1030 'gold' suit, which would become the standard model for the SR-71 pilots. From 1981 the S1031 'gold' suit was used by TR-1/U-2 pilots, and the SIO3OA high-altitude escape suit (modified from the S1030 in respect of the parachute harness attachments) was worn by the first four Shuttle crews.

At DCC, the women who worked on the S1030 and S1031 suits were Statia Banuskevich, Mary Mahassel, Dorothy Gilbert, Jane Michalak, Helen Hoyen, Marie Pignataro, Lillian Langevin and Gerry Simkus. Following the *Challenger* accident on 12 January 1986, NASA initiated a number of safety protocols, including a dictate that all future Shuttle crews would wear a Launch Entry Suit (LES). The contract for the S1032 LES – which would allow the astronauts to bail out of the Shuttle at up to 100,000 feet – was awarded to DCC, and fabrication of the suits began in 1988. The DCC seamstresses were Dorothy Gilbert, Mary Mahassel and Marie Pignataro, who would later work on the NASA S1035 Advanced Crew Escape Suit.

Mahassel later recalled: 'I started working at the David Clark Co in 1951 in another department. In 1973 I was transferred to the research and development department working as a stitcher, and was promoted to lead person of the stitching

Shuttle astronaut Cady Colman (right) presents Mary Mahassel with a Silver Snoopy award.

department in 1978. At one time I trained around a dozen stitchers making various suits ... In 1978, DCC fabricated the S1030 suits, and in 1981 the S1031 ... They were worn by the U-2 and SR-71 pilots. I did not get to meet any of the [Shuttle] crew-members that wore the SIO3OA suits.' In 2001, Shuttle astronaut Catherine (Cady) Coleman presented Mary Mahassel with a Silver Snoopy award, with the citation: 'Please accept our personal thanks for the outstanding support you have provided to the Space Shuttle program in your position as Stitching Lead for David Clark Company Inc., namely the SIO3OA Ejection Escape Suit, Anti-G suit, S1032 Launch Entry Suit, and S1035 Advanced Crew Escape Suit. Your dedicated contributions since 1979 have played a significant role in the success of this program. We who fly the Space Shuttle missions know that successful spaceflight depends to a great measure upon the excellent performance of individuals such as you. Your demonstrated commitment to excellence is indicative of the pride you have in producing a quality product for us. In appreciation, please accept the astronauts' personal award for professional achievement, the Silver Snoopy. Wear this pin with the knowledge that it is given only to those few we regard as the best in their profession. Your Silver Snoopy was flown on STS-103. Congratulations, and our best wishes for your continued achievement.' The award was a complete surprise, and was presented to her in the research and development area, where many employees were gathered. Mahassel retired from DCC in May 2002.

In 1980 Jean Alexander joined the Shuttle programme and began working with pioneering suit technicians Joe Schmitt and Al Rochford. She worked on the orbital flight-test missions STS-1 through STS-4, during which the two-man crews used the adapted SR-71 pressure garments for launch and entry in case they had to eject from *Columbia*. Following the loss of *Challenger* and the introduction of crew launch escape suits, she worked on suiting the crews from STS-26 onwards. Unlike the contractor suit technician who rotated every three to four years, the NASA suit technicians such as Alexander stayed for a career; but in the 1990s it became obvious that the NASA suit technicians would be replaced by contractor personnel, and so the role changed to training new contractor technicians to take over the role.[45]

During the continued development and testing of the Shuttle EMU, an unofficial endurance record was attempted. The record for a single EVA time inside a vacuum chamber (non-accumulative) was a little over 8 hours, and was held by a branch manager of the contractor supporting the Shuttle EMU programme.[46] This was not official NASA protocol, but was something that the test team and test subjects pursued. The record was not part of the official preparation for the use of EVA suits, though when the need arose to demonstrate that the suit could operate at the maximum optimum time, special permission had to be acquired based on test objectives and authorisation from the flight medical office and safety office to exceed a safety limit of 6 hours. There was, of course, the required pre-breathe period (at the time, 4 hours for women), depressurisation to a vacuum added another 30 minutes, and repressurisation back to normal pressure took another 30 minutes, totalling about 13 hours plus the few minutes needed to exceed the record. One of the female test subjects was so enthusiastic to try this test that she became the first women to exceed 13 hours in an EMU by a few minutes. She had declined the use of a fruit bar to chew on while in the suit, and chose only water stored at body temperature. As a result, she was exhausted when she emerged from the EMU, but was fully satisfied by her test run, and promptly ordered a pizza. The test and medical officers only allowed the run when the 8-hour-only limit in a suit was agreed, and the extra time was included for system start-up at vacuum and for system shut-down prior to chamber repressurisation.

Director of Johnson Space Center

The first woman to become Director of a NASA field centre was Carolyn L. Huntoon, who began working at JSC in 1968 and became a full-time life-sciences scientist there in 1970. She served as the sixth Director of JSC from January 1994 until August 1995, and afterwards took a senior executive post at the US Department of Energy. As Director for JSC she was responsible for more than 3,500 staff members at the Houston field centre, with a $3 billion annual budget. She became one of the most powerful women in NASA, and was the first scientist to head up the Houston operation after a long line of engineer-based administrators. Before working for NASA, Huntoon had gained a BS degree from North Western State University in Louisiana, an MS degree in 1966, and a PhD in physiology in 1968.

As a graduate student at Baylor College of Medicine in the mid-1960s she worked with her professor, who had a contract with what was then called the Manned

Spacecraft Center, studying spaceflight crew reactions to the stress as they prepared for flight on the Gemini programme. She enjoyed the work so much that she decided to try to work at NASA full-time after graduation in 1968, and worked on health issues for the Apollo astronauts before becoming a Principle Investigator on an experiment flown on Skylab. After Skylab she worked on the exchange of data with the Russians, on biomedical aspects of spaceflight as well as adaptation to and from long-duration spaceflight. By 1976–77, laws in the US were changing so that women could no longer be discriminated against in areas of employment. As a result, Huntoon was appointed to the selection board for the first group of Shuttle astronauts, including the first search for female astronauts by NASA. It was upon her suggestion that the applicants would not be asked any questions that had not been put to each of the male candidates; and in addition, any questions concerning their plans for marriage or starting a family were also not to be emphasised. This was an attempt to make the choice of candidates as fair as possible, relying on their educational and career experiences rather than personal influences. Huntoon never wanted to become an astronaut herself, but was happy to be involved in the selection of those who did. Her biggest disappointment was that the astronaut division remained at 90% men and around 10% women for many years, 'so there are still some issues here ... I don't know what the issue is ... If that is built into the selection process, or if it's just the way things are in our country.'[47] Huntoon rose to be Director of Space and Life Sciences at JSC in 1987, and was later Director of the Center – an appointment for which she needed to develop skills as a manager in overseeing engineers as well as scientists. She was told by the then NASA Administrator Dan Goldin that to work in an administrative position at NASA she need not be an engineer, but that managerial skills and a gregarious nature were required. From this, Huntoon admitted that she learned to not to be too quick to jump to conclusions and make decisions. She occupied a leading administrative role during a time of major change at JSC and in NASA during the period of transition from one political administration (George Bush Senior) to another (Bill Clinton), and in revising the objectives and organisation of the Space Shuttle and Space Station programmes, as well as in integrating with the Russians as full ISS partners and in developing the Shuttle–Mir Programme as a precursor to the ISS construction and operational phase. After serving as Director, Huntoon moved to Washington to work at NASA HQ, and subsequently accepted a position in the Office of Science and Technology Policy in the White House. She worked on science policy for the whole nation, and after two or three years she moved to the Department of Energy for another two years. When the administration changed again and George W. Bush became the new President, he asked her to stay until a replacement could be found. After more than 30 years of government service, Huntoon retired in summer 2001.[48]

REFERENCES

1 Jerri Truhill, interviewed by Nicola Humphries, Richmond, Texas, March 1999.
2 Frank Winter, *1928–1929. Forerunners of the Shuttle: The 'von Opel' Flights*, *Spaceflight*, **21**, 2, February 1979.

3 Willy Ley, *Rockets, Missiles and Space Travel*, Chapman and Hall, 1957.
4 Irene Sänger-Bredt, *The Silver Bird Story: A contribution to the History of the Development of the Aerospace Transporter – Eugen Sänger's favourite project*, *Spaceflight*, **15**, May 1973.
5 Nicolae-Florin Zaganescu, 'Dr Irene Sänger-Bredt: a life in astronautics', IAA-99-IAA.2.1.04.
6 Eugen Sänger and Irene Bedt, *A Rocket Drive for Long Range Bombers*, translated by M. Hammermesh, Santa Barbara, California, 1952.
7 Asif A. Siddiqi, *Challenge to Apollo: The Soviet Union and the Space Race, 1947–1974*, NASA SP-2000-4408, 2000, p. 33.
8 K. Gatland 'The Space Pioneers', in *The Illustrated Encyclopaedia of Space Technology: A Comprehensive History of Space Exploration,* Salamander Books, 1981, p. 20.
9 Frederick I. Ordway III and Mitchell R. Sharpe, *The Rocket Team*, Heinemann, 1979, p. 20.
10 Hanna Reitsch, *The Sky My Kingdom*, translated by Lawrence Wilson, Bodley Head, 1955.
11 R. Opitz, taped interview on Hanna Reitsch, 6 October 2003.
12 Hanna Reitsch, 'Flying the V-1 and Me163', presentation at SETP Symposium, September 1975.
13 Sheryll Goecke Powers, *Women in Flight Research,* monograph in aerospace studies, under the auspices of the NASA History Program: http://www.dfrc.nasa.gov/History/Publications/WIFR.
14 Beverly Swanson Cothren, NASA Oral History interview by Sandra Johnson and Rebecca Wright, Grass Valley, California, 15 June 2001.
15 Betty S. Love, NASA Oral History interview by Rebecca Wright, Palmdale, California, 6 May 2002.
16 Mary (Tut) Hedgepeth, NASA Oral History interview by Rebecca Wright and Sandra Johnson, Bakersfield, California, 12 June 2001.
17 K.G. Williams, *The New Frontier: Man's Survival in the Sky,* Heinemann Medical Books, 1959.
18 David M. Clark, *The Development of the Partial Pressure Suit,* David Clark Company Inc, Worcester, Massachussetts, 1992.
19 Ethlie A. Vare and Greg Ptacek, *Mothers of Invention*, 1988.
20 Correspondence from Bruce Hess, Staff Historian, Wright-Patterson Air Force Base, 2003.
21 Correspondence from Brigadier General Charles E. Yeager, 2003.
22 Correspondence from Joseph Ruseckas to Teresa Kingsnorth, 2003.
23 Milton Bracker, 'A Missile Pioneer Honoured in Death (Ashes of Dedicated Physicist were sent Aloft in Bomarc and Scattered in Ocean)', *New York Times*, 31 August 1956.
24 Richard P. Hallion, *On the Frontier: Flight Research at Dryden, 1946–1981*, NASA History series, NASA SP 4303, 1984, pp. 107.
25 Charles J. Donlan, 'The Legacy of the X-15', as presented in Proceedings of the X-15 First Flight 30th Anniversary Celebration, NASA Conference Publication 3105, 1991.

26 Correspondence from Darlene Feikema and Diane Leigh, University of Washington, to Teresa Kingsnorth, 2003.
27 A. Scott Crossfield with Clay Blair Jr, *Always Another Dawn: The Story of a Rocket Test Pilot*, Hodder and Stougton, 1961, p.265.
28 Betty S. Love, NASA Oral History interview by Rebecca Wright, Palmdale, California, 6 May 2002.
29 Correspondence from Betty Love, August 1991.
30 Harriet J. DeVries Smith, NASA Oral History interview by Rebecca Wright, Salano Beach, California, 7 November 2001.
31 Sheryll Goecke Powers, NASA Oral History interview by Rebecca Wright and Sandra Johnson, NASA Dryden Flight Research Center, California, 13 June 2001.
32 Bertha M. Ryan, NASA Oral History interview by Sandra Johnson and Rebecca Wright, Ridgecrest, California, 13 June 2001.
33 Steven E. Whitfield and Gene Roddenberry, *The Making of Star Trek,* Titan Books, 1991.
34 *Astronaut Selection and Training*, NASA Information Summaries, PMS-019 (JSC), December 1986.
35 Harold Sander and David L. Winter, *Physiological Responses of Women to Simulated Weightlessness: A Review of the Significant Findings of the First Female Bed-Rest Study*, NASA SP-430, 1978.
36 Barbara Rowes, *Housewives in Space*, *Omni*, **4**, 9, 1982.
37 Nichelle Nichols, *Beyond Uhura: Star Trek and Other Memories*, Boxtree, 1996.
38 Correspondence from Dr Kerry Joels, 2003.
39 Mitchell R. Sharpe, Alabama Space and Rocket Center, 'Make Way For The Women Astronauts!', *Spaceflight*, **18**, 12, December 1976.
40 Questions for Dr Ann F. Whitaker, questionnaire from Ian Moule, 25 March 2003.
41 C.S. Griner, M.H. Johnston and A, Whitaker, *The Concept Verification Testing of Materials Science Payloads*, NASA TM X-73320, June 1976.
42 Correspondence from Ann Whitaker, 1998.
43 NASA News JSC, 79–17, 19 March 1979.
44 NASA News, JSC, 80-047, 23 July 1980.
45 NASA Oral History Transcript, S. Jean Alexander, 23 June 1998; also 'Jean's Soft-Shoe Shuttle' in the *Daily Mail* (womens' section), 11 April 1984.
46 Skipper website.
47 Carolyn Huntoon, NASA Oral History interview, 5 June 2002.
48 The role that women play in the space programme has been a regular feature on the NASA web pages since the beginning of the ISS programme in the late 1990s. The *Behind the Sciences : Meet the People* section of the NASA Human Spaceflight website features a variety of workers interviewed about their roles at NASA, their backgrounds, and what or who inspired them in their careers. http://spaceflight.nasa.gov/shuttle/support/people.html.

Shuttlenauts

The advent of the American Space Shuttle system finally created an opportunity to broaden the selection of astronaut candidates to include skills other than high-performance jet flying, and also removed the requirement for holding a jet pilot qualification before assignment to an American spaceflight crew. In the selection to NASA between 1959 and 1969, the criteria had included test-pilot or jet-pilot qualification, or, for the 1965 and 1967 scientist-astronaut intakes, a requirement to graduate from a USAF jet piloting school as part of the basic training programme. For those women interested in flying in space, the chances of attaining such qualification were as difficult and remote as receiving a call from the Astronaut Office to fill a job vacancy, and it was not until well into the 1970s that American women finally began to make progress in both military and civilian commercial aviation.

In January 1973, Emily Howell Warner became the first female to pilot a commercial (Frontier) airline, and later the same year the US Navy announced that it would begin training women service personnel as pilots (but only as non-combatants), the first of whom were selected in 1974. That year the first female pilot in the US Forestry Service was Mary Barr, and in June 1974 Sally Murphy became the first woman to qualify as a US Army aviator. One of the problems in the advancement of female aviation was entry into military academies to gain higher educational qualifications. This was not open to women until 1976; but changes came when the USAF selected its first group of female aviation candidates in 1977, followed by the formation of the International Society of Women Airline Pilots in 1978. The Air Force began accepting women into the USAF Test Pilot School at Edwards AFB in 1974, but in the engineer class, not the aviation class. The first female graduate of the engineering course was Captain Leslie Halley Kenne (Class 74B), who during her career rose to the rank of Brigadier-General. It would be another fourteen years before the first woman pilot, Captain Jacqueline Parker, would graduate – the year before a certain Major Eileen Collins also made the grade.

SELECTING FOR THE SHUTTLE

The sequence of events in selecting America's first 'official' female astronauts began on 8 July 1976 – just four days after America celebrated its bicentennial, twelve days before Viking 1 made its historic unmanned landing on Mars, and on the seventh anniversary of the Apollo 11 Moon-landing. On that date, NASA announced that it would be accepting applications for a new group of astronaut candidates, to be trained to fly and operate the Space Shuttle. A total of thirty positions were to be considered: fifteen for pilot applicants, and fifteen for a new category called Mission Specialists. The closing date for applications would be June 1977, with all successful candidates being notified by December and instructed to report to NASA JSC in Houston for a two-year training and evaluation programme beginning on 1 July 1978. Upon the successful completion of this course, the astronaut candidates (Ascans) would be considered as NASA career astronauts and assigned further training in technical and support roles leading to assignment as flight crew-members.

In these announcements it was made clear that both minorities and females were encouraged to apply, as the Shuttle was promoted as a more relaxed approach to flying in space than that portrayed in the 'Right Stuff' era of Mercury through Apollo. The flight profile of the vehicle, its multirole crew and its diverse payloads required a broader group of qualified astronauts rather than jet pilots or scientists with jet-piloting skills. Early in the programme it had been decided that pilots would be required to 'fly' the vehicle and perform the duties of mission Commander, and therefore selection to this criteria was still reminiscent of earlier selections, with pilot candidates expected to replace the pioneering veteran astronauts who took America to the Moon.

Pilots were required to have at least a BS in engineering, the physical sciences or mathematics, and have at least 1,000 hours of command pilot time, with 2,000 hours desirable in high-performance aircraft, and preferably experience in flight-testing. (This severely limited female pilot applications for this first Shuttle-era selection.) They also needed to pass a NASA Class 1 physical examination (similar to a military or civilian Class 1 flight physical), including visual acuity of 20/70 or better uncorrected, correctable to 20/20 each eye, and a blood pressure of 140/90, measured in a sitting position. In addition, they had to have a standing height of between 5 feet 4 inches and 6 feet 4 inches.

The design of the Shuttle system allowed the opportunity to fly crew-members with little or no piloting experience, and with ascent and descent loads as little as 3 g (compared to 8 g on Apollo) the physical requirements for selecting career astronauts who would not pilot the orbiter could be more relaxed. Therefore, the selection of Mission Specialists focused more on academic and employment experience rather than piloting skills, though these would continue to be of an advantage.

Mission Specialists required at least a BS degree in engineering, biological sciences, physical sciences or mathematics, and preferably an advanced degree (masters or doctorate) or the equivalent employment experience. They had to pass a Class II physical examination, which had a greater latitude for non-standard vision

(vision-distance acuity – 20/200 or better uncorrected, correctable to 20/20 in each eye, blood pressure the same as for pilots) and hearing, and stand between 5 feet and 6 feet 4 inches in height.

The third category of Shuttle crew-member – the Payload Specialist – would be chosen as and when specific payload requirements required their participation. They required very little astronaut training, as they would mostly return to their previous careers or home agency shortly after the completion of the flight, and would not be considered NASA career astronauts.

The first Shuttle selection process

During the twelve months following the announcement, applications flooded into NASA, revealing a high interest in becoming an astronaut and flying into space, despite the lack of any firm plans for a space station or any hope of returning to the Moon in the near future. During 1977 – as television reports updated the world on the progress of the Space Shuttle *Enterprise* Approach and Landing Test Programme at Edwards AFB, Viking Landers sent regular reports from the surface of Mars, and the Voyager probes were prepared for missions to the furthest reaches of the Solar System – NASA announced, on 15 July, that more than 8,000 applications (including 1,544 women) for either the Pilot or Mission Specialist category had been received by the 30 June cut-off date. Some applicants had applied for both positions, and several tried to improve their chances by also putting their name forward for selection as a Payload Specialist on Spacelab 1.

With the extension of upper age limits and height regulations, as well as broadened criteria, a wide range of Americans expressed a desire to become NASA astronauts. The height limits were also stated to be dictated by the design of the Shuttle pressure suits they would be expected to wear for spacewalks, rather than by the limiting size of the spacecraft, as for earlier selections. Of the 6,348 most suitably qualified applicants, 668 were pilots. Of these, 147 were military pilots (no female applicants) and 521 were civilian pilots (of whom three were females). In the Mission Specialists applications, 5,680 were most suitably qualified, and of these, 161 were military applicants (three female) and a staggering 5,519 were civilians (including 1,248 females).[1]

All those who met the basic requirements were evaluated by discipline panels, which reviewed every one of the applications and found that only half met the basic requirements listed in the announcement. The military applicants had already been pre-screened and nominated by their respective parent service before being put forward for review by the NASA Astronaut Selection Board. All qualified application were re-reviewed, and from them, 208 individuals were invited to JSC for a week of interviews, orientation, and medical tests during July–September 1977. In this round, eighty Pilot applicants (76 military and four civilians – none of them female) were interviewed and tested; from the Mission Specialists applications, 128 were selected (45 military, of which two were female, and 83 civilians, including nineteen females). The 207 were called to JSC in ten groups of about twenty applicants each, as a mixture of Pilot and Mission Specialist applicants (with expanded details of the 21 female finalists):[2]

August 2 First group: twenty applicants.
August 15 Second group: twenty applicants.
August 29 Third group: twenty applicants, including eight women:

Nitza M. Cintron, 27, PhD; born in San Juan, Puerto Rico; attended high school in Santuree, Puerto Rico; currently at the Johns Hopkins University School of Medicine, Baltimore, Maryland.
Danielle J. Goldwater, 29, MD; born in West Haven, Connecticut; attended the New Haven High School; currently at Stanford Hospital, Stanford University, California.
Shannon W. Lucid, 34, PhD; born in Shanghai, China; attended high school in Bethany, Oklahoma; currently at the Oklahoma Medical Research Foundation in Oklahoma City, Oklahoma.
B. Tracey Sauerland, 29, MS and PhD; born in New Britain, Connecticut; attended high school there; currently assigned to the NASA JSC Space and Life Sciences Directorate, Houston, Texas.
Rhea Seddon, 29, MD; born in Murfreesboro, Tennessee; attended high school there; currently at the City of Memphis Hospital, Memphis, Tennessee.
Anna L. Sims (Fisher), 28, MD; born in Albany, New York; attended high school in San Pedro, California; currently at Harbor General Hospital, Torrance, California.
Victoria M. Voge, 34, Lieut-Commander USN, MD; born in Minneapolis, Minnesota; attended high school in New Brighton, Minnesota; currently stationed at the Naval Aerospace Medical Institute, Pensacola, Florida.
Milley H. Wiley, 31, PhD; born in Mineral Wells, Texas; attended high school there; currently at the Veterans Administration Hospital, San Francisco, California. (She subsequently flew as a Payload Specialist for Spacelab Life Sciences 1 (STS-40), in 1991 as Millie Hughes-Fulford.)

September 19 Fourth group: twenty applicants.
September 26 Fifth group: twenty applicants.
October 3 Sixth group: applicants, including one woman:

Sally K. Ride, 26; born in Los Angeles, California; attended high school there; currently at the Physics Department, Stanford University, Stanford, California.

October 17 Seventh group: twenty applicants, including eight women:

Kathleen Crane, 26, PhD; born in Washington DC; attended high school in Falls Church, Virginia; currently at the Scripps Institute of Oceanography, La Jolla, California.
Bonnie J. Dunbar, 28; born in Sunnyside, Washington State; attended high school there; currently at Rockwell International Space Division, Downey, California. (Selected in 1980.)
Joan J. Fitzpatrick, 27, PhD; born in Bayonne, New Jersey; completed high school there; currently at Colorado School of Mines Research Institute, Golden, Colorado.
Carolyn S. Griner, 32; born in Granite City, Illinois; completed high school in Winter Park, Florida; currently at NASA Marshall Space Flight Center, Huntsville, Alabama. (Participant in Spacelab simulations.)

Evelyn L. Hu, 30, PhD; born in New York; attended high school there; currently employed at Bell Laboratories, Holmdel, New Jersey,
Carol B. Jenner, 27, PhD; born in Washington; completed high school in O'Fallon, Illinois; currently at the University of Wisconsin, Madison, Wisconsin.
Mary Helen Johnston, 32, PhD; born in West Palm Beach, Florida; attended high school in Fort Pierce, Florida; currently at NASA Marshall Space Flight Center, Huntsville, Alabama. (Subsequently selected as a back-up Payload Specialist for Spacelab 3.)
H. Louise Kirkbride, 24; born in Philadelphia, Pennsylvania; attended high school in Upper Darby, Pennsylvania; currently at the Jet Propulsion Laboratory Pasadena, California.

October 25 Eighth group: twenty applicants, including one woman:

Jane L. Holley, 30; Captain, USAF; born in Shreveport, Louisiana; attended high school at Annandale, Virginia; currently at the USAF Tactical Fighter Weapons Center, Nellis AFB, Nevada.

November 7 Ninth group: twenty-three applicants, including two women:

Barbara J. Holden, 32, PhD; born in Los Angeles, California; completed high school in Lincoln, New England; currently at the Naval Weapons Center, China Lake, California.
Judith A. Resnik, 28, PhD; born in Akron, Ohio; attended high school there; currently at Xerox Corporation, El Segundo, California.

November 14 Tenth group: twenty-four applicants, including one woman:

Kathryn D. Sullivan, 26; born in Paterson, New Jersey; attended high school in Woodland Hills, California; currently a graduate student at Dalhousie University, Halifax, Nova Scotia.

After all these 207 were interviewed, examined and tested, 152 (including fourteen females, all MS, two military and twelve civilians) were found to be medically qualified, of which 149 (including all the females) were qualified medically and were still interested in continuing. In the 9 December 1977 issue of the JSC in-house magazine *Roundup* it was reported that the centre's Flight Medicine Clinic had been responsible for medical evaluations of all applicants (instead of the Lovelace clinic for Group 1, and Brooks AFB Medical Center for the other selections prior to 1978). This evaluation had four parts: a medical history (illness, injuries, surgery); a through physical examination; speciality evaluation (neurology, otohinolaryngology, ophthalmology); and special tests (treadmill, pulmonary functions, audiometer, body chemistry). The recommendations from the selection board are based on each applicant's educational qualifications, training and experience, as well as any unique qualifications and skills. According to the background details in the astronaut application package, 'because several hundred applicants fulfil the requirements, the final selection is based largely on personal interviews. Astronauts are expected to be team players and highly skilled generalists, with just the right amount of individuality and self-resilience to be effective crew-members.'[3]

Although the process of selection was progressing well, some of the administrative work was slowing down the procedures. On 12 December, NASA Administrator Robert A. Frosch held a meeting with the officials involved in the selection programme, but due to his involvement with agency budget activity he did not have time to review the data on the candidates. On 16 December he told JSC Director Christopher C. Kraft – who had supervised the selection process – that the review of final candidates would proceed early in the new year. On 19 December NASA officially reported that the formal announcement of the final group of candidates chosen for astronaut training would be delayed until January 1978. The same day, the *Washington Post* reported that Frosch was not 'crazy about the [up to] forty candidates'. About thirty of them were serving military officers, and the Administrator expressed some concerns on the significant military content, compared with the civilian contingent, of the Astronaut Office. He was also apparently not happy about having only two African-American pilots and only three female scientists on the short-list. A further review of applicants was therefore called for to determine whether the numbers in both categories might be increased. But no other African-Americans were selected, and five short-listed Pilots were moved to the next group to allow for increasing the number of female applicants in this first Shuttle-era selection.[4]

Twenty-nine new guys and six new girls

On 16 January 1978 NASA finally released the names of thirty-five new astronaut candidates – the first NASA astronaut selection in eleven years, since the second scientist-astronaut selection of 1967 (those in the 1969 group were transferred from the cancelled MOL programme), who called themselves the Thirty-Five New Guys. This was the first astronaut selection intended specifically to train as potential Space Shuttle crew-members, and consisted of fifteen Pilot candidates and twenty Mission Specialist candidates. There were twenty-nine male candidates, and the first women chosen as America's (and NASA's) first official female astronaut trainees: Anna Fisher, Shannon Lucid, Judy Resnik, Sally Ride, Rhea Seddon and Kathy Sullivan – all of whom were chosen as Mission Specialists. They were destined to become pioneers in America's quest into space.

At the time of the announcement, the Director of JSC was asked why there had been no minority or women candidates in previous astronaut selections.. He stated that up to 1967 there had been relatively few suitable applicants in either category, but added that with this group, 'The most rewarding thing is that there were large numbers of qualified women and minorities this time around ... We had no problems finding women and minorities who were qualified and highly motivated as to what they wanted to do.'[5] The following were NASA's 'original six' female astronauts:

Anna Fisher was a physician, and married. Her husband, William Fisher, had also applied, but was not selected (though he succeeded in 1980). During selection she used the name Anna Sims (although her maiden name was Tingle).

Shannon Lucid, born to American parents in China in 1943, was a post-doctoral fellow in biochemistry at Oklahoma Medical Research Foundation, and was the first mother (with three children) to be selected for astronaut training – a fact that was not lost to the media, although Chris Kraft pointed out that 'Shannon was selected because she had always wanted to fly in space. We took no account of her marital responsibilities because she asked us not to.'[6] Shortly after being selected, Lucid observed that during the late 1970s it was not harder for women to succeed than it was for men; it was society that made it more difficult: 'I always figured I would have a career and a family. The catch is that you have to find a man who feels the same way'.[7] According to Lucid, the family should be the first priority for both men and women, and she pointed out that many male astronauts have working wives as well as families. She also had a connection with one of the women from Lovelace's Women in Space programme. As Gene Nora Jessen recalled: 'One of my former flight students called to tell me that he had taught Shannon Lucid to fly. Since I had taught him to fly, this made me Shannon's [flight student] 'grandmother'. My former student explained this to her, and Shannon and I have laughed about our grandmother/granddaughter relationship!'[8]

Judy Resnik was an engineer at Xerox Corporation. At selection she stated: ' I don't feel like I'm a pioneer. I feel like I'm one of those persons selected to be an astronaut and it's a coincidence that I'm a woman. I think the six of us have always gone after the things we wanted; most of them are to a certain extent unusual. Some of us have had more obstacles than others, but we have overcome whatever was in our way, and this is another step in the pursuit of a continuing career.'[6] Resnik did not like the media attention. She did not think it was fair, because all the candidates – men and women – would go through the same training programme, and the women did not deserve all the public attention simply because they were women. On the day after the announcement of her selection she received almost a hundred interview requests and telephone calls.[7]

Sally Ride was a physics research assistant at Stanford University, and a talented former tennis player in junior rankings.

Rhea Seddon was a resident surgeon at City of Memphis Hospital. During selection she stated that the only thing that frightened her was 'that I might not do well. I don't think its particularly dangerous. We are all anxious to get started. We just need to find out where to begin.'[6] Seddon later added that all six women were focused on succeeding in their work, and none of them wanted to be considered as 'cute little girls NASA hired as stewardesses on the Space Shuttle.' She also said that they felt under pressure because they were first to be selected for astronaut training. They were in the spotlight, and were being observed by those who disagreed with their motives, by those who considered that whatever they did would refelect on women in general, and by other women aspiring to be astronauts themselves and who looked to this group to 'open the door' for them. She therefore felt a great responsibility. Reflecting on the domestic make-up of the group, Seddon pointed out that the average age of the six women was lower than that of the men. and while only two of the women were married, and only one had children, twenty-seven of the twenty-nine male candidates were married with children. 'That may say something about how women have to go about a career', she said.

Kathryn Sullivan was a post-graduate student in Earth sciences at Dalhousie University.

Highlighting the popular TV female detective series of the day, and under the direction of Chris Kraft, Director of NASA JSC, media headlines called the six new female astronauts 'Chris's Angels', who after training would 'count down to the loveliest lift-off of all, becoming heavenly bodies on Uncle Sam's payroll' – but said nothing about their academic attainments or the equality in the Astronaut Office which they were trying to portray.[9]

Upon selection to NASA and successful completion of the Ascan training programme, civilian candidates were expected to remain with the agency for at least five years, whilst military candidates would be detailed to NASA for 'a specific tour of duty' from their parent service, whereupon the option to remain at NASA or return to military service would be reviewed. One longstanding misconception about astronauts is that they receive huge salaries. However, salaries for the civilian astronauts were based upon the General Schedule pay-scales outlined by the US Federal Government at the time of selection for grades GS-11 ($21,800 per annum in 1978 dollars) through GS-14 ($33,800 per annum in 1978 dollars), and reflected each individual's academic achievements and experience. Salaries for serving military officers were based upon their current military rank. Though this has been refined over the years, the same basic philosophy applies to all current NASA career astronauts. Thus, the salary received upon becoming an astronaut at NASA was based upon past experiences and attainments and future promotions and academic achievements, and not, as is commonly assumed, in the form of large financial rewards for each flight into space. The early prospect of making regular trips into space rapidly deteriorated to the reality of only a few trips into space on the Shuttle – on average, two or three missions, each of a week to ten days over a period of 7–10 years. The rewards from flying in space were related to a personal career motivation, and were certainly not financial – at least not while an astronaut remained with NASA! The days of the famous *Time Life* contract, in which astronauts from the early selections were 'rewarded' for telling their 'official and personal' (hyped) stories to the one publisher, were long gone (and, in fact, began to dwindle in the late 1960s as the Astronaut Office expanded).

Follow-on selections, 1980–90

The process for selecting astronauts to fly on the Shuttle remained essentially unchanged over the next twenty-five years, though a change indicating a move towards future space stations and international operations began to appear from the 1992 selection.[10]

Class of 1980 – Group 9

On 19 May 1980 a further group of nineteen astronauts (eight Pilots and eleven

Mission Specialists) were selected for the Shuttle programme, and under an agreement with European Space Agency two male ESA Payload Specialists would train as European Mission Specialists with this group. Despite financial and technical difficulties at this time, the Shuttle's completion of a series of four to six two-man test-flights was expected to be followed by an increased launch rate of ten to twelve missions a year with crews of six or seven astronauts. With natural attrition due to retirement, the core of thirty to fifty astronauts would clearly not be sufficient to support such a rate, so the first of what was hoped would be an annual selection (a little optimistically predicted) was commenced in 1979. Applications as MS could now substitute an advanced degree for experience, although degrees in technology, aviation and psychology were not acceptable, as NASA required a broader natural science or engineering-based education of its candidates. (By 1985 this was amended to allow three years of related professional experiences in lieu of advanced degrees). Of the 3,465 applications for between ten and twenty openings, 121 candidates, including 21 women, were interviewed, but only two women were finally named to this group: Mary L. Cleave and Bonnie J. Dunbar, who had missed out on the 1978 selection. Five of the male military pilots selected (Blaha, Bridges, Grabe, O'Connor and Richards) had been finalists in the 1978 selection, but when NASA changed its mixture from a planned twenty Pilots and fifteen MS to fifteen Pilots and twenty MS, to allow more women into the programme, these five were deferred to the next selection.

Marsha Ivins inside the early Shuttle cockpit mock-up, *c.* 1975.

Class of 1984 – Group 10

In 1983 – the year after the Shuttle had been declared operational – NASA initiated what was hoped to be an annual selection of astronauts to respond to flight rates and natural attrition. There were 4,934 applications, but only six Pilot and six MS vacancies. Of the 128 people interviewed, twenty-three were women, and among the seventeen selected candidates (seven Pilot and ten MS) announced on 23 May 1984, three were women: Marsha S. Ivins, Ellen L. Shuman and Kathryn C. Thornton. This group was notable in that all its members were already employees of the US Government. Apart from the twelve serving military personnel, four civilians (including Ivins and Shulman at JSC) were already employed by NASA, whilst Thornton was employed at the US Army Foreign Science and Technology Center.

Class of 1985 – Group 11

With the increase in the launch rate, NASA's apparent bias in restricting new applications from industry or academia was clearly demonstrated in this selection. No applications were invited, and the agency merely reviewed 126 qualified civilian finalists from the 1984 process, inviting fifty-nine for interviews, and 120 military candidates, of which three were females. At the announcement of their selection on 4 June 1985, all thirteen (six Pilots and seven MS) were serving military officers or current NASA employees, including two women: Linda M. Godwin and Tamara E. Jernigan.

NASA Astronaut Group 12, selected in 1987.

Class of 1987 – Group 12

In 1985 NASA changed its selection process, stating that from 1 August it would accept applications from civilians on a rolling basis, and that those serving in the military would be nominated by their parent service once a year. The exact number of candidates selected would depend on the flight requirements at the time and the retirement of former astronauts. The new group would be named in the spring of 1986 (during a year expected to see more than fourteen Shuttle launches), and would commence Ascan training that summer. However, in January 1986 the loss of the Shuttle *Challenger* and seven astronauts – including Judy Resnik and Sharon C. McAuliffe – grounded all flights and seriously affected the use of the Shuttle (from the planned Vandenberg Air Force Base site) for polar orbits, its military and commercial payload programme, and delayed the deployment of several important scientific and planetary payloads for several years. Astronauts intended for selection in the spring of 1986 were finally named on 5 June 1987 (three Pilots and eight MS). Of the 2,061 applicants, 151 were interviewed, including twenty-three women. The two women selected in this group were N. Jan Davis and the first African-American woman selected by NASA for astronaut training, Mae C. Jemison.

Class of 1990 – Group 13

In 1988 the Shuttle was still grounded, the flight manifest was uncertain, there remained many astronauts who were in generic training from cancelled post-*Challenger* missions, several veteran astronauts had retired, and, there was a pool of unflown astronauts from selections since 1980. As a result, NASA amended its astronaut selection process for the Shuttle. Selections would hitherto be conducted on a two-year rolling cycle, with the cut-off date, for this intake, of 30 June 1989. Any applications received after that date would automatically be carried over to the next selection cycle. Using the now standardised selection criteria, around 2,500 applications were received by the official cut-off date; and of these, 106 applicants, including sixteen women, were interviewed. On 17 January 1990 – the day after the twelfth anniversary of the naming of the first Shuttle-era selection – twenty-three new astronauts were named. This was the second largest group selected for the programme; and reflecting a change in the make-up of the Astronauts Office, two-thirds were Mission Specialists. Of the seven Pilots selected, one was the first woman Pilot candidate, Major Eileen M. Collins, USAF; and of the sixteen Mission Specialist candidates, four were women, including the first Hispanic candidate Ellen Ochoa, as well as Captain Susan Helms, USAF, Captain Nancy J. Sherlock, US Army, and Janice E. Voss.

The new focus on educational backgrounds rather than test or operational flying skills became evident with the Group 14 selection in 1992. This was also the beginning of a new era, to train future crews to construct and crew the large space station that had been in planning for a decade.

New roles for new astronauts

With the first selections of astronaut trainees in what was to be known as the Shuttle era (approximately covering the astronaut Classes of 1978–92), two main roles of specialist emerged: Pilot astronauts and Mission Specialists. These categories remain, and though from the 1994 selection the focus was more on Shuttle–ISS operations than on purely solo Shuttle operations, their roles have remained essentially the same as when the first Shuttle era selection was announced in 1978.

Pilot astronauts

This category has been the mainstay of the NASA astronaut selection process since the selection of seven men for Project Mercury in 1959. At that time, military jet and test-piloting skills were required, and with civilian test-pilots being acceptable with the second selection in 1962 this also effectively ruled out any women candidates, who would not acquire sufficient high-performance experience. By the 1963 selection, test-piloting skills were dropped, although always remained desirable. This allowed two women to apply, but they were not among the thirty-two finalists; and again, in 1966 six women pilots applied but were unsuccessful. In 1969 the selection was comprised of transfers from the recently cancelled USAF Manned Orbiting Laboratory (MOL) programme; and once again there was no opportunity for women to enter the NASA astronaut programme. With the Shuttle, prospects improved, but though Pilot applicants from the 1980 selection could have a degree in biology, the minimum logged flight time had to be in jets. The first successful female Pilot candidate was Eileen Collins, selected in the Class of 1990. Between 1959 and 2004, 148 Pilot-category astronauts were named by NASA, of whom three (Collins, Still and Melroy) are female – a reflection of the difficulties encountered by women not necessary to attain the suitable qualifications (or desire?) to apply to NASA but in progressing through the selection process as a Pilot. Since 1990, however, piloting skills have been less significant than scientific or engineering skills – reflected in the number of female Mission Specialists selected since 1978. Perhaps with the advent of the new Crew Exploration Vehicle within the next two decades, more women Pilots may be accepted for piloting roles – which may include the chance to land a vehicle on the Moon or, one day, on Mars.

Since 1978, Pilot astronauts have served as Space Shuttle Commanders and Pilots, while a few have occasionally served as a Mission Specialist on their first flight. The role of the Commander during the mission includes the onboard responsibility for the vehicle, the crew, the success of the mission, and the safety of the flight. The Commander may also serve as shift Commander on multi-shift scientific missions, and act as back-up during important phases of the mission such as the handling of Remote Manipulator System.

Six pre-1969 astronauts were assigned as Commander on their first Shuttle missions due to past spaceflight experiences and assignments within NASA (or, with Engle, as a former X-15 pilot with the USAF), while six of the seven former MOL astronauts each flew first as Pilot (Peterson only flew once as an MS) before attaining a command of their own on their second mission. From 1978, *all* Pilot candidates have completed at least one mission as a Pilot (or, in a couple of cases, MS2/Flight Engineer) before assignment to their own command – in some cases flying as a Pilot more than once before taking the left-hand seat. By 2004 only Eileen Collins had attained the status of Shuttle mission Commander on her third mission, and is scheduled to lead the post-

Columbia return-to-flight mission as Commander of STS-114 in 2005; whilst both Still and Melroy have each flown twice, but only as Pilots. Pilot astronauts also usually handle support roles in EVA (originally as EVA crew-members, although this was changed to Intravehicular Activity (IVA, or IV) support roles early in the programme) and rendezvous and docking missions, flying the orbiter during periods of undocking, and flying around from a space station (Mir or ISS) to acquire experience for when they dock to a station as Commander on a subsequent mission

Mission Specialists

In 1965 and 1967, astronauts were selected under a 'scientist-astronaut' category that in the mid-1970s evolved to become Mission Specialists. Mission Specialists work in cooperation with the Commander and Pilot, and have responsibility for the coordination of a number of Shuttle-based areas: Shuttle primary systems and sub-systems; crew activity planning; use of consumables; and experiment and payload operations. Their training is reflected in detailed studies of the onboard systems of the Shuttle orbiter, in addition to operation characteristics, requirements and objectives of the specific mission, and any supporting equipment and procedures for each of the experiments, payloads or hardware on their assigned mission. Mission Specialist assignments on a Shuttle flight have varied with the requirements of the specific mission, and safety and operational requirements.

Mission Specialists are assigned numerical designations from MS1 to MS5, depending on the mission to be flown. Each MS has a specific role, and except for MS2 these roles are interchangeable, depending on the individual crew-member, the Commander, past experience, and the mission design. Essentially, however, they are as follows:

MS1 Usually the occupant of Seat 3, on the flight deck to the aft and right of the Pilot, who assists in the ascent phase or descent phase of the mission by monitoring displays and checklists. This is also the position that most Payload Commanders have been assigned, though this is not always the case. On some descents this position is occupied by MS3.

MS2 Also known as the Flight Engineer, the third member of the orbiter crew, who assists the Commander and Pilot in the ascent and descent phases of the flight, supporting calls and actions on the flight deck. Seat 4 is located in the centre aft position on the flight deck between the Commander on the left and the Pilot on the right.

MS3 Seat 5 on the mid-deck next to the side hatch, with responsibility (since 1988) for activation of the slide pole in the event of an emergency escape, and for assisting the returning space station crews in stepping in, and post-landing assistance. During some descents, MS3 occupies the Seat 4 position on the flight deck, and MS1 completes the entry phase on the mid-deck. Therefore, MS1 is normally 'ascent phase' trained, and MS3 'descent-phase' trained, though again this is not a firm rule, as on three-person ISS crew exchanges, MS3, 4 and 5 have been mid-deck-seated ISS resident crew-members for both ascent and entry phases.

MS4/MS5 Seats 6 and 7 (and 8 when flown) on the mid-deck, which can also be occupied by Payload Specialists (PS1 or PS2) or three ISS crew-members.

In addition, Mission Specialists also perform others functions such as the following:

RMS Operation of the Shuttle's Remote Manipulator System (robotic arm) is normally handled by a Mission Specialist astronaut working as the primary RMS operator from the aft flight deck of the Shuttle orbiter. One or more other crew-members are also trained as RMS operation back-ups. RMS operators are also an integral part of the EVA team.

EVA Training for, and operation of, contingency, unplanned and planned spacewalks is handled by Mission Specialists working in teams of two (EV1 and EV2, or EV3 and EV4 when necessary). If more than one EVA is scheduled, teams of two astronauts normally rotate alternate excursions outside, providing support and back-up on the days that they do not perform EVA to spread the work-load and offer a redundancy to ensure mission safety and success. Intravehicular (IV) crew-members can include the Pilots, but Mission Specialists also act as EVA photographer or choreographer.

Prox Ops During periods of rendezvous and docking, or operations in close proximity to other large objects (Prox Ops), Mission Specialists support the Commander and Pilot by handling the 'flying' of the orbiter and the Primary RMS operator, and by performing visual cueing, laser ranging tasks, and photodocumentation – thus lending more pairs of eyes to a tricky operation.

Science support On missions involving a significant amount of scientific research and investigation, Mission Specialists support the collection of data by acting as a 'science crew' rather than an 'orbiter crew' who maintain the Shuttle systems (normally the Commander, the Pilot and MS2). In the 1990s a new role – Payload Commander – was introduced for significant scientific payloads in preparation for space station missions, in which the changing role of Mission Specialists on scientific research missions would see the role of Science Officer introduced.

Seats to spare?

During the first four orbital flight tests, Shuttle crews consisted only of Commander and Pilot. On STS-5 and STS-6 two Mission Specialists were assigned, while the first members of Group 8 flew from STS-7. A 5–7-person crew became the norm, with only one mission (STS 61-A in 1985) flying a maximum eight-person operational crew. (The Shuttle can, in theory, support ten crew-members for short-term descent in a rescue/recovery role, although this has never happened). In 1995, three Mir crew-members were returned by a five-person Shuttle crew. Since 1990, most missions have featured seven crew-members.

Early studies evaluated repetitive flights of crews if the Shuttle attained the launch rate originally planned; but this was never really adapted. On 15 May 1976 – two months before the call for the first Shuttle astronaut candidate was issued – NASA's Associate Administrator for Manned Spaceflight, John F. Yardley, informed the press in Paris, after a three-day meeting with ESA officials, that the Shuttle was planned to carry out two hundred missions with crews of up to seven astronauts, including women astronauts and representatives from Europe, on space science missions of up to a month. NASA's plans envisaged over seven hundred Shuttle flights in the twelve years from 1980, with a fleet of five orbiters completing a total of sixty flights per year (each orbiter flying one mission a month). A core NASA crew of three astronauts (Commander, Pilot and MS) could be supplemented by other Mission Specialists or up to four Payload specialists, who could expect to fly one or two dedicated missions – whereas a NASA career astronaut might expect to fly a maximum of twenty or thirty separate missions!

By early 1980, NASA planning documents illustrated a crew planning chart which revealed that when Shuttle launch rates exceeded twenty launches in one year, a whole crew would be recycled intact within a year. This manifest of twenty-two launches would require eleven whole crews. These documents were used to assist in justifying the further selection of astronauts in 1980 before any of the previous thirty-five candidates had even been assigned to a flight. By the end of the following year the Shuttle had flown two missions, and a new flow chart had been produced based on STS planned schedule models that revealed projections for attrition of veteran astronauts with an increase of the estimated rate of launches. With two launch-pads available from KSC, one under planning from Vandenberg, and four or five orbiters available, it revealed that during the 1980s and 1990s a sustained intake of astronauts would be required to support all planned operations and to take into account natural attrition and programme changes.

No names were assigned to any of the flights, nor were crew positions identified. It was merely a planning document to show the management of the increasing number of personnel in preparing, supporting and managing, as well as flying, so many missions, if everything took place as planned. It was estimated that by Fiscal Year 1987, 50% of pre-1980 astronauts – including the recently selected Group 8 candidates – would have left the programme. All this, of course, depended on the funding of the programme, the delivery of new orbiters, the development of a fully commercial launch manifest, science payloads, support of the military using the Shuttle as a major launch vehicle, international interest, a space station research facility, operational use of both launch sites, an effective and sustainable crew training flow, and a large and sustained budget, supported by proven results and returns from the investment. One of the interesting revelations in this document was that MS candidates would be 'teamed together' in pairs, flying their entire careers together on 5–8 missions to maximise past experience and reduce between-flight training loads.

NASA remained optimistic throughout the early years of the programme (until 1986), despite serious delays and setbacks to this planning schedule. In May 1984 it announced that STS crews would remain intact as 'units' for future mission assignments as mission rates increased, instead of previous assignments on an individual basis. This would help alleviate bottlenecks in the limited Shuttle mission simulators, and require generic proficiency training for teams of Commander, Pilot and MS2 Flight Engineer, with MS1 supporting the flight from the flight deck, and liaison with other crew-members on a science crew, or specific payload requirements. The four would also be able to perform contingency EVA operations if required, and include a primary RMS operator. As a test-case, Bob Crippen would evaluate the shortest time required for ascent–decent training, having experienced three missions to date (STS-1, STS-7 and STS 41-C) and being assigned to a fourth (STS 41-G).

This idea, although not followed rigidly, was adopted for most of the missions throughout 1984–90, as some payloads were delayed and the core crews completed generic training. In November 1983 a four-person core crew was also assigned as a DoD launch-ready stand-by crew, and in May 1984 two four-person crews were named to deploy two planetary missions in 1986, before cancellation due to the *Challenger* accident and the introduction of a NASA policy to revert to a five-person minimum assignment.

As the Shuttle programme evolved over the ensuing twenty years it became apparent that there would be far more astronauts than available flight-seats; but it is also interesting to note that *every* member of the NASA astronaut selection from 1978 (Group 8) through the 1995 (Group 15) completed at least one flight into space before leaving the agency. As long as the candidate completed the Ascan training programme and remained at NASA to support the programme, a flight into space, though never guaranteed, was at least probable (excepting serious injury or death, as in the case of Stephen Thorne shortly before the end of his Ascan training in May 1986), though a second flight was not always a certainty. From the mid-1990s this expectation of a flight was no longer guaranteed upon completion of Ascan training, as there were far more astronauts than there were seats available.

Dedicated back-up crews were also terminated after STS-3, as there remained a qualified pool of suitable trained astronauts to replace a single crew-member and have minimal impact on any crew-training and scheduled flight (as happened on various occasions), although crews were used to support others of similar nature, and occasional back-up crew-members were assigned on an individual basis. In the late 1990s, regular back-up crews were reintroduced for resident ISS crews, but not for Shuttle core crews.

For the women as well as the men, securing a seat on a flight crew remained as much a mystery as it had for those selected in the 1960s. A combination of dedication, team-work, individuality, the so-called 'astro-politics', the pecking order, the influences of military hierarchy in the Astronaut Office, compatibility, demonstrated skills, flight rate, programme funding and scheduling, and the ever-present luck ... all played a part in getting an astronaut off the ground; but not until after graduating from the astronaut 'grubby school' – the Ascan training programme.

ASCAN TRAINING

The six women of the 1978 selection and their twenty-nine male colleagues embarked on an Astronaut Candidate (Ascan) training programme that initially extended over two years but was amended when it was found that it could be completed in twelve months (although it has sometimes been re-extended over two years, depending on the size of the intake and the demands on limited training resources) This programme is completed by all candidates *prior* to being classified as a NASA astronaut. The basic Ascan programme has changed very little since 1978, except for additional language studies and courses on the ISS and international programmes. It is expected that the course on the Space Shuttle will be gradually phased out as that programme is wound down over the next decade.

Most people assume that an astronaut's life is spent flying in space, and have no idea what goes into preparing a person for spaceflight on the Shuttle. The women's training role is as close to being identical to their male colleagues as possible, and they certainly do not ask for special treatment, reinforcing the desire of all astronauts to blend into the 'team'. When selected by NASA, trainees are fully aware of their status as role models, but there is a desire to focus on the 'team effort' rather than the individual, looking to hard work and dedication plus a little luck, and the opportunity to succeed in their chosen careers. They do not, however, like to be identified by gender or ethnic background – especially when named to a 'crew'. NASA is a government agency, but the core of NASA – those who work at the field centres – belong to the 'NASA family', and at the astronaut level the 'flight crews' are often seen as a 'family unit', with all the advantages and disadvantages that such a unit can bring.

But what makes 'an astronaut'? The following is a description of astronaut candidate training from the late 1980s, but it can be applied to any group between 1978 and 2000, and to any member, male or female. It is worth summarising what an astronaut has to undergo *before* sitting on the launch pad and heading into space.

One of the questions most frequently asked is: 'What qualifications do you need to become an astronaut? Do you need to be a pilot or a doctor, an engineer or a scientist?' In the 1960s it was certainly true that military jet piloting skills were essential, and this continues to dominate applications for Pilot astronauts for the Shuttle. However, the advent of the Mission Specialist opened up the opportunities to a broad range of non-pilots.[1]

The first Ascans

In July 1978 the thirty-five new astronauts reported to NASA JSC in Houston, Texas, for their training and evaluation programme, setting the pattern for all subsequent selections through to 2004. By August 1979 it was clear that they would all complete their basic training programme early, and on 31 August 1979 they all 'graduated' from Ascan school to take up assignments in the Advanced Training Branch to prepare them for assignments to Shuttle flight crews. In the ensuing two decades, almost 250 individuals have been selected for Ascan training, and almost all have graduated and progressed to at least one spaceflight each. The two exceptions were Stephen Thorne (Class of 1985), who was killed in an off-duty aircraft crash in May 1986, at the end of his Ascan training programme, and Patricia C. Hillard Robertson, who was also killed in an off-duty flying accident in May 2001.

Patricia C. Hilliard Robertson (*b.*1963) was, in June 1998, a member of the seventeenth NASA astronaut class, and as the Ascan training programme wound down she was given ISS technical assignments as a support crew-member for the second expedition crew. By profession she was a qualified doctor specialising in aerospace medicine, and prior to joining NASA she was working as a medical officer in the Medical Science Division at the Life Sciences Directorate. She had become a certified flight instructor and holder of a commercial pilot's licence. She was talented and capable, well-liked, and was widely expected to be assigned to a long-duration space station crew – probably as a NASA Science Officer – within a few years. On 24 May 2001 she died from injuries sustained in the crash of a small private aircraft two days earlier. She was aged 38, and was the first female NASA astronaut to lose her life without making a flight into space.

Training to survive

The first phase of the new astronaut training programme focuses on preparing for crew training on the T-38 training aircraft used by NASA astronauts to travel around the continental United States. Those pilots who have already flown T-38s receive refresher training for the aircraft, while those with little or no jet-pilot training take a course to qualify them as a T-38 crew-member. For most of the women who have been selected to the NASA astronaut programme, this T-38 crew-member training familiarisation was the course in which they participated.

Ascan Training Phase 1: Orientation and flight training

- JSC orientation by centre management representatives.
- Pilots: T-38; physiological training (if required); water survival training (if required); T-38 flight checkout.
- Mission Specialists: T-38 ground school to facilitate transition from passenger to crew-member; physiological training; survival training; ejection; parasail training; communications; navigation; cross-country flying; crew coordination; T-38 systems; T-38 familiarisation flights.

Mary Cleave (selected in 1980) recalled the surprise that civilian women were being trained as T-38 crew-members in a class where most were military test-pilots: 'There was all sorts of 'Ooh, there's girls here' kind of thing, so that was very, very, different. When we started flying [T-38s] it was not a world where women went.'[11] By 1981 Rhea Seddon, who was single at selection, had met and married fellow astronaut Robert Gibson. In July 1982 she gave birth to their first child, and during her pregnancy, using her knowledge as a medical doctor, she was able to argue that women could still fly through the first stages of pregnancy. Prior to this the USAF grounded pregnant pilots, but Seddon was able to demonstrate – to the amazement of military pilots – that this was unnecessary.

Survival-school training in the Shuttle era is quite different from the survival training of the Apollo era. In the 1960s, because Apollo missions from the Moon could conceivably land anywhere in the Earth's equatorial regions, most of the survival training focused on water egress training and jungle and desert survival courses. On the Shuttle, the crew has some control over where the vehicle lands, and so survival training was amended accordingly. The USAF Water Survival School at Holmstead AFB, Florida, trained those who had not received similar training while serving in the US armed forces (including most of the women candidates). This was followed, a few weeks later, by an ejection survival course at Vance AFB, in Enid, Oklahoma, which featured training for a simulated ejection from a T-38 cockpit, parasailing, parachute training, and descent into water and onto land.

Aircraft operations continue from the start of Ascan training until departure from the active flight list or in the Astronaut Office support roles that some veteran astronauts perform after their last spaceflight, such as weather coordination flights in support of launch and landing operations. These are completed in the smaller T-38 training aircraft for cross-country flights and proficiency flying, or in the KC-135 for zero-g parabolic flights. Many astronauts, including the Mission Specialists, also log time in sports planes, sailplanes or gliders, as well as having parachute experience. Kathy Sullivan – a professional geologist as well as an astronaut – logged time as a crew-member in NASA's high-altitude WB-57F (modified U-2) Earth Resources aircraft.

For some of the women, parasailing and parachute training were not set up for people of their size. During a windy day in Florida, Mary Cleave was completing paragliding during Ascan training, and after cutting the lines, as she was meant to do, she continued to ascend – towards Miami, leaving behind the horrified Air Force trainers and supervisors, who were thinking that they would have to report an astronaut lost at sea. There was much relief when she descended as the winds shifted, and she landed safely. After this occurrence, two sizes of parachute – regular and extra small – were requisitioned for training; but fortunately, none of the women have had to eject from a T-38 on a standard parachute. Cleave also recalled that in

Ascan training: a break during parachute training instruction Group 8, 1978. Left–right (women only), Resnik, Ride, Sullivan, Seddon, Fisher, Lucid.

the early days of training she was expecting more 'action stuff' as well as a large amount of bookwork. For her inaugural flight in a T-38, instructor pilot Bud Ream had been instructed not to do anything that would make the her sick. He therefore followed the instruction; but Cleave pleaded for aerobatics and promised that she would not be sick, as she had waited all her life to fly in a high-performance jet. But it would be more than six months before she and the other candidates experienced the 'action stuff' that she craved.[11]

Technical training

The next phase of the Shuttle-era selections was the technical training programme.

Ascan Training Phase 2: Technical Training

- Shuttle systems: Guidance, Navigation and Control (GNC); Data Processing Systems (DPS); instrumentation; electrical and environmental systems (EPS and ECLSS); tracking techniques.
- Flight operations: aerodynamics (ascent and entry profiles); orbital mechanics (Prox Ops and rendezvous).
- Mission operations: flight overview (ascent, orbit and entry) and planning; payload integration; payload carriers (Spacelab, PAM and IUS); payload deployment (including RMS); EVA systems (including scuba diving); ground support roles; mission rules.
- Manned spaceflight concepts: guest lecturers (including former astronauts); Shuttle design; space station configurations (which in the 1990s expanded to include Russian programmes such as Mir and the development of Freedom into the ISS).
- Sciences: space physiology; medicine; Earth observation and photography;

Ascan wilderness training, Group 12, 1987. Back to camera, at centre, Davis; at rear, Jemison.

astronomy and star identification; planetary science; space physics; atmospheric science; life science; materials science; geology; oceanography; meteorology.

Technical assignments

Once the second phase of astronaut training is completed, Ascans move on to a variety of technical assignments in the Astronaut Office, providing 'on the job training' as well as participation in a number of simulations, meetings, contractor visits, public appearances and support of flight activities in various roles, coordination of numerous test and development activities, presentations and participation in test and evaluation environments, and media support roles.[12] Some of the leading mission support roles in which Ascans have been involved prior to first flights have included:

KSC launch support team Activities at the Kennedy Space Flight Centre, at Cape Canaveral, Florida, preparing payloads and orbiters for launch; supporting launch-day preparations (such as count-down demonstration tests and emergency egress simulations), and activities on launch day and during post-landing procedures. This team is generally known as the 'Cape Crusaders', and includes astronauts who configure the flight-deck displays and switch panels for launch, establish communications between the orbiter and launch teams, support the crew on entry into and exit from the orbiter (the Astronaut Support Person), and serve as an End of Mission Exchange (EOM) crew, configuring the orbiter for safe transport back to the Cape (should the landing occur elsewhere) or to the Orbiter Processing Facility.

Capsule Communicator CapCom – the astronaut in Mission Control, Houston, who talks directly to the on-orbit crew and is the point of contact between ground teams and the crew in space. Working in shifts (Ascent/Entry; Orbit 1; Orbit 2; Orbit3/Planning) around the clock, these astronauts also pass on messages to and from the families, and act as the voice of the crew in the interpretation of real-time flight activities on non-critical issues. A separate team of CapComs in the ISS Flight Control Room now handles Space Station communications while the Shuttle flight controllers and CapComs operate from a separate control-room facility.

During a recent interview,[13] Ellen Ochoa (selected in 1990) pointed out that the role of Ascans in supporting missions while awaiting their first or next mission was as important as flying the mission itself: 'You know that if you are not training yourself, you are basically helping other people do it successfully, and as you get here [inside the NASA organisation, and particularly the Astronuat Office] you realise that there are a lots of ways to contribute that are really important. I have worked as a CapCom in Mission Control, which is a really fun and challenging job, because you have to learn that flight, not in as much detail as the crew-members on it, but enough to really do your job well. You are working with the whole flight control team, and that is also very interesting and challenging. In everything that you do, you see how it impacts the actual missions that are going on and the success of those missions, so in general the support jobs are very interesting and challenging and the people are very motivated. You are a member of the team; whether it's being part of the crew or part of the team looking at one aspect of that mission, you are always a member of a team.'

Abort site support Astronauts can be deployed to contingency abort landing sites during ascent, where they act as coordinator for the emergency landing should one occur (but which to date has never happened). Transatlantic abort sites in Spain and on the west coast of Africa are supported by Contingency Action Centres and Mishap Representatives in case of major malfunctions or abort scenarios. These are based at KSC, JSC, and the Dryden Flight Research Center next to Edwards AFB in California.

Shuttle avionics laboratory This is the location at JSC where the avionics software for Shuttle flights is evaluated and verified. 'Crews' of Pilots and Mission Specialists test flight or payload software, or 'fly' various scenarios. During stand-down time between scenarios, participants swap 'war stories', and really get to know the people working behind the scenes on every flight.[11]

WETF A team of EVA-trained astronauts remains on stand-by to enter the water

Tammy Jernigan during JSC WETF training for EVA operations at JSC.

tanks in support of any contingency EVA that may occur during flight, or to assist in completing EVA tasks. They also work to develop new EVA techniques and training regimes.

Family escorts Astronauts are assigned to escort families of the flight crew during launch and landing, to advise on mission development. They are also the first point of contact should anything go wrong, and provide the all-important 'one of our own' link between the crew and close family members.

Working with veteran astronauts on technical assignments was also a daunting challenge initially. Cleave recalled that because the toilet had not worked correctly during STS-1, Chief Astronaut John Young gave her her first personal technical assignment of fixing it. She tried to explain that she was more accustomed with the other end of the system than with the plumbing end, but Young, recalling her background in microbiological ecology, civil and environmental engineering and research work in a water research laboratory, replied: 'You're a smart girl. You'll figure this out.' Young knew that the new astronauts were there because of their ability to solve problems. They were part of the astronaut team, and were there because of their skills and experience, not because of their gender or ethnic background. Cleave fixed the problem, and in doing so realised that the system actually assisted her in understanding the aerodynamics and aviation engineering class: 'My god! This aerospace [technology] is just [like] an inside-out sewer pipe', to which she could relate from her earlier engineering career. However, when she shared her interpretation of state-of-the-art jet engineering, 'None of the test-pilots were at all impressed with this great visualisation of their [spacecraft] as an inside-out sewer pipe.'[11]

Pilot-pool training

The next stage of training is assignment to the Pilot pool advanced courses for Ascans (and crew-members recently off a flight assignment to maintain their proficiency prior to assignment to a new crew), with more detailed familiarisation with the various systems of the Shuttle. These include:

Avionics systems Guidance, Navigation and Control (GNC); Data Processing Systems (DPS).

Orbiter systems Caution and Warning (CandW); Electrical Power System (EPS); Environmental Control and Life Support System (ECLSS); Communications and Instrumentation (COMM/IN); Propulsion (PROP).

Crew systems Photography/Closed Circuit Television (Photo/CCTV); habitability (galley, waste management, personal hygiene, stowage, sleep stations, crew equipment and clothing); emergency procedures.

Flight operations Ascent, orbit and entry (including pre-launch embarkation and post-landing egress of the orbiter crew module).

Technical assignments for the first women astronauts

By the end of 1979 and the completion of their Ascan training programme, the Group 8 astronauts had been assigned their first various technical duties in the Astronaut Office. Most of the Office was supporting the launch of the first flight and the series of orbital flight tests planned to qualify the Shuttle system for operational use. Members of Group 8 were assigned to these flights in support roles, and received technical assignments in the development of future payloads and flight equipment. The six women astronauts were given the following assignments:

Anna Fisher RMS operations, EVA equipment and procedures, tile repair, medical operations for both STS-1 and STS-2, and duties in SAIL.

Shannon Lucid Spacelab 1, crew training, SAIL, Flight Simulation Laboratory duties, Large Space Telescope (later renamed Hubble Space Telescope), proximity operations and rendezvous, Shuttle pallet satellite (SPAS).

Judy Resnik Spacelab software, payload software, RMS software, RMS operations, Power Extension Package (planned to extend the orbital lifetime of the Shuttle), 25-kW power module and training issues.

Sally Ride RMS operations, support crew for both STS-2 and STS-3, and point of contact for visual operations, RMS procedures and the orbit Flight Data File for those missions. Also assigned as CapCom for STS-2 and STS-3 and as an STS-1 T-38 chase-plane crew-member for the landing phase. Early in her training, Ride

Kathy Thornton participates in emergency pad egress training at KSC.

demonstrated an uncanny skill in operating the robotic arm of the Shuttle during simulations. According to one training technician this was possibly due to her hand–eye coordination developed through competitive tennis.[14]

Rhea Seddon Food systems, Spacelab 3 and Spacelab 4 (a life sciences mission), payload software, CRT display verification, orbital medical kit, medical checklists, medical operations support for STS-1, STS-2 and STS-3, SAIL duties.

In a recent interview,[15] Seddon supported the process of assigning astronauts to technical assignments between their flights: 'It's very good that [NASA] sometimes assigns you to things that you know about, and they sometimes assign you to things you don't know anything about, so you make a contribution in some areas and you do on-the-job training in other areas. The food systems were certainly not the most popular job to be assigned to, but I had an interest in [nutrition aspects] and I enjoyed it, so it was a nice fit.' She also explained that the EVA suit was one of the less enjoyable things that she evaluated: 'They could never get one to fit me, as I was too small, and they questioned whether they should invest millions of dollars into redesigning the suit to fit all sorts of women's size and shape differences, so I

On 1 July 1979, Kathy Sullivan flew as a crew-member of the NASA WB-57F reconnaissance aircraft, piloted by Jim Korkowski, flying out of Ellington AFB, near JSC, Houston. During the 4-hour mission they flew to 63,300 feet, during which time Sullivan operated a colour infrared camera and multispectral scanning equipment for 1.5 hours. The flight set a female aviation record.

was never a candidate for wearing the suit or doing EVAs.' EVA requires considerable upper body strength, and Seddon stated that she would never have been very good at it. Kathy Thornton (selected in 1984) was also a small person, but she worked out a lot to build her upper body strength to make up for lack of physical size, and became a successful spacewalker. Seddon added: 'I think I could have done that too, but it certainly wasn't at the top of my list of where I wanted to make my contribution. I wanted to do research.' She also always had problems with the huge launch and entry suits introduced after the *Challenger* accident. She found them really awkward, and always referred to the operation as 'suit wrestling'. She managed, 'but it was not my favourite thing to do.'

Kathy Sullivan OASTA-1 and OSS-1 payload packages, EB-57F crew-member, flight software verifications (Versions 16 and 18), STS-2 T-38 chase-plane crew-member for the landing phase, and vehicle integration at the Cape.

Although these assignments would continually change during their careers and would depend upon flight assignments, as new female astronauts came into the programme they received similar assignments – a further step in securing that all-important first selection to a flight crew.

Shuttle crew training

Ascans who complete the entire training course, technical assignments and pool training become eligible for crew assignment. Exactly when they join a crew depends on flight-seat availability, the launch manifest, the objectives of the mission, the availability of more experienced astronauts, their own eligibility, and a little luck. Once assigned to a crew, specific flight-crew training becomes priority, and the Ascans move towards the top of the list for simulator time. The flight-specific crew-training programme encompasses the following:

- Commander and Pilot training in the Shuttle Training Aircraft (STA), practicing approach and landing profiles until a few days prior to launch
- Crew Systems refresher courses
- Flight operation refresher courses, and any mission-specific profiles for the mission
- Flight-specific training
    - ✦ Payload Deployment and Retrieval Systems (PDRS)
        - ✧ Operations (including RMS)
        - ✧ Payload deployment
        - ✧ Payload retrieval
    - ✦ Carriers
        - ✧ Spacelab (modules and pallets)
        - ✧ Payload Assist Module (PAM) (phased out by the 1990s)
        - ✧ Inertial Upper Stage (IUS)
        - ✧ Space Hab (from the 1990s)
    - ✦ Attached Payloads
        - ✧ GAS canisters

    - ✧ Hitch-hiker attachments
    - ✧ Mission-specific packages
  - ✦ Mid-deck experiments
    - ✧ Student experiments
  - ✦ Mission-specific experiments
    - ✧ SAREX (ham radio)
  - ✦ Proximity Operations/Rendezvous:
    - ✧ Specific commercial satellites
    - ✧ Shuttle Pallet satellite
    - ✧ Science satellites (Solar Max; LDEF; HST)
    - ✧ (Since 1994, docking training with Mir and then the ISS)
  - ✦ Extravehicular activity
    - ✧ EVA hardware and procedures
    - ✧ Contingency operations and scenarios
    - ✧ Planned EVAs for specific missions (such as HST servicing)
    - ✧ Manned Manoeuvring Units (MMU in early 1980s, and SAFE from the early 1990s)
    - ✧ Space station construction techniques and operations
  - ✦ DTOs/DSOs: a programme of
    - ✧ Detailed Test Objectives (normally engineering-based)
    - ✧ Detailed Supplementary Objectives (more experiment-based)

After a crew has completed its mission and any post-flight requirements such as debriefing, readapting to the 1-g environment, technical report writing, and public tours and appearances, it is disbanded. Individuals will either re-enter the pool training group for reassignment to a new crew, retire from active flight status to enter a management or administrative role, return to their parent service, or leave the agency to pursue other careers and interests outside government service.

Astronaut Office branch office assignments

The NASA Astronaut Office (CB) is part of the Flight Crew Operations Directorate (FCOD) – one of several directorates at NASA JSC in Houston, Texas. The Aircraft Operations Division is also part of FCOD. The Astronaut Office has several branches, which change depending on programme requirements and future planning. There is a Chief of the Astronaut Office as well as a Deputy Chief and Branch Chiefs, under whom Lead astronauts are assigned to organise other astronauts to relevant technical assignments, support roles, training and coordination, and Astronaut Office point-of-contact roles. Even from the early days of the programme it was obvious that no single astronaut could learn all aspects of a mission, and so multi-role skilled back-up team assignments strengthens the Astronaut Office and the flight crews. Although the branch or department names and roles change slightly over the years, these examples from the Shuttle era of the 1980s provide an impression of the work that astronauts were assigned on the ground.

- Operations and Training Branch
  - ✦ Operations: CapComs, crew equipment, Flight Data File

    - Training: Shuttle mission simulator
- Mission Safety Branch: crew, hardware, systems and procedures
- Mission Development Branch
    - Deployment of payloads and upper stages
    - RNDZ/Prox Ops
    - EVA/EMU/MMU
    - Science and technology payloads
    - Space medicine issues and hardware
    - Payload accommodations (PDRS, Spacelab, pallets, payload communications and data)
    - Mission integration
- Systems Development Branch
    - Software, improved GPCs, payload operations control, orbiter experiments, landing site aids, auto land, Shuttle training aircraft
- Test Branch
    - Shuttle Avionics Integration Laboratory (SAIL)
    - KSC Support Branch ('Cape Crusaders')
- Crews in training (including astronauts with some organisational roles as well as crew training requirements)

First female crew-members on the Shuttle

Early in 1982, the Group 8 Ascans, including the first six women astronauts, were ready for assignment to their first missions. The Shuttle Orbital Flight Test programme was nearing completion, and the crews for the first two 'operational flights' were nearing the final stages of training. It was time to announce the crews for the next few missions – and for the first time, women would be assigned to fly on an American spacecraft in orbit.

REFERENCES

1 David J. Shayler, 'The Thirty-Five New Guys: The First 10 Years', *Orbiter*, No. 35, January 1988, AIS Publications.

2 NASA JSC News Release Nos. 77-42 (29 July); 77-44 (11 August); 77-46 (25 August); 77-52 (15 September); 77-54 (26 September); 77-56 (3 October); 77-59 (12 October); 77-66 (20 October); 77-70 (4 November); and 77-75 (11 November).

3 Astronaut Selection and Training: NASA Information Summaries, NP-1997-07-006, JSC, July 1997, p. 3.

4 *Washington Post*, 19 December 1977, postscript to lead article.

5 'NASA announces selection of 35 Shuttle astronauts', *Washington Post*, 17 January 1978.

6 Carlos Byars, 'Prospective astronauts visit space Center', *Houston Chronicle*, 1 February 1978.

7 Donna Miskin, 'Women and minorities star in the new astronaut class', *Washington Star*, 19 November 1978.

8 Correspondence from Gene Nora Jessen, 9 June 1998.
9 Brian Vine, 'Chris's Angels: America's first ladies of space', *Daily Express*, 15 March 1979.
10 David J. Shayler, 'NASA Astronauts', in *Who's Who In Space*, ISS (third) edition, by Michael Cassutt, Macmillan Library Reference, 1999. pp. 1–17.
11 Mary L. Cleave, JSC Oral History interview by Rebecca Wright, Washington, 5 March 2002.
12 Alcestis Oberg, *Spacefarers of the '80s and '90s: The Next Thousand People in Space*, Columbia University Press, 1985, pp. 35–52 – a useful account of the 'astronauts' job'.
13 AIS telephone interview with Ellen Ochoa, NASA JSC, 2 March 2004.
14 Ref. 12, p. 45.
15 AIS telephone interview with Rhea Seddon, Vanderbilt University, 21 April 2004.

Sally and Svetlana

On 18 June 1983, Sally Ride became the first American woman to fly into space, as a crew-member of the STS-7 Space Shuttle mission. Since then, 35 women have flown on American Shuttle missions, including four who lost their lives in the two accidents of 1986 and 2003. Ride, however, was not the second woman in space. On 19 August 1982 the world was informed that a new Soviet crew was on the way to Salyut 7, including female cosmonaut Svetlana Savitskaya as Cosmonaut Researcher.

At the time of writing (2004), the American space programme is focusing on the return to flight of the Shuttle system – hopefully during 2005. As part of the restructuring of the US manned space programme following the loss of *Columbia*, the only purpose of future flights of the Shuttle will be to complete the construction of the International Space Station before the Shuttle is retired after 2010. To support this effort, a cadre of astronauts, including several women, has been assigned to planned missions to the ISS via the Shuttle, and a further group awaits assignment. For almost all the female space explorers who have followed the trail of Tereshkova, their access to space has been via the US Shuttle system (only three women have ridden into orbit onboard a Soyuz spacecraft), but after the loss of *Columbia* and the plans to ground the Shuttle from 2010, the new 2004 astronaut class will probably be trained on a new vehicle for access to space, as well as receiving instruction on the Russian Soyuz spacecraft that has been the only other way for people to enter space for over forty years. (In October 2003, however, the manned debut of the Chinese Shenzhou spacecraft offered a third alternative, and in early 2004 there were indications that the Chinese are to select a group of female takionaut trainees for Shenzhou training.) The achievements of these female space explorers at various space stations since 1982 are covered in a separate chapter. Here we review the assignments and activities of the women who flew in space between 1982 and 1985, including the 'race' to place a second woman in space twenty years after Tereshkova flew.

BLAZING A NEW TRAIL

In April 1982 – a little over two weeks after *Columbia* returned from her third mission in space (STS-3) – discussions were being held at JSC to select suitable candidates from the 1978 astronaut selection for assignment to STS-7 and STS-8. The first four (originally six) Shuttle missions were part of the Orbital Flight Test programme, and the fourth and last crew was in the final stages of training for a June launch. The crews for STS-5 (the first 'operational' satellite deployment mission) and STS-6 (the first flight of *Challenger* and the deployment of the important Tracking and Data Relay Satellite to improve orbital communications with the Shuttle) had been named earlier in March, though the crews had all been in training supporting OFT missions for several months. Another crew was in training for STS-9, which was carrying the first Spacelab research mission. Most of these assignments were filled by pre-1978 members of the Astronaut Office, so the first opportunity for any of the Group 8 astronauts to fly in space would be on the satellite deployment missions of STS-7 and STS-8.

The first assignments

All thirty-five candidates were technically qualified to fly on either mission, and each had demonstrated a good public presence. To reduce the candidates to a suitable short-list, an Advisory Group on Shuttle Crew Selections met and reviewed the qualifications of the whole group and their Ascan training performances.[1] Several key figures contributed to the review, including George Abbey and Bob Crippen, who would be assigned as Commander of STS-7. The Pilot candidates had already been selected, and so the Advisory Group focused on the Mission Specialist candidates for the two missions. At this time, a database for Space Adaptation Syndrome was being collated for all of the candidates and the active veteran astronauts – the start of a Medical Division protocol to assemble such data for all astronauts. This would result in the assignment of medical doctors to the crews of STS-7 and STS-8 to study first-hand the effects of adapting to spaceflight in the first few hours and days of a mission on the Shuttle, which would have implications for mission planning.

It was recognised that with STS-7 one of the female candidates could receive a flight assignment, and as such would receive notable media attention as the 'first American woman in space'. It was also recognised that the opportunity to fly the first 'ethnic minority' astronaut was available on STS-8, which would also generate additional media interest. The candidates short-listed for STS-7 were:

Judith Resnik	Extremely well qualified on the RMS, but not so comfortable with public affairs issues.
Sally Ride	Extremely well qualified on RMS systems and operations, but more experienced than Resnik on orbiter systems. Possessed a very good public presence.
Rhea Seddon	Limited experience on either payload support or STS-7 mission specific training. She was also pregnant, and was soon

	disqualified from selection for the first female astronaut flight.
Kathy Sullivan	Possessed significant experience in support training roles.
Shannon Lucid	No significant comments provided.
Anna Fisher	Experienced in support role training, and described as an 'outstanding public presence'.
John Fabian	The only male candidate for STS-7. Described as a leader, well organised, and extremely well qualified on proximity operations and RMS operation.

The STS-8 candidates were:

Dale Gardner	Extremely well qualified on STS software.
Ron McNair	Limited RMS training, but an 'outgoing and outstanding public presence' (African-American).
Guion Bluford	Limited RMS experience (African-American).

Interestingly, Ellison Onizuka (Hawaiian-American) is not listed here; and Fred Gregory (African-American) was a Pilot candidate and not eligible.

All six female candidates were short-listed for STS-7, but none for STS-8. This was reflected in the announcement of 19 April 1982,[2] naming the four-person crews for STS-7 and STS-8 as:

STS	*Commander*	*Pilot*	*MS1*	*MS2*
7	Crippen	Hauck	Fabian	Ride
8	Truly	Brandenstein	Bluford	Gardner

With this announcement, Sally Ride became the first American woman to be assigned to a US flight crew, bringing access to space for women a significant step closer.

The Soviet response

News of a woman being assigned to a Shuttle flight crew increased media interest in the event, generating numerous articles and TV documentaries on the progress and prospects of women in space. From the time of their selection in 1978, all through their training, and following the results from the first flights of the Shuttle system, it became obvious that the first American woman would follow Tereshkova and become the second female to fly into space, on an early Shuttle flight. The Soviets, however, had other ideas.

With the news that NASA would be sending a female explorer into space, the Soviets, after years of ignoring the opportunity, decided that they were not going to be beaten by the Americans and would launch a Soviet woman into space in the early 1980s, before the Americans could do so on the Shuttle. The Soviet space station programme had developed to the point where resident crews were flying for up to six months, and short-term visiting crews were staying for about a week to replace the ageing Soyuz spacecraft that launched and returned the crews. It was therefore inevitable that if a flight of a Soviet woman was to take place quickly, then a short mission to the Salyut station would be the obvious goal.

The Soviets had decided to develop a semi-permanent manned space station in Earth orbit from the early 1970s – partly in response to losing the Moon race to the Americans in 1969. Instead of first developing a reusable space Shuttle, like the Americans (though one was on the drawing board, and part of the long-range goals but would not be ready in time to surpass the American Shuttle), expanding the space station programme became the Soviet aim. The first station – Salyut – was launched in April 1971, and the Soyuz used as a crew ferry was an integral part of the system – a profile that has continued to the present day on the ISS.

Overcoming various difficulties and setbacks, the Soviets gradually increased orbital duration from three weeks to more than six months by 1980. The lifetime of the station also increased from a few months on the first Salyut to several years by the time Salyut 6 was operating a decade later. The improved Soyuz could only carry up to three cosmonauts at that time, so research opportunities were limited by crew-time for maintenance and housekeeping duties. The spacecraft also had a limited orbital lifetime, even when docked to the station in dormant mode, so the Soviets needed to periodically exchange the spacecraft with a fresh and reliable rescue and return vehicle for the resident crew, should it be needed. From Salyut 6, therefore, a series of short visiting missions was devised to deliver the new Soyuz and return the older one. Research time was still limited; but it afforded the opportunity to double the station crew for a week or so, and fly a third person as cosmonaut-researcher.

Once the Societs knew that America was selecting non-career astronauts from the US and Europe to fly as payload specialists on Shuttle missions, they again took the initiative by establishing a series of missions by cosmonaut guest researchers from Communist Block countries, flying with a Soviet Commander on a visiting mission to Salyut 6. This station was due to be replaced by Salyut 7 in 1982 and by a larger modular station in the mid-1980s, and as part of the new station planning it was decided to recruit a new group of female cosmonauts in order to train and fly them before the Americans could put their first female astronauts into space onboard the Shuttle.

A varied selection

The idea of selecting a new group of female cosmonauts for a Soviet mission had actually been in the mind of NPO Energiya Chief Designer Valentin Glushko for some time. In 1974, Glushko had taken over the reins of the former OKB-1 Koroylov Design Bureau that had placed the Soviets in the lead in the space race in the 1960s, but had floundered after the death of its inspirational designer in 1966. Some fifteen years after the selection of the first Soviet female cosmonauts, the programme, but not the ideology, had changed. The 1970s had seen a rise in the professional skills of Soviet women in both science and engineering, and there was therefore a large resource of talent to fill positions in numerous technical fields. Unlike the lack of male counterparts following the Second World War, however, there was by now a balance of numbers between Soviet men and women, so that fewer Soviet women reached higher-level positions or made inroads to leading technologies, such as the development of the space programme. In addition to completing a full working week alongside their male colleagues, Soviet women also kept the home and family, and looked to pioneering achievements by Soviet women – such as the flight of Tereshkova – to improve their own fortunes.

In 1978, the case for more Soviet female cosmonauts was not helped by comments by two former cosmonauts, now in key administrative roles within the manned programme. Georgi Beregovoi – pilot of Soyuz 3 in 1968, and now Director of the Gagarin Cosmonaut Training Centre – stated that 'social differences' between the sexes added to the risks of flying a mixed-sex crew or an all-female crew, and that women were much more emotional and could be easily upset. Director of Cosmonaut Training Vladimir Shatalov – a three-time space veteran – tried to play down the lack of female cosmonauts as a gesture to prevent undue harm and the discomfort of long spaceflights.[3]

However, within twelve months NASA had selected a new group of astronauts, including six women, for the Space Shuttle. Not wishing to fall behind the US in space prestige, Glushko initiated a new selection process; but instead of the publicity skills of the first female cosmonaut group in 1962, this time the individual's specialisation in her field was a criterion. The selection would lead to what was hoped would be the second woman in space before the Americans could orbit one of their female astronauts on the Shuttle, followed by a 'space first' by flying an all-female crew to a space station. Their professional and educational qualifications were studied, as was their commitment as Communists and Soviet citizens. Recommendations from the local Communist youth party, the Komsomol, the local trade union organisation, and from the local branch of the Communist party, together with a degree of ethnic background checking, were all considered to ensure that mostly 'Russian state' candidates applied. As this was a civilian selection process, the applicants applied for the role of engineer-tester, rather than pilot-cosmonaut as in the Air Force selections. But when members of the first group of female cosmonauts tried to apply, they were informed that they were too old. Tereshkova was now a 'national treasure', and would not have the chance to fly again. As a result, the upper age limit for candidates was set at 33.

Several hundred applicants were screened and, partly in response to concerns over their ability to perform their tasks in space, a large group was submitted for medical testing, beginning in May 1979, followed by a short course of basic cosmonaut training (academic, wilderness, flying and simulator courses) in preparation for assignment to crews. This course was completed between December 1979 and June 1980. Two of the identified women who were not selected were Svetlana Beregovkina (*b.*1950), a doctor of the IMBP, and Lyudmilla Sviridova, an NPO Energiya engineer. On 30 July 1980 the women were approved by the Joint State Commission (GMVK) for enrolment as cosmonaut candidates in the team. Unlike their American counterparts, however, their names were not publicly announced, and their training remained out of the limelight. The ten new cosmonaut trainees included five women physicians from the Institute of Medical and Biological Problems (IMBP) – the leading Soviet institute in the field of space medicine – who would be early candidates for long-duration missions to Salyut stations. There were three female engineers – two from NPO Energiya and one from the Leningrad Mechanical Institute – and one representative from the Soviet

Academy of Sciences. The tenth candidate was a test-pilot from the Yakovlev aircraft factory:

Galina V. Amelia, 26, a physician on the staff at IMBP.
Yelena I. Dobrokvashina, 32, also a staff physician at IMBP.
Yekaterina A. Ivanova, 30, an engineer at the Leningrad Mechanical Institute.
Natalya D. Kuleshova, 25, an NPO Energiya engineer who told her father about her application, only to hear his opinion that women had no place in the cosmonaut team.
Irina D. Latysheva, 26, a radio engineer on the staff of the Moscow Institute of Radiotechnology and Electronics (MIRZ) within the Academy of Sciences.
Larisa G. Pozharskaya, 33, an IMBP physician.
Irina R. Pronina, 27, an engineer from NPO Energiya, who learned that a new selection of female cosmonauts was being sought from her father Rudolf, an engineer at the Baikonur cosmodrome.
Svetlana Y. Savitskaya, 32, an aeronautical engineer and test-pilot for the Yakovlev design bureau.
Tamara S. Zhakarova, 28, a physician at IMBP.
Olga Klyushnikova, 26, also a physician from IMBP. She was approved by the GMKV and completed some initial general spaceflight training, but was not formally enrolled in the IMBP team and was later medically disqualified.

The nine remaining group members then completed a programme of general Soyuz–Salyut training, leading to the award of the Cosmonaut Certificate that signified their graduation from the cosmonaut training programme. However, some of the members did not receive their certificate for several years, with the final one being presented (to Latysheva) as late as 1991 – eleven years after selection.[4] Being selected for cosmonaut training – or, indeed, astronaut training – does not guarantee a spaceflight, and not even official inclusion in the team ensures a trip into orbit, as members of the first female selection could testify. On more than one occasion the new team were told during training that the programme was to be suspended, and that they should consider themselves on stand-by until they were needed for a spaceflight, working at their normal positions and receiving occasional spaceflight training courses to retain their proficiency.

When Sally Ride was named to the crew of STS-7 in April 1982, the urgency of launching a woman into space first was realised. The Soviets responded by announcing that they would orbit a second woman (then unidentified), although when and for how long was not revealed. The programme to prepare for that flight was already in progress. The female group was training for a visiting mission to the new Salyut station, and the requirement to launch the mission in late 1982 (to beat the Americans) required an abbreviated training programme and a short mission. Flying and engineering skills would be an advantage over the scientific or biomedical skills required for a long mission, and in light of this, Svetlana Savitskaya was assigned as Cosmonaut Researcher to a Soyuz crew commanded by veteran cosmonaut Leonid Popov, with space rookie Alexandr Serebrov as Flight Engineer. The back-up crew was Vladimir Vasyutin, Viktor Savinykh and Natalya Kuleshova.

Also included in the group were two other female candidates, Irina Pronina and Yekaterina Ivanova, though they were not formally assigned to a crew. It soon became clear that Kuleshova's preparation was not adequate for the demands of specific mission training, and she was replaced by Pronina. Although it was not the hoped-for space spectacular of an all-female crew, this mission would fly to Salyut 7 and place the second female in orbit almost a year before the Americans. The rest of the women in the group would have to wait for a new assignment – and it would be a long and fruitless wait, as only Savitskaya would ever fly into orbit. Her back-ups also realised that she was the obvious choice for this mission, and had an all-female crew flown, she would have been the mission Commander.

SVETLANA, SOYUZ AND SALYUT

Savitskaya came from an aviation background. Her father was Marshal of the Soviet Union Yevgeny Saviksky – twice Hero of the Soviet Union, a Second World War air ace, and a former Deputy Chief of the Air Defence Forces. As a young girl Savitskaya apparently preferred playing with model aeroplanes than with dolls, and dreamed of flying high-speed aircraft. As a teenager studying at an aero-technical school, she qualified as both a pilot and a parachutist before entering a course of aeronautical engineering at the Moscow Aviation Institute. After graduation in 1974 she served as a DOSAAF association flying club instructor, participating in sports flying. In 1970 she travelled to the Hullavingston Air Show in the UK as a member of the Soviet National Aerobatic Team for the sixth World Aeronautics Championships, where she became World Champion for all-round flying and was feted in the British Press as 'Miss Sensation' for her skills. In 1974, Savitskaya entered the Ministry of Aviation Productions test-pilot school at Zhukovsky Air Force Engineering Academy, prior to being employed at the Yakovlev design bureau, going on to set many speed records in Soviet aircraft, and earning a Master of Sport rating with more than five hundred parachute jumps. She also claimed 'I can both knit and sew', and found housework a relaxing pastime after a hard day's work flying![5] Fluent in English and a skilled swimmer and ice-skater, she had set sky-diving records from over 14,200 metres at the age of only 17. Her background was completely different from Valentina Tereshkova's or Sally Ride's.

For Savitskaya the path to space would be by R-7 launch vehicle as a crew-member of the Soyuz T-7 ferry mission – the latest variant of a spacecraft that had been in service since 1967.[6] After a short flight to the Salyut 7 station, the three-person visiting crew would join the resident crew for a week-long programme of medical and scientific experiments before returning to Earth in the older Soyuz T-5 spacecraft, leaving the resident crew to complete their record-breaking mission of 211 days and return in the fresher Soyuz T-7 spacecraft. Salyut 7 had been launched in April 1982 to replace the ageing but highly successful Salyut 6, and the resident crew was the first of a planned series for the new station. Savitskaya's Soyuz T-7 was the second visiting mission to the station that year, and although she was a pilot by profession she was also an engineer, trained to perform the role of Cosmonaut

Researcher during the mission. But despite the science and the achievement, some aspects of the mission did not go as smoothly as intended.

Soyuz T-7: a mission to Salyut

Before the mission, Savitskaya expressed her life-long dream of flying into space. As a pilot she regarded the Soyuz and Salyut as she would a new aircraft (without wings), and would strive to master the new equipment, set new records, and attain higher speeds. The prospect of a non-stop flight for one week also fascinated her as a new experience for a pilot, along with the speed of 28,000 km/hr while flying around the world in 90 minutes. Also new for her was the programme of experiments and scientific research: 'My profession as a pilot is being enriched by a new profession: that of Cosmonaut Researcher,' she stated.[7] Savitskaya also said that she was aware that the prospect for further flights into space by female cosmonauts rested upon the success of her mission. In her book *Almost Heaven* (published in 2004), author Bettyann Holtzmann Kelvles suggested that this was also a comment on the apparent lack of support from Tereshkova, who apparently had not visited Savitskaya during training. She was also not present at the launch, which would have added to the media coverage of this new flight into space by a woman.[3]

Operationally, Soyuz T-7 (call-sign Dnieper) was launched on 19 August 1982 from the Baikonur cosmodrome and docked to the aft docking port of the Salyut 7 space station the following day. After opening the hatches and transferring into the Salyut, the joint crew completed their programme of scientific research and experiments. For Savitskaya this entailed participating in a programme of experiments deigned to evaluate female adaptation to spaceflight. It had been more than nineteen years since Tereshkova had flown, and all biomedical data, apart from hers, had been gathered from male astronauts and cosmonauts. 'The time had come to get information about women as well,' stated IMBP Professor Oleg Gazenko.

Experiments to perform

The programme of the three cosmonauts included twenty-five experiments that were primarily medical and biological in nature, designed to compare the results of each cosmonaut. Popov was a veteran cosmonaut who had spent six months on Salyut 6 in 1980 and a short visiting mission to the same station, while Serebrov was a first-time space traveller. Savitskaya would provide a new set of female data for direct comparison with her male colleagues' data throughout the duration of the mission.

In his diary entry for 19 August – the day of the launch of Soyuz T-7 – Valentin Lebedev noted that visitors upset the routine for a main crew on the station, even though they are a welcome relief of new faces, fresh supplies, mail, and news from home: 'I don't like it when our settled routine is interrupted, even though I think life will be easier with this crew. They won't disturb us as much as I expect. Popov and I were in the same training group for three years [in a crew for a cancelled long-duration mission to Salyut 6], and we know each other very well.'[8]

During one radio communication, Savitskaya commented on life onboard the station: 'It is very clean and comfortable here. We've launched our technical, biological, medical and astrophysical research programmes. The equipment is in ideal order. Breathing presents no problem onboard this spacecraft. With the module space of more than 100 cubic metres, there's a lot of room here. There is enough room for five crew-men aboard this lab.'[5]

Using the French echograph instrument while exercising on the station's treadmill, Savitskaya's pulse rate was recorded at 120 bpm, which compared with 110–132 bpm during a similar Earth-bound exercise. Comparable recordings were also taken when the cosmonaut was not exercising. Another experiment, called Braslet (Bracelet), placed an Earth-simulated load on her cardiovascular system by restricting the flow of blood to her upper body. Koordinatsiya (Coordination) studied how eye movement initiated motion sickness in space, while another experiment monitored this sensation and its prevention. Other experiments completed by the cosmonauts involved observations of the Earth's atmosphere and deep space objects, and focused on biomedical investigations and technological experiments.

During their stay onboard Salyut, the Soyuz T-7 crew exchanged their personal form-fitting support couches with those in the Soyuz T-5 Descent Module, which had been docked to Salyut 7 since May and was reaching the end of its useful design lifetime. As their mission drew to a close, they packed away their exposed film cassettes, experiment results and personal belongings into Soyuz T-5. On 27 August 1982, Soyuz T-5 undocked from the station and began its return to Earth, landing just over three hours later 70 km north-east of Arkalyk after a flight of 7 days 21 hrs. Initial reports from the landing site indicated that all the cosmonauts were feeling well and had suffered no negative effects from their short trip into space. Initial Western reports indicated a highly successful mission, in contrast to Tereshkova's ordeal. For a while it appeared that the chances of flying other Soviet female cosmonauts, with varying specialities in science, engineering and medicine, looked bright. But were they?

No special privileges

Beregovoi – a one-time doubter of such flights – now reported that one positive return from the mission was that the mix of a male–female crew was apparently better than that of an all-male crew. He also reported: 'Women are more accurate and painstaking, especially in experiments that require fine and punctual work. In this situation, a woman cosmonaut may prove her worth to advantage.'[9]

Savitskaya's training programme had been the same as that of her male colleagues, but initial articles in the Western media continued to reflect the difficulties facing women in the Soviet regime. Although it was said that Soviet women were equal to the men, the reality was quite different, and it was several years before details of Savitskaya's flight emerged, via a number of sources.[10]

According to Valentin Lebedev's diary, Savitskaya initially remained in the Soyuz while her two colleagues transferred to the Salyut. Wondering where she was, they looked back to see her brushing her hair, and only when she had finished did she move over to the main space station compartment. Savitskaya was welcomed onboard with a bunch of flowers that had been grown onboard the Salyut. As Lebedev wrote: 'We presented Sveta with an *arabidopsis* plant, with small fragile flowers on it. We explained that this plant was the first to complete its entire cycle of development from seed to seed in space. It seemed appropriate to give these first space-flowers to the first woman on our station as a symbol of human settlement in space.'[11] It was portrayed as a joke in the Russian press, but it had a deeper meaning in Russian culture. But for their first meal together, she was presented with an apron, because although she was a pilot and a cosmonaut, to the rest of the crew she was first and foremost a woman and therefore a hostess (cook) for them. According to Lebedev's diary she was agreeable to this role – but probably grudgingly, because in Russian tradition, giving a young woman an apron before she has a pretty dress suggests that her housekeeping skills are of more importance than her appearance! Each day, she spent some time in the Soyuz to 'make herself beautiful', and during the course of the mission Lebedev grew to like her, and realised that she was a determined woman.

More revelations surfaced from resident crew Commander Berezovoi's letters to his wife (which were auctioned at Sotheby's in 1993). He commented about the age-old sailing superstition that a woman onboard a ship is a bad omen, and that he was worried about having a mixed crew in such a small station. There were no separate sleeping berths or toilet facilities, so Savitskaya had to sleep strapped against the wall of the station in her sleeping bag or in the Soyuz, and had to use the same personal hygiene facilities (a chemical toilet and rudimentary shower cubicle) as her male colleagues, although it had a curtain. With the ever-present noise of fans and equipment within a volume not much larger than the inside of a railway carriage, living amicably with crew-mates is essential, as going outside to 'cool off' is not an option. Berezovoi and Lebedev described their difficulties alone in the station during their 211 days – sometimes not talking with each other for days, and working at opposite ends of the compartments. The arrival of a visiting crew upset the routine, and with a woman on board Berezovoi was additionally worried: 'I will say nothing of Savitskaya. She is a woman, and that says it all. It will not be easy.'[12] For her 'personal needs', Savitskaya stayed in the Soyuz ferry while her male colleagues used the station facility, but she took her rest periods with the rest of the crew in the main station compartments.

As the joint flight progressed, Berezovoi – who was in overall command of the group of cosmonauts during their stay on the station – noted that mistakes were being made by the visitors, although in TV and radio broadcasts he said that 'we stay ever cheerful and lie a lot.' Berezovoi admitted that the presence of a woman on the station severely limited their freedom and complicated daily life. He was also very frank about the lack of team-work between Savitskaya and Serebrov, who were 'like a cat and dog' at each other, with Serebrov voicing his concerns about his female colleague and Savitskaya doing very little more than was expected of her from the flight programme. Although veteran cosmonaut Popov got on well with his colleagues, the resident crew was clearly relieved when the end of the mission approached and they said goodbye to their visitors. According to Lebedev, hosting a visiting expedition was very tiring, and it would take two or three days to return to the normal operating routine after they departed – a problem that continues with current station operations.

After the flight

During post-flight press conferences, Savitskaya naturally attracted the most interest and questions; but she was not happy when the questions focused on how she thought being a female affected her performance in space. As the American female astronauts would discover on the early missions, the fact that a woman was on the crew was more of a headline than the mission or its achievements. To the astronauts and cosmonauts (except, perhaps, one or two of the Russians), being a man or woman was not an issue. They were assigned because of their talents, skills and achievements, and were a crew, not special individuals. At a time when female careers and equality were of global interest, the Soviet promotion of a 'normal Soviet woman' who could fly in space was being countered by articles on the flowers, the apron and the gap between men and women in the Communist state.

Savitskaya tried to promote the results of the flight with the medical and biological data obtained, and hyped up the potential for many more flights by female cosmonauts; but she was found to be a difficult woman to interview, or even arrange to interview. She quickly became known as an Iron Lady, and on more than occasion was compared to the then UK Prime Minister, Margaret Thatcher. Then, in April 1983 the German press agency DPA featured a report by a German physician who had colleagues on the T-7 medical team. They had suggested that there was an attempt to make Savitskaya pregnant during the mission. This raised a number of media articles around the world, but it has never been substantiated by evidence or with other reports. Overall, the mission was a success, and plans were being formed to fly another woman on the next mission to Salyut 7 as part of a long-duration flight.

Meanwhile, Savitskaya, in contrast to her lack of charm in the post-flight press conference, later commented on the necessity of flying women on space missions. This was supported by at least one Russian medical source,[9] which continued to hype the skills of female cosmonauts in typically Soviet fashion. Savitskaya supported mixed crews, stating that they would perform better on longer trips in space, and that there were some things that women do better than men, such as in the precise work of biotechnological research in which 'the delicate fingers of a [highly qualified] woman cosmonaut are better suited.' Ophthalmologists also reported that a woman's eyes are better in determining shades of colour than those of men, 'hence it takes a woman cosmonaut to locate deposits of mineral resources, and it takes pedantic women cosmonauts to blaze a trail in astrophysics. Women cosmonauts can mix the purest alloys, grow crystals and test new machinery and, simply, do anything that demands a high degree of concentration, punctuality and discipline. Women cosmonauts have proven [by 2001] to be such excellent workers that the question of whether they should crew spaceships is no more. This does not mean that every single woman should fly. Cosmonautics is the lot of highly qualified people.'

Savitskaya firmly believed she had performed well on her mission – better, in fact, than her male colleagues – and proposed new flights of other women cosmonauts. Though not identifying them in person, it was clear that the Soviets were indeed intended to upstage Sally Ride's flight in the summer of 1983 with another female cosmonaut mission – hopefully a long-duration mission to Salyut 7. But although the crew was assigned, and completed training over six months, the mission did not take place, leaving the launch-pad clear for Sally Ride to become the first American woman to fly into space.

RIDE, SALLY, RIDE

As the launch of STS-7 approached it was difficult to find a vacant hotel room along the 'space coast' – the eastern seaboard of Florida near the towns of Cocoa Beach and Titusville. The interest was not so much in the mission, but in the crew, which included America's first woman in space. As the crowds prepared to witness the launch on 18 June 1983, billboards adorned the long straight roads leading to the space centre displayed 'Ride, Sally, Ride!' Twenty years earlier, on this same day, Valentina Tereshkova was in orbit approaching the end of her three-day mission onboard Vostok 6. Now America would at long last have its own first lady of space.

Sitting next to Sally

In May 1983, during one of the last pre-flight crew press conferences, a reporter asked Sally Ride if the male colleagues she was flying with felt neglected, as the attention was focused on her and not them. Ride typically deflected the obviously irritating question by asking why the question had been directed to her. Pilot Rick Hauck's response reflected the feelings of the crew about the triviality of such questions: 'If anything, sitting next to Sally, I'm sure my picture will appear in more places than I need it to appear.'

In the months since her selection to the flight, both NASA and Ride herself constantly tried to play down the fact that a woman had been assigned to an American spaceflight crew for the first time. Inevitably, however, media headlines resorted to tiresome wordplay such as 'Sally's Ride', and reporters continued to focus their attention more on whether she would wear make-up in orbit rather than her role on the mission.

Ride was flying as MS2 (Flight Engineer) – a role that included supporting Commander Bob Crippen and Pilot Rick Hauck during the ascent and entry phases of the mission. In an abort situation during ascent, she would provide the Pilots with the data they needed to either bring the orbiter to a safe landing or reach orbit. In addition, her experience as an RMS operator and her skill in manipulating the device were prime qualifications for her selection to the flight. Prior to her assignment on STS-7, Ride had served as a CapCom during STS-2 (November 1981) and STS-3 (March 1982), and was impressed by the descriptions of Earth from those crews, replying, 'When do I get my chance?'

For five decades, crew selection has essentially remained privileged information available only to those within NASA Flight Crew Directorate, but a good educational background, the ability to learn new things, and be a team player, are important skills when it comes to consideration for a flight crew position. Ride said that she thought the idea of flying in the small confines of the Shuttle orbiter for a week was 'No big deal. I didn't come into the space programme to be the first

[American] woman in space. I came in to get a chance to fly as soon as I could.' Above all, she had gained the respect and trust of her fellow crew-members, and, like them, was regarded as just one of the crew – a point that became obvious to her when, as she observed, 'Crip even stopped opening doors for me.'

Preparing to fly

Once selected to a flight crew the pace of training increases, depending on the availability of mission simulators, the time to launch, the complexity of the mission, and the progress of the payload and hardware towards the launch date. Although every mission's training programme is different, an impression of what crews assigned to a Shuttle mission in the 1980s had to complete before launch can be gained from official STS-7 training documentation.[13]

Crew training for the STS-7 astronauts began with support for the STS-5 crew in their integrated simulations. This required a high load in training hours until the STS-5 simulations were complete, followed by a reduction in hours (while the STS-6 crew used the simulators) until their own software came on line in the simulators and their training loads increased considerably. The crew organised themselves, with Crippen and Hauck attending all the training lessons related to non-payload issues, and most of those with payload issues. MS1 John Fabian and MS3 Norman Thagard, added to the crew in December 1982 to study adaptation to Space Motion Sickness, attended only payload and orbit timeline lessons, while Sally Ride's role as MS2/FE necessitated her attendance at almost all of the lessons.

Crew systems training began in July 1982, and was completed with a final lesson on 8 June 1983. The training included Flight Operations for ascent, orbit and entry; Crew Systems training; Contingency EVA operations (support role for Ride); Payload Assist Module training; Proximity Operations training; Payload Deployment and Retrieval System training; Integrated Simulations (with the assigned flight control teams working in MCC Houston and the crew in the Shuttle Mission Simulator) held between 22 April and 14 June 1983; and then integrated training (again with the flight control teams). A 58-hour long-duration simulation, held four weeks prior to launch, was determined to be 'an excellent way to put the whole mission together and thus enable the crew to gain an insight and information that would not have been acquired otherwise.' The crew concluded, post-flight, that the mission would not have progressed as smoothly as it did without this simulation. In the post-flight debriefing regarding their training, Ride commented that the preparation and training that the crew had with the experiment and training hardware enabled her to feel comfortable with the equipment and systems in flight. She also found helpful the Crew Equipment and Integration Test and the In-Flight Maintenance walk at the Cape, inside *Challenger*, to determine where items were stowed. Observing the STS-6 crew CETI/IFM process also added to her knowledge of what equipment and systems were behind which panels to assist in housekeeping chores during her mission. Training is not only about operations and contingencies, but also about familiarisation and repetition. And practice makes perfect.

The training also included reviewing various workbooks and written documentation, simulators, spacecraft and hardware mock-ups, loose equipment, and other training devices. The following summarises Sally Ride's hours spent in training for her first mission as MS2 on STS-7:

Training load

Course	*Planned (hrs)*	*Actual (hrs)*
Ascent integrated simulations	63	49
Orbit integrated simulations	132	160
Entry integrated simulations	61	40
Ascent flight operations	108	112
Orbit flight operations	65	49
Entry flight operations	136	124
Crew systems	81	78
EVA operations	10	9
Orbiter support systems	40	35
PAM/SUSS systems	49	69
Payload	33	30
PDRS	53	54
SPAS-01	88	87
Prox ops	52	42
Totals	971	938

Integrated simulation

Simulation type	*Planned (hrs)*	*Actual (hrs)*
Ascent aborts	23	11
Orbit Day 1	14	15
Orbit Day 2	14	8
Orbit Day 3	12	11
Orbit Day 4	12	12
Orbit Day 5	24	22
De-orbit preparations	6	15
Entry	17	6
58-hour simulation	0	49
Orbit Day 6	12	5
Totals	134	154

Payload training

Training task	*Training (hrs)*
Pam-D briefing	4
Palapa-B1	4
Anik-D	4
MLR/GAS	2
OSTA-2	4
CFES	16
Total	34

'Definitely an E ticket'

Following a smooth and trouble-free countdown, *Challenger* was launched on its second mission at 07.33 EDT on 18 June 1983, with the Public Affairs Office commenting, 'Lift-off of STS-7 and America's first woman astronaut, and the Shuttle has cleared the tower.' After 8 min 32 sec (which seemed very slow to Ride), the main engines were shut down, and PAO commented, '*Challenger* has delivered to space the largest human payload in the history of mankind: four men, one woman.' America had its first woman in space and safely in orbit. The view impressed Ride, who commented: 'You spend a year training just to know which dials to look at, and when the time comes, all you want to do is look out the window. It's so beautiful.' She called the ground, comparing the ascent to the most exciting ride at Disneyland. According to Ride, a Shuttle launch was 'Definitely an E ticket' – the special pass for the most exciting rides in the famous amusement park.

Later, during the post-flight technical crew debriefing in July 1983, Ride commented on her first ascent into space.[14] It was a much milder ride than she had expected – even milder than the simulator had prepared her for, but she did not get any surprises, and thought that the simulator was an excellent replication of the ascent profile, with the vibrations not as severe as she expected. She was surprised by the brightness of the SRB separation and the feeling of deceleration at main engine cut-off.

With the crew safely in orbit, it was time to unstrap and take off the launch entry helmets. (Pressure garments for launch and entry were not used on the Shuttle between November 1982 and January 1986.) This was the first time that the sensations of orbital flight became apparent to Ride: 'I felt uncomfortable during the first two hours in orbit. I was very uncertain how to move around the cockpit, how to get from one place to another. Making keyboard inputs was difficult, opening lockers was difficult, and it took me quite a bit longer than I was expecting it to take. It was a good thing that the post-insertion timeline was relaxed, because there was a real strong learning curve in just doing the simple things that you don't even think about, like opening and locking lockers, and stuffing things into lockers and pulling things out. It took a lot longer than I was prepared for it to take, but after three hours on orbit I felt comfortable moving around and felt that I could do anything that I was supposed to do. By Day 2, I felt very comfortable in orbit.'

Sally Ride: from sport to space

Sally Ride could have been a professional athlete instead of a professional astronaut, but she chose studies and then space over sport as a career. At first she enjoyed softball and football, but at the age of ten she began playing tennis, and in her teens became a nationally ranked amateur player and captain of her high-school tennis team. She pursued her academic career at Stanford University but continued playing tennis, and her talent was seen by tennis pro Billie Jean King, who suggested that she should turn professional. However, Ride also loved mathematics and physics, and wanted to continue studying for her PhD in astrophysics, electing to devote her time to science instead of sport. By the time she completed her thesis in 1978 she had been selected for astronaut training.

Her role on STS-7, in addition to duties as Flight Engineer, was to support the deployment of the communication satellites Anik C-2 for Canada and Palapa B-1 for Indonesia, as well as the deployment and retrieval of the Shuttle Pallet Satellite by RMS and a number of onboard experiments.

'Isn't science wonderful?'

About eight hours into the mission, Ride assisted Fabian in preparing for the deployment of the first commercial satellite, Anik. This was deployed by spinning the satellite to even out the thermal levels before ejecting it by explosive charges and springing it out of the payload bay. When it reached the position to fire the onboard motor to boost the satellite to a stationary orbit forty-five minutes after deployment, it was at a safe distance from *Challenger*. The next day the second satellite, Palapa, was deployed using the same method, with Ride again assisting Fabian from the aft flight-deck controls of *Challenger*. During the second day, Ride also assisted Fabian to power up and check the SPAS and the RMS in preparation for deployment and retrieval of the pallet satellite. Before the day was over, she had also activated the Monodisperse Latex Reactor material-processing experiment and several materials experiments on the OSTA-2 package in the payload bay. During the third day she again assisted Fabian from the aft flight-deck, running the SPAS in the payload bay. Despite some overheating problems, several onboard experiments on the pallet were activated.

During the fourth day (and also on the sixth day), Ride activated and operated the McDonnell Douglas CFES experiment package, designed to evaluate the use of an electrical field to separate biological materials – a step towards developing space-manufacturing of medicines for potential commercial marketing purposes. During

Sally Ride trains on the RMS simulator at JSC in preparation for STS-7.

the post-flight debriefing she commented that it took about an hour to become accustomed to working with the device on the mid-deck – especially the hand-holds, which she had not trained with on Earth. After a couple of hours, operation became much easier, and by the second half of Day 4 and during Day 6 she found the operation very easy to accomplish. She also noted that had she been called upon to activate and operate the system on Day 1, shortly after entering orbit, she would have found it very difficult. During the operation she also hosted live TV broadcasts in which she explained the facility and its operation, prompting CapCom to observe that she would make an excellent science tour guide on the Shuttle. She replied: 'Isn't science wonderful?'[15]

Flight Day 5 would be busy, with the release and recapture of the SPAS twice. This was to provide some important firsts for the Shuttle programme, and was a valuable experiment in proximity operations. Knowledge gained here would be used in later satellite servicing missions, and in later years would be built upon with the docking of the Shuttle to a Russian space station and, more recently, in the constructing of the ISS. Again, Ride had an important role in this phase of the mission, in assisting Fabian and leading the use of the RMS. The operation would include the first incidence of a payload being released from the RMS, the first demonstration of the ability of the orbiter to keep station and manoeuvre in close proximity to a second object, critical experience for imminent satellite servicing operations planned for the Shuttle, a demonstration of the Ku-band rendezvous radar system, and the first RMS retrieval operation.

Despite concerns about overheating, the SPAS operations proceeded well and included the operation of the material processing experiment MAUS-2 (MAUS-1 operated while in the payload bay) during a two-hour run while in free flight. At one point, SPAS ventured 300 metres from the payload bay of *Challenger*, and used its onboard TV cameras to produce spectacular first images of a Shuttle in orbit, with its cargo bay doors open, serenely orbiting over a cloud-covered Earth. The crew realised the opportunity to identify the images with their mission, and dutifully configured the RMS to a figure 7, representing their flight number. These new images of human space exploration in the Shuttle era became very popular in books and articles over the coming years. Ride captured the satellite for the last time, and lowered it into the bay after a very long and successful day's operations.

Day 6 was essentially a 'wrap up' day, but due to deteriorating weather at KSC (the planned landing site, using the Shuttle Landing Facility for the first time) the mission was eventually diverted to land at Edwards AFB in California, where on Day 7, after a flight of 6 days 2 hrs 24 min, it achieved a textbook landing on Runway 15. It was a highly successful mission; but then the crew received some bad news. Cold beer awaited them – but it was in Florida, 3,000 miles away! Furthermore, Ride's family lived only 100 miles from Edwards AFB – but they had travelled to Florida to watch the landing.

At Edwards AFB, a small crowd gathered with posters proclaiming 'A woman in space today, equality tomorrow'; but for Ride, the sense of excitement overpowered any sense of 'destiny' that her mission might have encouraged. Upon landing, the crew congratulated each other enthusiastically, and during the post-landing speeches

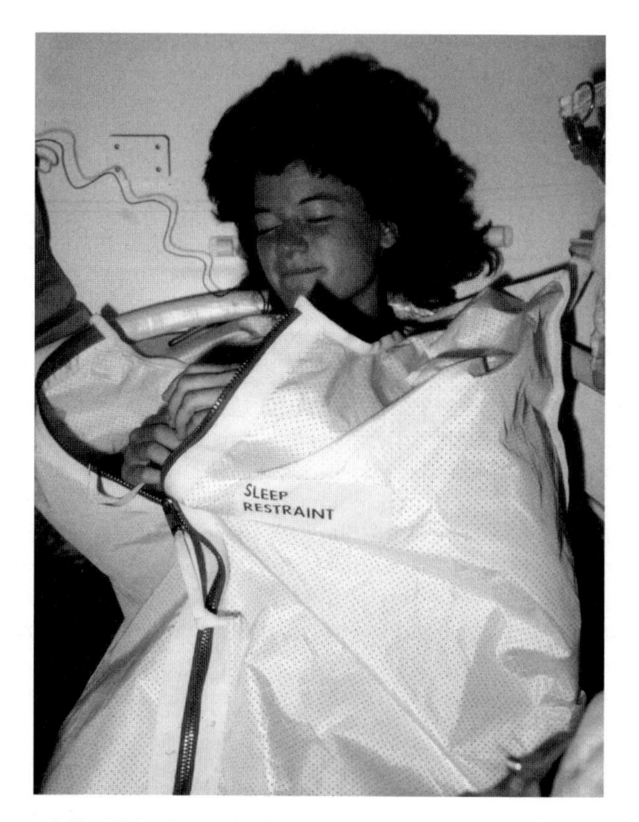

Sally Ride sleeps during mission STS-7 in June 1983.

Ride emphasised her enjoyment of the adventure: 'The thing that I'll remember most about the flight is that it was fun. In fact, I'm sure it was the most fun that I will ever have in my life.'

During the post-flight debriefing she commented on the habitability of the Shuttle. For the first two nights she slept in one of the blue Shuttle sleeping bags, attached vertically at the head and feet ends to the mid-deck lockers. But though she slept well, it was not that comfortable, so for the third night she used the bag but unpinned it, and curled up in a ball, floating over the teleprinter. This was also uncomfortable, so on the fourth night she unpacked a white Apollo sleeping bag (which was more capacious) and attached it to a hand-rail on the airlock. She slept well in the foetal position, with her head in the mid-deck and her feet in the airlock, and used that location for the rest of the mission. She also described using the exercise treadmill every day, and though it did not feel like running at all, she thought that psychologically it was a good thing and increased her heart rate. The restraint straps worked very well, but it seemed to exercise different leg muscles to those she normally used when jogging on Earth.

Back in training

Shortly after completing her mission Ride was assigned to a second mission for 1984. It was the same mission on which Kathy Sullivan would perform the first EVA by an American woman, and hopefully by any female; but in Russia, Svetlana Savitskaya was also being prepared for a return to space – once again to beat the Americans on both counts, as the first woman to fly twice, and the first to perform EVA. For now, however, America could claim success with both the STS-7 mission and the performance of Sally Ride – the first of many women who, over the ensuing twenty years, would fly on the Shuttle.

FOLLOWING SALLY

For more than twenty years American women had been denied access to space, but this was changed by the selection of six female astronaut candidates in January 1978, followed by Sally Ride's mission on STS-7. Two decades had separated the historic flight of Tereshkova and the missions of Savitskaya and Ride, but now there were other women who hoped to follow their trail into space. For some it would be a long and frustrating wait – and some would never leave the launch pad, feel the power of rocket launch, experience the freedom of orbital flight, marvel at spectacular views of Earth, or face the challenge of life after spaceflight.

Female shuttlenauts

The road to and from space has been difficult and at times tragic, but the will to attain orbital flight remains. For more than twenty years the Shuttle has provided access to space during short missions for both NASA astronauts and a variety of specialists, scientists and passengers.

As Sally Ride orbited the Earth onboard *Challenger* during the STS-7 mission, the other five women of her selection and the two chosen in 1980 were in various stages of training, or fulfilling flight support roles leading to their own first flights into space. All seven would make one flight each by the end of 1985. Following the loss of *Challenger* and its crew (including Judy Resnik and Christa McAuliffe) in 1986, Shuttle flights did not resume until 1988, with the first women crew-members flying in 1989. Between then and 2003, a further twenty-eight female astronauts flew into space onboard the Space Shuttle. In the twenty years between STS-7 and STS-107, thirty-six women flew onboard the Shuttle on more than ninety separate missions, several flew together on the same crews, and four of them completed Shuttle-based EVAs. But before this expansion occurred, each of the other five original NASA female astronauts selected in 1978 attained orbital flight in missions between July 1984 and June 1985. Judy Resnik, Kathy Sullivan (with Sally Ride on her second mission) and Anna Fisher flew in 1984, while Rhea Seddon and Shannon Lucid orbited the Earth for the first time in 1985. Sullivan and Ride flew on *Challenger*, and the others flew on *Discovery*.

A voyage of discovery

During the process of selecting crew-members for STS-7, Judy Resnik was deemed to have very good skills in operating the robotic arm, and when the crew for STS-12 (the maiden flight of *Discovery*) was being formed, these skills would prove decisive in her assignment to this early mission. The crew for STS-12 was named on 4 February 1983, with Resnik being assigned as MS3 to assist in the operation of the RMS and the deployment of a prototype solar sail designed to evaluate power generation in orbit. On 9 September 1983, NASA changed the designations of Shuttle missions, and STS-12 became STS 41-D – denoting the planned fiscal year of launch (1984), the launch site (KSC – 1) and the sequence in that fiscal year (fourth – D).

Resnik had a little more difficulty in getting off the ground than had Ride on STS-7. STS 41-D was originally scheduled for 25 June 1984, but was delayed for a day because of a faulty back-up computer, cancelling the launch just thirty-two seconds before lift-off. The following day the count-down passed seven seconds and the engine ignition sequence was started. Engine number 3 and engine number 2 fired, but before engine number 1 could ignite, a failure in number 3 aborted the launch. Only four seconds remained before ignition of the twin SRBs would have commited them to lift-off. The cut-off came as a surprise to the crew, but having gone through countless simulations their training prepared them for the sequence of actions to make *Discovery* safe on the pad. As they worked inside the orbiter, a fire was reported around the engine compartment of *Discovery*, triggering water-spray nozzles on the pad to extinguish any remnants of a blaze. Some forty minutes later the crew was ready to leave the orbiter, but with thousands of gallons of water having deluged the pad they decided not to use the slide wire baskets and instead descended the tower to the crew transfer van. In the process, they became extremely wet. After an engine change in the VAB, *Discovery*'s payload bay was reconfigured to include elements from the 41-D mission and the now-cancelled 41-F mission to maintain the launch schedule. At the third attempt, on 29 August, the launch was again cancelled by computer problems, but on 30 August Judy Resnik finally became the second American woman to enter orbit.[16]

During ascent, Resnik occupied the mid-deck near the side hatch window, and later reported that at staging (SRB separation), she saw a big flash and a lot of spray, with the window immediately clouding up and debris passing the window all though the first stage of ascent. At main engine cut-off, several loose washers and nuts (missed in pre-launch cleaning) now began to float around.[17] Shortly after achieving a safe orbit, Resnik commented on the view: ‘The Earth looks great,’ she exclaimed and when President Reagan called the crew and asked Resnik if the flight was all she hoped it would be, she replied, ‘It certainly is . . . and I couldn't have picked a better crew to fly with.’

‘Hi Dad’

Judy Resnik was born in Akron, Ohio, into a Jewish family. In 1966 she graduated from high school and went on to gain a BS and PhD in electrical engineering. She worked as an engineer for the Xerox Company prior to selection to NASA for astronaut training. Her speciality was the RMS, and she worked on Shuttle computer software prior to being named to the 41-D crew. Typically thorough in all she attempted, Resnik had undertaken a strict regime of diet and exercise to help in her selection to the astronaut programme, and had gained a private pilot's licence. She even visited former astronaut Michael Collins in his Washington office in the National Air and Space Museum, and fellow Ohioan, Senator John Glenn, to help her prepare to join NASA, as she was worried about not being selected. It was clear that Resnik never enjoyed the public relations part of the astronaut role. She resented articles mentioning her divorce and her Jewish origins, and regarded them as intrusions into her private life; and when she appeared in public she found the task daunting. She never wanted to be labelled the first Jewish astronaut, nor even a female astronaut – just 'an astronaut'; and she must have been relieved when she was not selected to be the first American woman in space. She was pleased to be assigned to a flight shortly after Sally Ride, which took her off the rota of general public appearances so that she could concentrate on crew training: 'I am just pleased to have a flight assignment. I feel that I am very fortunate to be any woman in space . . . any person in space.' She loathed the public duties, but adored the training programme, and when finally in orbit the normally quiet and reserved astronaut thoroughly enjoyed herself, experiencing the thrill of launch and weightless flight, and coming to terms with the challenge of anyone in space with long hair (known as a 'bad hair day'). Photographs of her in flight show a smiling astronaut and a personal message to her father, 'Hi Dad' – the family support ever important to her both in training and on a mission.[18]

A well-behaved solar array

During the first few orbits, Resnik's job was to relocate items of equipment from the mid-deck to the flight deck and vice versa, while her four fellow astronauts on the flight deck concentrated on configuring the vehicle for orbital flight. The crew decided that Payload Specialist Charlie Walker would remain in his seat – essentially keeping out of the way – while Resnik went about her tasks. He supported her by holding items or reading off checklists, and this worked so well that Resnik recommended that future Payload Specialists should remain in their seats on the mid-deck as long as possible, and not venture into the crowded flight deck until after the completion of all tasks.

The first task on the mission was the deployment of the communication satellites, but Resnik's main role was the deployment of the large solar-array wing, as a flight test for potential use in supplying additional electrical power in large space structures. The ultra-light array was stored folded (like an accordion) inside a box only 18 cm across, but when the fibreglass mast was extended at 3.8 m/s, it could be raised to 31 m above the payload bay. With Resnik controlling the deployment, the first extension began during Flight Day 3, to a height of 22 m, or 70%. Some of the panels stuck – probably due to prolonged storage on Earth – but they did not hamper the deployment. After just nine minutes the wing was extended to the required 22 m. 'It's up and it's big. It's very steady and stable,' exclaimed the astronaut, who then retracted the structure to ensure that it would fold correctly.

Once this had been achieved, she redeployed it to 22 m again to test its rigidity when the orbiter thrusters were fired. Some swaying was observed, but on the whole the wing remained very stable. The next day, deployment to 22 m was repeated, after which the wing was fully extended to 31 m, and further dynamic tests were completed. The swing at the top of the array was only 50 cm each way, which was much less than predicted before the flight. After the flight Resnik said that the experiment was very well behaved, recalling that the flight activities were very close to the numerous ground simulations that the crew had performed, and that operation in orbit was much smoother and quicker than in the simulations.

Ice-busters

During STS 41-D, one problem caused a little discomfort. During Flight Day 4, the waste water dump outlet on the side of the orbiter was found to be clogged with a chunk of ice, which was inspected by the TV camera on the RMS. The crew turned *Discovery* to face the Sun to try to melt the ice; but this did not work, and so plans were formed to try to remove it with the end of the RMS, without damaging either the arm or the orbiter. A back-up plan for an EVA was also considered. However, the RMS approach worked, and the crew earned the soubriquet 'ice-busters'. On the ground, Sally Ride helped devise the RMS procedures for commander Hank Hartsfield to carry out, while Resnik observed from the restricted view of the orbiter flight deck side windows. One major incentive for clearing this blockage was the restricted use of the orbiter waste management systems, forcing the crew to use the back-up system – the Apollo-era plastic bags! Though Resnik used the bags, she was critical of their design for female use and suggested amendments for future flights. During post-flight debriefing she observed, 'The Apollo bags, if you're female, don't work. Period!' During the flight, the crew were full of praise for the Apollo astronauts, who had to use these devices, and nothing else, for twelve days. But they also commented, 'You do not want to hear what Judy has to say!'

After a week in space, *Discovery* landed on 5 September, with Resnik seated on the flight deck in seat 3, having exchanged places with Mike Mullane. Looking through the overhead flight-deck windows, she commented on the sights of re-entry: 'It was just like a giant orange fireball . . . the duration and frequency you see with lightning, only every time you saw a flash, it was this orange fireball.'[17] During the crew post-flight press conference, Resnik reiterated the ease with which the array was deployed and stowed, and what benefits might be expected from such technology, especially in a future space station programme. As usual, the media also asked irrelevant questions about her full head of hair floating off her shoulders: 'For any woman in the future who might be going up into space, any advice on hairstyles?' Her curt reply – 'No advice' – soon ended that topic.[19]

'Don't fall in love with your mission'

As Resnik became the second American woman in orbit, the other four female astronauts from the first group were already assigned to subsequent flights. As they were completed, all six were quickly recycled back into training for their second missions. The pair from the 1980 selection were performing support roles in the

Shuttle programme, and would also soon receive their first flight assignments as the mission launch manifest picked up in 1985.

Anna Fisher had been named to a Shuttle crew (41-G) under the command of Rick Hauck on 2 September 1983, but by November of that year, launch schedules had changed, and the crew was assigned to STS 41-H – until August 1984, when they were again reassigned to STS 51-A, which they finally flew in November 1984.

Rhea Seddon had been named to the crew of STS 41-E on 21 September 1983, under the command of Karol Bobko. This crew was reassigned to mission 41-F in November 1983, until the delays with *Discovery*'s maiden launch forced the combination of payloads and the reassignment of Bobko's crew to a later mission, designated 51-E. In March 1985, continuing problems with the TDRS forced NASA to rethink crewing and missions that summer, and the 51-E crew was therefore again reassigned to STS 51-D, which they finally flew in April 1985. At the same time as preparing for her first mission, on 2 February 1984 Seddon had been named to the first Spacelab Life Sciences mission (SLS-1), manifested as STS 61-D (formerly Spacelab 4). She continued to train for this mission over the next seven years until it finally launched in June 1991.

Kathy Sullivan had received her assignment to the STS 41-G mission on 17 November 1983, and was in training for the first EVA by a woman. She was joined on the flight by Sally Ride on her second mission – another female spaceflight 'first'. The Commander of the mission would be Bob Crippen, who had commanded Ride's first trip into space onboard STS-7.

On the same day, NASA announced that Shannon Lucid would fly onboard STS 51-A under the command of Dan Brandenstein. Lucid was the sixth and final member of the first female selection to receive a flight assignment, but over the next decade she would amass more spaceflight time than all the other five members of her selection put together. By August 1984 she and her fellow crew-members had been reassigned to mission 51-D, and in March 1985 they moved to 51-G, which they flew in June 1985.

Typically during this period (1983–85), the common rule, when learning of a flight assignment in the Astronaut Office, was: 'Don't fall in love with your orbiter, don't fall in love with your crew, or even your mission – it's prone to change!'

KATHY, ANNA, RHEA AND SHANNON

The next of the six to fly was Kathy Sullivan, onboard the thirteenth Shuttle mission in October 1984. NASA had announced that Sullivan would complete an EVA with Dave Leestma, which they hoped would be a 'first'. However in July 1984 Svetlana Savitskaya became the first woman to fly in space twice, and also completed an EVA, thus stealing the limelight from the American mission three months later.

First American EVA

Launched on 5 October 1984 and landing eight days later, the mission of STS-41G carried five men and two women – at that time the largest crew ever launched into space on one vehicle. The primary payload of this mission was a package of instruments designed to photograph and record radar images of the Earth's surface for later use in mapping resources, interpreting geological features, and studying the use of land across the globe. This objective pleased Sullivan, as it was her primary responsibility in the experiment package, and was close to her specialist field as a professional geologist. In addition, she would be stepping outside to become the first American woman to experience walking in space – all on her first mission!

Kathy Sullivan

Kathy Sullivan had been selected with Sally Ride for astronaut training in 1978, and both were assigned to the 41-G crew. But it was not the first time they had worked together. Many years earlier, while at elementary school in California, Sullivan and Ride were classmates for a short time. Following graduation from the University of California at Santa Cruz in 1973, Sullivan moved to Canada and pursued advanced studies for a PhD in geology from Dalhousie University. She then participated in a variety of oceanographic expeditions under the auspices of the US Geological Survey, during which she explored the Mid-Atlantic Ridge, the Newfoundland Basin, and fault zones located off the coast of Southern California. She also found time to climb some of the highest mountain peaks in America and Europe, and learned five languages. After operating the OSTA package on STS 41-G, she was finally assigned the role that she wanted as an astronaut: operating scientific instruments in space. When Sullivan applied to NASA she was not sure what NASA wanted from her, based on her experience in oceanography, but her brother said that she was a female scientist who stood a good chance of selection; and he was right. (He also applied, but was turned down). Now, not only would she fly in space and operate scientific instruments, she would also assist in observing the oceans from space and would perform an EVA. Her dedication to her career and NASA left little time for family life for the unmarried astronaut. As she said shortly before launch: 'We all have different goals in life, and I am getting close to mine. There just isn't room for anything else. It's the anticipation of what it's like to walk in space which has dominated my whole being for the last year.'[20]

STS 41-G and the first US female EVA

The launch on 5 October was nominal, and nine minutes later the crew was in orbit and configuring *Challenger* for the mission. It was Sally Ride's job to operate the RMS and deploy the Earth Radiation Budget Satellite (ERBE) on the first day. Problems with the Ku-band antenna in tracking the TDRS arose on the first day, and with a second problem in latching the Shuttle Imaging Radar (SIR-B) antenna shut, the planned EVA was delayed from Flight Day 5 to Flight Day 7.

The EVA was planned to simulate the refuelling of a satellite from the Shuttle to increase its orbital lifetime. Sullivan and Leestma would install a special access valve in the mock-up satellite fill panel to allow them to make the correct plumbing connections from a supply rack to the 'take-up' tank. Once back inside the orbiter they would transfer 85 kg of hydrazine from the supply tanks to the take-up tanks by remote control. The system was installed on a Mission Peculiar Experiment Support

Structure (MPRESS) located in the payload bay, and included a separate set of plumbing connections independent of the EVA work area for a number of fuel transfers before and after the EVA operation. Sullivan assisted Leestma in testing the fuel transfer during Flight Day 2.

During Flight Day 3, plans were formed to allow the EVA crew to complete their EVA on Flight Day 7 after SIR-B operations had been completed, and have Sullivan and Leestma manually latch both of the antennae and stow the Ku-band antenna. In preparing for the EVA, the two astronauts were assisted by pilot Jon McBride as IV crew-member during Flight Day 6.

As the moment to exit the airlock approached on 11 October 1984, Commander Bob Crippen informed Mission Control that Sullivan was 'champing at the bit to go outside.' Leestma exited first, followed by Sullivan four minutes later. On exiting, Sullivan exclaimed that EVA 'feels really great ... I love it,' as she became the first American woman to walk in space. Working in the payload bay, the two astronauts successfully completed their tasks, with Sullivan acting as a 'plumber's assistant' by passing Leestma the tools and photographing his actions for later evaluation. The fuel tests would be completed once the astronauts had returned to the crew compartment after the EVA. For 3 hrs 30 min the two worked in the payload bay and successfully stowed the SIR-B antenna and Ku-Band dish, and also conducted their assigned Orbital Refuelling System tasks. The success of the operation increased confidence that future in-flight contingencies could be solved by EVA and in refuelling satellites on orbit. During the EVA, Sullivan performed a 'fancy dive' position in moving about the bay that would have earned her high marks in an Olympic diving competition! The mission ended with a successful landing at the Shuttle Landing Facility at the Cape on 13 October.[21]

Two up and two down

Since her childhood, Anna Fisher had wanted to be an astronaut; and on 8 November 1984 she achieved her dream by becoming the fourth American female in space as MS onboard *Discovery* on the STS 51-A mission. And as she had a 1-year-old daughter, Kristin Anne, she was also the first mother in space. Her week in space was once of the most exciting times of her life; but it was also one of the most difficult, as she was not with her daughter. Including the pre-flight quarantine and post-flight activities, she was separated from Kristin for about two weeks. But Fisher was also determined to show her daughter, as she grew up, that it was normal to have astronauts for parents (Fisher's husband Bill was selected in 1980), and to realise that as a female, any job was open to her.[22]

The objective of the mission was to deploy two communication satellites and, by means of EVA, retrieve two others (Palapa and Westar) that were stranded after deployment from STS 41-B the previous February. Although they were deployed without incident, the Payload Assist Motors on each satellite had failed to boost them to a usable orbit, and they were rendered useless. During the summer of 1984 a plan was devised to send the STS 51-A crew to retrieve both satellites and bring them back to Earth for possible refurbishment and later relaunch. Anna Fisher's role on the mission was as launch and entry Flight Engineer (MS2) – assisting the

Commander (Rick Hauck) and Pilot (Dave Walker) during ascent and descent – and also as primary operator of the RMS, supporting the EVA operations by fellow MSs Joe Allen and Dale Gardner. Flight Director Jay Greene recalled that on past missions the Shuttle had deployed satellites, rendezvoused with satellites, repaired satellites, and retrieved and returned free-flying pallet satellites to Earth; but this was the first time that all this would be undertaken during one mission.

Anna Fisher

Anna Fisher decided upon a career as an astronaut when she was young. She had a keen interest in mathematics and science at school, and in reading books with female leading characters. She was twelve years old when the first people began flying in space, and she decided that she would like to fly in space herself. There were no female astronauts at that time, but there were women in medicine and several (such as Dee O'Hara) involved in space medicine. Anna therefore decided to become a doctor – not confined to the Earth, but on a space station or on a Moon base. Despite the apparent lack of prospects for female astronauts, she pursued her medical career through college and graduated with a BS degree in chemistry, after which she worked on a medical degree. By the time she was completing her medical internship in 1977 she had been studying for ten years after graduating from high school; and by then NASA was looking for female astronaut applicants. She never gave up her dream of becoming an astronaut, and in January 1978 her determination allowed her to become one of the first American female astronaut candidates. Her husband Bill (also a medical doctor) also applied for selection in 1978, but was not successful. However, in 1980 he reapplied, and was selected. He subsequently flew on one mission and conducted two EVAs in 1985. The Fishers were therefore the first married couple (independently) selected for astronaut training. (Tereshkova and Nikolayev were married in 1963 *after* both of of them had flown in space, whereas the Fishers were married in 1977 *before* either were selected for astronaut training.)

After Ascan training, Anna Fisher was assigned a number of support roles in the Astronaut Office, including specialising on the RMS, serving as a 'small' test-subject for outfitting the Shuttle Extravehicular Mobility Unit (EMU), and evaluating contingency Shuttle EVA procedures including closing payload bay doors by means of EVA. Prior to STS-1 she was also involved in demonstrations of methods for repairing the thermal protection system on orbit. In SAIL, she worked on ascent and entry procedure verification and on RMS operating issues, and also supported the flight software for STS-2 to STS-4. For STS-5 to STS-7 she worked at KSC, supporting the verification of payloads carried on those missions, and on STS-9 she served as CapCom for the first Spacelab flight. For the first four Shuttle missions, flown under the Orbital Flight Test programme, she was assigned to the team of support astronauts as a medical physician in a helicopter at either the primary or back-up landing sites, ready to render emergency medical aid. This was part of her involvement in the development of crew rescue procedures. Prior to this assignment the Astronaut Office had sent her to attend an Emergency Medicine Conference in San Diego, which included lectures on the movement of injured persons out of confined places (such as the orbiter crew modules) and the procedures to treat skin injuries from toxic fuels (such as those carried on the Shuttle).

Satellites for sale

At the time of insertion into orbit, *Discovery* was 27,336 km behind Palapa, the first satellite to be retrieved; but the Shuttle's orbit was slightly lower, and they were catching up by several hundred kilometres on every 90-minute orbit. On Flight Day 2, the first satellite – Anik D2 – was deployed from the payload bay, followed by the second satellite – Syncom 4-1 – on Flight Day 3. The crew prepared for the EVA operations as the Shuttle gradually closed in on Palapa, and the final rendezvous was achieved on Flight Day 5.

The first EVA took place on 12 November, with Allen flying the Manned Manoeuvring Unit out from the orbiter to retrieve the satellite by using a specially developed 'stinger' device (a shaft with a wheel-shaped handle at one end and a stabilisation guide-ring attached). This was inserted into the satellite's PAM rocket nozzle to capture the satellite, which did not incorporate hand-holds or grapples. Allen then returned to the payload bay of *Discovery*, and worked with Gardner in attaching an RMS grapple fixture on a metal brace on top of the satellite so that Fisher could use the robot arm to grab the satellite and allow the MMU astronaut to release it. An A-frame should have been bolted to the base of the satellite to fix it in the bay for return to Earth, but when this was tried in flight it was found that the upper RMS brace could not fit because a 'feed horn' used for signal transmission was protruding from the satellite an ⅛ -inch too far. This was not the case on the ground mock-ups or in any of the documentation used for preparing the flight hardware for the capture process. The only solution, therefore, was to have the RMS attach to the grapple fixture and the two astronauts manoeuvre the satellite, holding it in place while the A-frame was locked in place.

With Palapa in place, the orbiter was manoeuvred for rendezvous with Westar two days later. Meanwhile, a new plan was devised to ease the retrieval of the second satellite, which was of the same design as the first. This time, while Gardner flew over to the satellite, Allen would mount the Multiple Foot Restraint on the end of the RMS and grasp the omni-antenna atop Westar as a hand-hold, instead of first cutting it away, as during the recovery of Palapa. Acting as a human A-frame, Allen would hold the satellite while Gardner stowed the MMU and then fitted the adapter, allowing Fisher to safely berth the second satellite. Everything proceeded remarkably well, and the astronauts even had time to pose with a 'For Sale' sign over the two satellites. The next day was a rest day, during which they prepared for entry and landing on the following day. The crew was delighted at recovering the satellites so easily, despite potentially mission-threatening problems.

An exhausting exercise

For Fisher, her first mission was demanding. Since Walker, the Pilot, was assigned to assist the EVA crew in their preparations, Fisher was assigned to assist Commander Hauck during the final stages of the rendezvous with both satellites, working together to complete forty-four burns of the Orbital Manoeuvring System and the

Reaction Control Systems to reach the satellites. The EVAs accomplished on the mission were 'a physically and mentally exhausting exercise for all five crew-members.'[23] Fisher and Hauck were involved in rendezvous manoeuvres from six hours before each EVA started, and were permanently occupied until after the end of the EVA period – a working day of about twelve hours. By the end of post-EVA activities, all five astronauts were extremely tired. During the EVAs themselves, Fisher operated the RMS, took photographs, and shared the responsibility for monitoring the orbiter systems with Hauck – all of which allowed little time for rest or meal breaks. The recommendations from the crew after experiencing these difficult working procedures on STS 51-A resulted in changes to EVA protocol on future missions.[24] The RMS performed well during the entire mission, and was crucial in supporting the EVA programme, the altered EVA timeline resulting from real-time planning on EVA-1, and amendments to the flight plan for EVA-2. Fisher had trained as primary RMS operator for the mission, and completed the entire training syllabus, including extensive contingency scenarios with partial systems failures in the arm. Her previous training and support of other flights helped both her preparation for the flight and her understanding of how to overcome in-flight problems and revised operating procedures.

Orbital surgery

Medical doctor Rhea Seddon became the next woman in space, with the launch of the delayed STS 51-D mission on 12 April 1985. This was the sixteenth Shuttle mission, launched on the twenty-fourth anniversary of the first manned spaceflight by Yuri Gagarin and the fourth anniversary of the first Shuttle flight. This was also a satellite deployment mission – and another mission that did not proceed according to plan. The first comsat, Anik C-1, was successfully deployed, and although the second, Leasat 3, was deployed correctly from the Shuttle by the crew, its sequencer failed to initiate the deployment of the antenna, the spin-up of the satellite, and the ignition of the perigee kick motor. To allow the astronauts to perform an unscheduled EVA, the mission was extended by two days in order to devise a plan for triggering the sequencer prior to the Shuttle's return home. During these preparations, Seddon's former skills as a professional surgeon would prove useful.

Rhea Seddon

On 5 May 1961, fourteen-year-old teenager Rhea Seddon sat in front of the television and watched Alan B. Shepard become the first American in space. One day, she thought, women will also travel into space. Two years later she was impressed with the flight of Valentina Tereshkova, but soon realised that this was not a swift advance for women in space, but more of a test mission in which the cosmonaut survived. Though she dreamed of becoming an astronaut, the path was just as difficult for American females as for Russians, and so she pursued a career in medicine. She did not think her grades were good enough for medical school and concentrated on a nursing programme rather than a doctorate in medicine, enrolling at Vanderbilt University. But after just one year, her desire and determination to become a doctor was too great, and she enrolled at the University of California at

Berkeley, completing her medical degree in 1973 at the University of Tennessee College of Medicine. After surgical internship and three years as a general surgeon in Memphis, Tennessee, Seddon was looking towards extending her specialisation in surgical nutrition to earn a PhD when she learned that NASA was looking for new astronauts, and that for the first time, women candidates could apply.[25]

Girl Scout camp

When Leasat failed to operate correctly it was decided to allow the crew to try to activate the faulty switch lever on the side of the satellite to begin the programmed sequence and send the satellite into its operational orbit. The plan to achieve this required the crew to fabricate a snare device and attach it to the RMS during an EVA. With the EVA crew back in the airlock, the RMS/snare configuration could be placed next to the slowly spinning satellite in order to trip the switch as it passed by. This became known as the 'fly-swatter'. With limited materials to work with, ingenuity was the order of the day, and, working with team on the ground, the astronauts gathered a metal swizzle stick reach mechanism, plastic book covers, a Swiss army knife, duct tape and scissors. This reminded Seddon of activities at a Girl Scout Camp, and as a professional surgeon, Seddon was tasked with fabricating the fly-swatter (and a back-up device resembling a Lacrosse stick) with the assistance of PS Jake Garn. The process included some stitching to hold the devices together: 'We were measuring with tape measures and cutting and pasting and wondering what in the world this thing was going to look like when we finished.' The task also generated extra work and updates from the ground, including masses of paper on the Shuttle teleprinter to update the flight plan.

According to Seddon, the crew had great fun cutting and pasting, playing arts and crafts, and putting together a book from a roll of paper – the first time a full checklist book had been uplinked to a crew.[26] In the weightless environment, Seddon had to use her surgeon's skills as well as others that were not used in operating theatre procedures. She narrated the procedure during the post-flight crew debriefing press conference slide presentation: 'This is the surgeon at work. You can't tell I've got a pair of scissors in my mouth, which is usually a 'no-no' in the operating room. I have a knife in one hand and I'm slicing into a plastic book cover, beginning the work on making the fly-swatter. I got to use scissors, scalpel, bone saw and needle and thread, so, they made me feel like a real surgeon working up there.'

There was some doubt as to whether the RMS would be effective in using such primitive tools to activate the switch. There were three good hits at the switch, as revealeded on the film taken by the crew, and their own observations, before the allocated time for the separation manoeuvre in case the crew had activated the switch. The Shuttle had to be moved away from the satellite before the internal rocket motors activated. The switch was flipped, but it failed to activate the full deployment sequence, and so the process was abandoned in favour of a second mission to attempt an EVA 'rewiring' later in the year (successfully flown as STS 51-I in August 1985). Photographs indicated broken wires on both the fly-swatter and lacrosse-stick devices, prompting Seddon's crew to celebrate her skills on hitting the '3 Wire' (on an aircraft carrier), an achievement for a naval aviator, or, in Seddon's case, a Navy wife (she was married to navy test-pilot and astronaut Robert Gibson). During post-flight evaluation, Seddon reported that the instructions to make the snare devices were excellent, and she found it very useful to talk with colleagues (including Sally Ride, who helped evaluate the sequencing of RMS manoeuvres) on the ground and to use the onboard TV to show what the crew had built and to receive feedback. She had more confidence in the fly-swatter instructions, which were written down, than in the Lacrosse instructions, which were verbal and were received, in her opinion, too rapidly to be written down. The crew recommended that future instructions to a crew would be more suitable in written form than in verbal form.

In addition to fabricating the snare devices and operating the RMS, Seddon also participated in evaluating the Toys In Space educational experiment, demonstrating how familiar toys performed differently in space. According to Seddon, the Slinky would not 'slink' at all, and she became very efficient at collecting jacks in the game jack-stones, by carefully not letting the floating ball hit the floating jacks. In another use of her former career as a surgeon, Seddon was responsible for the medical experiments and investigations on the flight, including the flight echo experiment, in the first American use of an echocardiograph device in space. During the mission she used the device on herself and three of her crew-mates. STS 51-D ended with a landing at the Cape on 19 April.

A most successful international mission

Shannon Lucid became the sixth American woman in space (and the last of the original group selected to fly in space) with the launch of STS 51-G on 17 June 1985. A decade later, on her fifth mission, she would make the longest flight by a woman in spaceflight history; but her first mission was a short, one-week satellite deployment mission. The eighteenth mission of the Shuttle programme was described as one of the most successful missions to date, with an on-time launch and the fulfilment of all of its pre-flight objectives. The crew successfully deployed three satellites for Mexico, Saudi Arabia and the US telecommunications company AT&T. The crew included two Payload Specialists, Frenchman Patrick Baudry and Saudi Prince Sultan Salman Al-Saud, and this international flavour of Lucid's first mission was a feature of both her space career and her personal background.

Shannon Lucid

Shannon Lucid was born in Shanghai, China, where her parents were missionaries. During the Second World War the family was interned by the Japanese for a year, and after returning to the US and settling in Oklahoma she completed her schooling and pursued a PhD in biochemistry at the University of Oklahoma. At the same time as her academic studies, she worked at a variety of jobs, including teaching assistant, science laboratory technician, chemist, and research and graduate assistant, as well as raising a family and finding time to earn a pilot's licence.

After selection to NASA and Ascan training, she worked in SAIL and supported several missions. Immediately after her first spaceflight she served a tour as CapCom in Mission Control, Houston. For STS 51-G, Lucid's role included the deployment of X-ray equipment to take photographs of the centre of the Milky Way, operating the RMS, and operating the medical equipment onboard the Shuttle to gather biomedical data on the crew throughout the flight. She also assisted the two Payload Specialists with their experiments and research objectives.

Into orbit ... by a nose

For the launch of STS 51-G, Lucid would be seated in the third seat on the flight-deck behind the pilot, John Creighton (also on his first spaceflight), while rookie Steven Nagel (MS2/FE) was seated between Creighton and Commander Dan Brandenstein. This flight would see the one-hundredth American astronaut enter space by rocket; but with three rookies flying into space on the same mission, which of them would be number 100? In a light-hearted report from space, the crew indicated that since Nagel was seated slightly forward of Lucid, he was the hundredth American in space ... the length of his nose deciding it, as Lucid was seated a few centimetres further back on the flight deck![27] Lucid thought the ascent was more or less as she had imagined it, with a good view out of the flight-deck windows. 'The only surprise on the ascent was that I heard a metallic thump, and then the ET sep [external tank separation] occurred.'[28]

Lucid was responsible for the deployment by RMS of Spartan-1 – the free-flying pallet satellite carrying X-ray recording equipment. Spartan flew free from *Discovery* for forty-five hours, and recorded data for twenty-four hours over sixteen orbits of Earth before being recaptured by the crew using the RMS. In May 1985, a month before launch, the crew had visited KSC to go through system procedures with the flight unit and the Goddard Space Flight Center support staff prior to close-out for flight. Lucid used an RMS controller to operate the flight hardware by turning the Spartan on and off and simulating a few malfunctions by linking a hand controller to a small computer that was programmed the same as the RMS hand controller on *Discovery*. She found this very useful training for using the real RMS on orbit. The operation in space proceeded very smoothly, and Lucid thought the flight operations were a lot easier than the simulations. But she also acknowledged that without working on the ground simulators, operating the RMS in space would be much more difficult because of the complexity of the multiple positions in which the arm could be placed.

Mission accomplished

Discovery landed on 24 June 1985, ending the first flight of the sixth and last of the first group of American female astronauts to fly into space. Of the six, Ride had made two flights and Sullivan had accomplished an EVA. All six would resume training for further missions, and new astronauts were now joining the original six and the two female astronauts of the Class of 1980. The Ascan Class of 1984, completing their training programme, included three female candidates: Marsha Ivins, Ellen Shulman (later Baker) and Kathy Thornton. Then, just a week before the launch of STS 51-G, NASA announced its eleventh group of astronauts, including two more women: Linda Godwin and Tammy Jernigan. By the end of 1985 NASA had thirteen women in various stages of flight-readiness and training, and several women from the US, Europe and Japan were in training as Payload Specialists for science missions. It had been twenty years between the flights of Tereshkova and Ride, and now in two short years the opportunities for females to fly into space had increased due to the availability of the Shuttle. The prospect of orbital flight had even extended to the selection of non-scientists for a one-off mission in 1986. Several teacher candidates were being evaluated for a flight in Earth orbit, including Sharon Christa McAuliffe and Barbara Morgan – who were soon to write their own chapter in human space exploration.

REFERENCES

1 JSC FCOD Memo, 14 April 1982, in the Cliff Charlesworth file collection, NASA History Archive, University of Houston, Clear Lake; also David J. Shayler, *The Shuttlenauts, 1981–1992*, vol.2, 'Flight Crew Assignments', Astro Info Service Publications, December 1992.
2 NASA News (JSC), 82-023.
3 Bettyann H. Kelvles, *Almost Heaven*, Basic Books, 2004, p. 86.
4 Bert Vis, 'Soviet Women Cosmonaut Flight Assignments, 1963–1989', *Spaceflight*, **41**, 11, November 1999, 474–480.
5 'Russians in Space', *Voice of Russia* radio service, 2001.
6 Rex D. Hall and David J. Shayler, *Soyuz: A Universal Spacecraft*, Springer–Praxis, 2002.
7 Gordon R. Hooper, *The Soviet Cosmonaut Team*, vol. 2, 'Cosmonaut Biographies', GRH Publications, 1990, pp. 254–258.
8 Bert Vis, 'Soviet Women Cosmonauts', in catalogue of Sotheby's Russian space history auction, New York, 11 December 1993, Lot 140; Valentin Lebedev, *Diary of a Cosmonaut: 211 Days in Space*, Bantam Paperbacks, 1990; Bettyann H. Kelvles, *Almost Heaven*, Basic Books, 2004, pp. 86–89.
9 Hooper (ref. 7) p. 255.
10 Lebedev (ref. 8) pp. 188–189.
11 Ref. 8, p. 191.
12 Sotheby's, 'Russian Space History', New York, 11 December 1993.
13 STS-7 Post-flight Training Report, Mission Operations Directorate Training Division, TD225/A192, 13 October 1983, NASA LBJ Space Center, Houston, Texas.
14 STS-7 Technical Crew Debriefing, Mission Operations Directorate, Training Division, NASA LBJ Space Center, Houston, Texas, July 1983.
15 John Pfannerstill, '*Challenger*'s Return to Space', *Spaceflight*, **26**, 2, February 1984, 82–88; and **26**, 3, March 1984, 130–135.
16 David J. Shayler, *Disasters and Accidents in Manned Spaceflight*, Springer–Praxis, 2000, pp. 137–139.
17 STS 41-D Technical Crew Debrief, Flight Crew Operations Directorate, JSC-20079, October 1984, NASA JSC.
18 Joanne E. Bernstein and Rose Blue with Alan Jay Gerber, *Judith Resnik, Challenger Astronaut*, Lodestar Books, 1990.

19 STS 41-D Post-flight (Crew) Press Conference, 12 September 1984, NASA JSC, Houston, Texas, transcript.
20 Michael Parry, 'Off to take a giant step for womankind', *Daily Express*, 4 October 1984.
21 Additional information on the flight of STS 41-G can be found in Henry S.F. Cooper, *Before Lift-Off: The Making of a Space Shuttle Crew*, Johns Hopkins University Press, 1987.
22 Carole S. Briggs, *Women In Space: Reaching the Last Frontier*, Lerner Publications, 1988, pp. 45–48.
23 STS 51-A Flight Crew Report, Astronaut Office Memorandum CB-85-048, 28 March 1985.
24 David J. Shayler, *Walking in Space*, Springer–Praxis, 2004 pp. 265–267.
25 AIS interview with Rhea Seddon, 21 April 2004.
26 STS 51-D Post-flight press conference transcript, 23 April 1985.
27 STS 51-G air-to-ground commentary.
28 STS 51-G Technical Crew Debrief, JSC 20649, Flight Crew Operations Directorate, August 1985.

Shuttle specialists and passengers

By the end of 1985, the launch manifest for the Shuttle over the next decade was looking increasingly busy, with at least four orbiters flying from two pads at the Kennedy Space Center in Florida and, starting in 1986, from the single military launch pad from Vandenberg AFB in California. With around a hundred NASA astronauts in various stages of assignments or training and a new astronaut selection due in 1986, most of the 1978 selection were being assigned to a second or third flight, and those selected in 1980 and 1984 were receiving their first assignments. The Ascans of 1985 were well into their year-long training programme. In addition, the first 'part-time' astronauts – the Payload Specialists – had begun flying in 1983 and were soon to be joined by the first Space Flight Participant (a teacher), followed shortly after by the second (a journalist).

FROM THE PLANET EARTH

As 1986 dawned, the prospects of a flight on the Shuttle by NASA astronauts, international scientists, USAF DoD Manned Spaceflight Engineers and Space Flight Participants looked promising, with an almost monthly launch rate manifested. But on 28 January 1986, just seventy-three seconds after the launch of the twenty-fifth Shuttle mission (STS 51-L), a huge explosion led to the destruction of the *Challenger* orbiter and the deaths of its seven crew-members, including Judy Resnik and schoolteacher Christa McAuliffe. The whole programme was grounded for thirty-two months. When the Return-to-Flight mission (STS-26) restored confidence in the Shuttle system and flights resumed in September 1988, the launch rate slowly increased once again. By 2003, the Shuttle had changed its role from a satellite deployment vehicle and science research platform to a space station construction and logistics vehicle. In those fifteen years, dozens of new astronauts flew on the Shuttle, including a growing number of female astronauts from the US and several other countries.

Then tragically, for the second time in seventeen years, a Shuttle vehicle was lost, claiming the lives of seven more astronauts. The loss of *Columbia* and the crew of

STS-107 (including Kalpana Chawla and Laurel Clark) not only grounded the Shuttle again, but also seriously threatened the continual occupation and expansion of the International Space Station. The resulting inquiry recommendations led to the decision to retire the Shuttle fleet after completion of the station around 2010, replacing it with a new Crew Exploration Vehicle. The CEV would not only service the ISS, but would hopefully pave the way for a return to the Moon by 2020 and an infrastructure for manned flights to Mars by 2030. But in the period 1985–2003, although flawed and plagued by delays and setbacks, the Shuttle proved to be a versatile vehicle, and served as an access to space for dozens of space explorers, including many women.

A laboratory for space

In the early 1970s, as NASA developed the Space Shuttle System, the European Space Agency developed the Space Laboratory to fit inside the payload bay, offering either pressurised work facilities, unpressurised pallet support structures, or a combination of both, for a wide range of scientific research fields on missions of a week to ten days.[1] These crews included a number of Payload Specialists – non-career astronauts chosen for a specific flight, experiment or payload, who would complete a minimal spaceflight training programme before assignment to a specific flight crew and then return to their former posts shortly after completing the mission. The specialists could come from academia, scientific research, engineering companies, the military, foreign countries supplying a major Shuttle payload, or leading contributors to stated mission objectives.

Spacelab 1

The first call for Payload Specialists came in 1977, in both Europe and the US, to fill the role of PS on the first Spacelab mission (Spacelab 1). Each of the twelve member countries of the European Space Agency conducted a national selection process and submitted up to five candidate nominations to ESA head office by September 1977. A further three would be selected from the staff of ESA, resulting in fifty-three candidates for the single-seat flight opportunity. On 2 September, forty-nine candidates from respective member countries were evaluated, with six finalists joining four from ESA staff. In the first European space explorer selection process, several female candidates nominated, but none were selected. One of the applicants was Anny Levasseur-Regourd (French, *b.*1945), who was working as an engineer at CNES. She went on to hold prominent positions in the Union Astronomique Internationale (UAI), the International Space Science Institute (ISSI), and COSPAR. Most recently (2004), she was working as a lecturer at the University of Paris.[2]

On 22 December, ESA named the four European finalists for Spacelab 1 as Nicollier, Merbold, Ockels and Malerba. All four would eventually make spaceflights, with Merbold becoming the first European astronaut to fly onboard a US space mission (STS-9/Spacelab 1). In the American PS candidate selection announced on the same day by NASA, Ann Whitaker (*b.*1939) was named as a semi-finalist for Spacelab 1, but in the 1 June 1978 announcement of the finalists, she was unsuccessful in gaining a flight seat. She has since gone on to receive a PhD in materials engineering from Auburn University, Alabama (1989), and later (2004) served as Director of the Materials and Processing Laboratory at NASA's Marshall Space Flight Center.

Spacelab 2

At the same time as the selection process for Spacelab 1, there was also a selection process for Spacelab 2 – a pallet-only astro/solar physical mission. On 19 April 1978, eight scientists (six American and two British) were identified as candidates for two PS seats on the mission. The one female candidate was American Diane Prinz (*b.*1938), who was a research physicist at the Naval Research Laboratory in Washington, and was working on optics and computer software for the Spacelab 2 mission. She became a semi-finalist in August 1978, and continued supporting the development of the payload until being named as back-up PS on 8 June 1984. She served as a scientific CapCom, and provided mission support from the Payload Operations Center (POC) at Marshall Space Flight Center during the mission. Following the mission, she was a leading candidate for assignment on the Sunlab solar research mission planned for STS 61-L in 1986, but due to the loss of *Challenger* the mission was cancelled, and Prinz lost her chance to fly in space. She returned to NRL until 2000, when ill heath forced her retirement. Sadly, she died of a lymphoma on 12 October 2002, aged sixty-four.

Spacelab 3

On 8 June 1983, four PS candidates were named from seven finalists for Spacelab 3 – a life and materials science laboratory mission. Included in the finalists was metallurgist Mary Helen Johnson (*b.*1945), who had previously worked at MSFC on the 1974 all-female Spacelab ground simulation. She was a candidate for the first NASA Shuttle selection in 1978, but was not selected. With her own materials processing experiments onboard Spacelab 3, she qualified for consideration as potential PS on that flight, but when the crew announcement was made Johnson was named as a back-up PS, serving as science CapCom and support crew-member from the POC at Marshall. After leaving Marshall, Johnson worked for the Tennessee Space Institute at Tullahoma.

Spacelab 4

In January 1984 the candidates were announced for Spacelab 4 – a life sciences laboratory mission. All were American, including Millie Hughes-Fulford (*b.*1945), a biochemist who in 1978 had tried to join NASA but without success. In spring 1985 she was nominated as back-up PS for Spacelab 4, and was scheduled to fly onboard the second Spacelab life sciences mission. Due to delays resulting from the loss of *Challenger*, one of the original PSs for the first life science flight (Robert Phillips) was medically disqualified from the mission by the time it was rescheduled in 1989, due to his age (he was 60) and health problems. On 2 November 1989 he was replaced by Hughes-Fulford, the only qualified PS replacement available. She later flew as a PS on STS-40 in May 1991.

Manned spaceflight engineers

While NASA's female astronauts were preparing for flights as Mission Specialists and several American female scientists were in training for various Spacelab missions, the US military was also preparing a cadre of officers to fly on classified DoD Shuttle missions, accompanying military payloads and experiments. The idea of flying career military officers was initiated in the late 1970s, resulting in three MSE selections and a variety of special selections of military observers and scientists before the programme of dedicated military Shuttle missions was terminated in the early 1990s (although several NASA astronauts have continued a number of military man-in-space experiments since then).[3] This included observations from STS-31 in 1990 by NASA astronaut Kathy Sullivan – a geologist and oceanographer. with the rank of Lieut-Commander in the Naval Reserve.

Although some names from the first selection were leaked after their selection in August 1979, their identities were not officially revealed until October 1985. In that selection, fourteen officers (twelve USAF and two USN) were selected, although none were female. In the second selection of fourteen candidates in September 1982 there were two female officers chosen, followed by a third in the final MSE selection of September 1985. In addition to these three officers, a fourth was selected under the Weather Officer In Space programme. Although fully trained as MSE candidates and essentially as Payload Specialists with NASA, none of them flew in space, and they returned to their military careers after the cancellation of the programme.

Maureen LaComb (*b*.1956) had earned a BS degree in radiation health physics and an MS degree in computer science before entering active duty with the USAF prior to selection to the MSE programme in September 1982. In November 1985, she was assigned as a military PS for the planned Shuttle Spacelab laboratory mission Starlab, dedicated to the Strategic Defense Initiative Organisation (SDIO). It would have been a nine-day evaluation of tracking and pointing technology, but in August 1990 the programme was cancelled and LaComb returned to other USAF duties at the Consolidated Space Test Center at Onizuka AFB in Sunnyvale, California.

Katherine Roberts (*b*.1954) gained a BS degree in physics and an MS degree in space physics and engineering. She joined the USAF on active duty in 1976, subsequently completed three years as an orbital analyst at NORAD, and worked at the National Security Agency. At the time of her selection to the MSE programme in 1982 she was a satellite intelligence officer in Special Projects. She was assigned to the first KH-11A payload (a series of imagining reconnaissance satellites called Lacrosse) in November 1985, but during 1986, in the recovery from the loss of *Challenger*, plans to launch polar orbiting Shuttle missions from Vandenberg were abandoned. The payload survived and was launched on STS-27 in December 1988, and for a time Roberts was considered for a PS seat on that mission, until NASA announced that no PS would fly on the first five post-*Challenger* missions (STS-27 was the second return-to-flight mission) and she lost her chance to fly in space. Returning to her USAF career, Roberts continued in the Space and Missile Systems Division in various roles, including space operations, acquisition of space systems, and staff roles including Vice Director of Space Operations at US Strategic Command in

2002. In 2004 she was promoted to Brigadier-General, and she now serves as the Deputy Commander for Command, Control, Communications, Computers, Intelligence, Surveillance and Reconnaissance Enterprise Integration.

Theresa Stevens (*b*.1960) graduated from the USAF Academy in 1982 with a BS in operations research, and received her first assignment as a Shuttle flight controller seconded to NASA JSC in Houston, Texas. In April 1986 she was selected for the USAF MSE programme. Unfortunately, this was shortly after the *Challenger* accident which grounded all manifested Shuttle missions. By the time flights resumed, the DoD had remanifested many national security payloads on to expendable launchers, and senior Air Force officials decided to phase out the Manned Spacecraft Engineer programme, closing the office in July 1988. With that decision, Stevens was assigned to the Woomera Air Station in Australia, after which she returned to take up a post at the Pentagon.

Carol Belt (*b*.1953). In addition to the MSE programme, the military initiated a number of other associated programmes. One of these was the Weather Officer In Space Experiment (WOSE) to fly professional meteorologists and provide expert assessment and interpretation of the conditions in the atmosphere from the vantage point of low Earth orbit. Carol Belt was selected in March 1988 for the primary PS position for a flight in 1989, but the mission was never formally scheduled and remained unflown. In 1992 the programme was cancelled. Belt had received a BS degree in meteorology in 1975 – the same year that she entered the USAF – and subsequently earned an MS degree in 1981 and her PhD in 1984. She logged more than 500 hours as an aerial reconnaissance officer, flying onboard WC-130 Hercules aircraft gathering data from typhoons in the Pacific and other weather phenomena, and other assignments. She was a finalist in the first WOSE selection of 1985, and when the programme was cancelled in 1992 she decided to retire from active duty.

International Shuttle candidates

With the expansion of manned spaceflight opportunities on both American and Soviet spacecraft, and with the proposed (but later cancelled) French Hermes space shuttle and European Columbus space station module, several nations around the world embarked on national astronaut selection programmes.[4]

Canada

With a major input in the Shuttle programme due to the Remote Manipulator System, Canada was the first nation to take the opportunity to select its own astronaut team. In June 1983, a call for applications for 'Space Team Canada' resulted in more than 4,400 responses. From these, nineteen semi-finalists were medically examined, and six finalists were announced on 5 December 1983. They included Roberta Bondar, who would fly on the STS-42 mission in January 1992. By 1991 the original contracts for the first six astronauts were expiring, and a new selection process was initiated. This time there were 5,300 applicants, with fifty-three short-listed in October 1991 for medical tests and evaluations. Four new astronauts, including Julie Payette, were named on 8 June 1992. In August 1996 Payette was selected to join the NASA Class of 1996 (Group 16) in their Ascan training programme to qualify her for assignment on Shuttle flights as a Mission Specialist. Payette flew to the International Space Station onboard STS-96 in May 1999.

Roberta Bondar (*b.* 1945) earned several university degrees between 1968 and 1977, including a BS in zoology and agriculture, an MA in experimental pathology, a PhD in neurobiology, and a Doctorate in medicine. She continued her medical career after joining the Canadian astronaut programme, lectured at the University of Ottawa, and taught at Ottawa General Hospital. She worked as a back-up in 1984 and 1985 for a subsequently cancelled 1987 Canadian mission, and was selected as an International Microgravity Laboratory (Spacelab) candidate in 1989, being named as a prime PS in March 1990. After her flight in 1992 (STS-42), Bondar left the astronaut programme to return to her medical career full time and to lecture on her experiences as an astronaut.

Julie Payette (*b.*1963) earned a BS in electrical engineering in 1986 and an MS in applied science in 1990. She followed a career in the computer industry prior to selection for astronaut training. This proved valuable, as her early technical assignments reflected human–computer interaction in space research. She has also obtained her commercial pilot's licence in addition to military flying credentials, and has studied Russian, Spanish and Italian. She has logged more than 120 hours in the reduced gravity aircraft programme, and has also qualified as a deep-sea diver. She would fly on STS-96 in 1999 to the ISS on an early logistics mission.

Indonesia

With an Indonesian Palapa satellite scheduled for deployment from the Shuttle in 1986, such a major payload qualified the customer to provide a national PS to assist and monitor the deployment of the satellite and perform mid-deck experiments. In October 1985, four candidates were submitted to NASA for consideration, resulting in a prime and back-up candidate being selected. The primary candidate was Pratiwi Sudarmono, who was assigned as a PS onboard STS 61-H in June 1986 before its cancellation as a result of the *Challenger* accident.

Pratiwi Sudarmono (*b.*1952) earned a medical degree in 1976 and an MS degree in microbiology in 1980, completing her PhD in generic engineering and biotechnology in 1984. Prior to selection as a PS, she was a lecturer in microbiology at the University of Indonesia, and she returned to that institution after the cancellation of her mission in 1986.

France

France had nominated candidates (including one female candidate) for the ESA Spacelab 1 mission in 1977, and was negotiating with the Soviet Union to send one of its citizens to a Russian space station on a short visiting mission. After the success of this 1982 flight to Salyut 7 by Jean-Loup Chrétien, it was decided to create a national team of space explorers for assignment on US or Soviet missions, looking towards crewing the Hermes mini space shuttle then under development. On 9 September 1985, seven finalists, including Claudie Deshays, were named from 700 applicants.

Unlike some of her male colleagues Deshays was never assigned to a Shuttle mission, but she flew to Mir and the ISS onboard Russian vehicles. The French were also interested in a dedicated French Spacelab (Spacelab F) mission, possibly flown under the proposed Spacelab Life Sciences 3 mission, but this did not materialise.

Germany

Following the success of Ulf Merbold on Spacelab 1 in 1983 and Spacelab D1 (STS 61-A) in 1985, the West German Federal Aerospace Research Establishment was completing the selection of a second cadre of West German astronauts at the time of the *Challenger* accident. This selection for Spacelab D2 was postponed until 3 August 1987, when the five finalists named included Renate Brummer, who served as a back-up PS to the STS-55 (Spacelab D2) mission in 1993, and Heike Walpot, who served as a crew interface coordinator for the Spacelab D2 mission.

Renate Brummer (*b*.1955) was a professional meteorologist who was a research associate at the University of Boulder. She had also completed her 1986 PhD studies

The 1997 German astronaut selection for Spacelab D2, including Renate Brummer (rear right) and Heike Walpot (front left).

in the United States on the effects of mountains on the weather, and was studying the prediction of tornadoes. As well as holding a private pilot's licence, Brummer had originally qualified and worked as a secondary school teacher of mathematics and physics. Had she flown onboard Spacelab D2 instead of serving as a back-up astronaut supporting the flight from the German control centre at Oberpfaffenhoffen, near Munich, she could have become the first school-teacher to orbit the Earth. After the Spacelab D2 training group was disbanded in November 1993, Brummer left the German astronaut team and returned to the US to take up a post with the National Oceanographic and Atmosphere Administration (NOAA) in Boulder, Colorado.

Heike Walpot (*b*.1960) was a qualified physician, a former member of the German national swimming team, and a competitor at the 1976 summer Olympic games in Montreal. Her astronaut assignments following basic training included acting as German crew representative for the Hermes space plane project. In 1992 she was assigned to the D2 team, and in April 1995 was selected to train for a possible long-duration flight to Mir. She therefore began lessons in the Russian language. But she was not selected as either a prime or back-up crew-member, and left the German astronaut team to return to the Aachen University medical department.

European Space Agency

Following the selection of the first European Shuttle Payload Specialists for Spacelab 1 in 1977, the European astronaut team consisted of three male candidates. During most of the 1980s, European representatives on manned spaceflights came from national teams, but in June 1989 ESA finally began a selection programme to expand the group of European astronauts for flights onboard Hermes, the American Shuttle and the Russian Soyuz/space station programmes. At that time there were two European Spacelab missions planned (Spacelab E1 and E2), as well as deployment and retrieval operations of the EURECA free-flying platform. All thirteen ESA member states were invited to submit three to five candidates each, and Canada was also invited to participate. Sixty applicants were screened, with the plan to select ten finalists. Some of the candidates were already members of national teams, and some had already flown in space, including the UK former Juno (Mir) crew-member Helen Sharman and CNES rookie Claude Deshays. One of the four Irish candidates was Dr Deirdre MacMahon (*b*.1960). The selection was for six Laboratory Specialists (LS) for assignment to space station or Spacelab missions, and four Space Plane Specialists (SPS) for training on Hermes, but prior to naming the finalists, ESA was forced to scale back its plans. On 15 May 1992 only six new astronauts were announced, including Marianne Merchez from Belgium.

Marianne Merchez (*b*.1960) studied for a doctorate in medicine and took a course in aerospace medicine before entering general practice in 1985. She also became a qualified airline pilot, attending Belgium's Civil Aviation School and serving as a co-pilot on commercial flights. Following selection by ESA she completed preliminary cosmonaut training at TsPK, but withdrew from possible consideration for the planned 1994 and 1995 ESA mission to the Russian station, and instead moved to the United States to join her husband and fellow ESA astronaut Maurizio Cheli at JSC in Houston. When she was not selected for Mission Specialist training with a view to flying on the Shuttle to the ISS, she resigned from the ESA astronaut programme.

Japan

On 7 August 1985, Japan announced the selection of three candidates to train for a Japanese Spacelab Mission (Spacelab J) planned for 1988. From 533 applications received for the single flight-seat, the three candidates would complete a NASA PS training programme before the finalist was named to a crew in May 1987, with the other two serving as back-ups. The trio included Chiaki Naito (later Mukai). Following the *Challenger* accident the mission was postponed, but the team continued generic payload training until the summer of 1990 when the crew announcements were made. On 16 July 1990, Mukai was named with Takao Doi as joint back-up to Mamoru Mohri, who flew the Spacelab J mission as STS-47 in September 1992.

Chiaki Mukai (*b*.1952) pursued a medical career, gaining a medical doctorate in 1977, subsequently working as a surgeon and cardiovascular surgery instructor, and then gaining a PhD in physiology. Following her assignments on Spacelab J, she flew as a PS onboard the sixteen-day STS-65 (IML-2) mission in July 1994, served as alternate PS for the sixteen-day Neurolab Spacelab flight of STS-90 in April 1998, and flew a second mission as PS for STS-95 in October 1998. (She also wrote the Foreword for this book.)

Spaceflight participants

In the late 1970s an advisory panel to NASA, in an attempt to broaden the appeal and understanding of manned space exploration to a wider audience, began to consider assigning 'civilian passengers' to fly on Shuttle missions. Eventually, NASA agreed to support such a venture, and a selection programme was conceived: the Space Flight Participant (SFP) programme. By 1984 the first group to be considered for this programme were journalists, and a selection process began. However, in August of that year President Ronald Reagan announced that the first 'ordinary citizen' in space would be a primary or secondary school teacher. As a result, the Teacher-in-Space project was initiated, while the Journalist-in-Space project became the second programme in the series. Overall this was well received – except, of course, by some of the journalists, as they had covered the space programme from its outset. But space education was also a long-established aim of NASA and the space community, and it was argued that future engineers, scientists, administrators, space explorers and policy-makers would be taught by school-teachers. A flight in space by a teacher might inspire young students to support the space programme in their adult years. As these programmes were being pursued, NASA also invited VIPs from the Senate (Jake Garn) and Congress (Bill Nelson) to take controversial seats on Shuttle missions. Due to launch delays, however, both politicians flew into space before the teacher was launched in January 1986.

The announcement of the SFP programme generated a host of applications from members of the film and music industries, and poets and artists. In the UK press in November 1985, actress Haley Mills – daughter of Sir John Mills – was identified as being interested in flying on a three-day Shuttle flight around 1996, hoping to become the first actress in space. Nothing evolved from these applications.

Teacher-in-Space programme

The Council of Chief State Schools received 10,463 applications for the TIS programme. All fifty US states, plus US territories, the Department of State and Defense overseas dependent schools, and the Bureau of Indian Affairs, were eligible for applicants. On 3 May 1985, 114 elementary and secondary level teachers were named as candidates, with two from each state (except from the Bureau of Indian Affairs, and New Mexico). There were fifty-three female teachers in the group, and after further consideration the final ten candidates announced on 1 July 1985 included six female teachers:

Kathleen Beres, a biology teacher at Kenwood High School in Baltimore, Maryland.
Judith M. Garcia, a French and Spanish teacher at Jefferson School for Science and Technology in Alexandria, Virginia.
Peggy Lathlaen, a teacher of gifted children at Westwood Elementary School in Friendswood, Texas.
Christa McAuliffe, a social studies teacher from Concord High School, Concord, New Hampshire.
Barbara R. Morgan, a second-grade teacher at McCall-Donnelly Elementary school in McCall, Idaho.
Niki M. Wenger, a teacher of gifted children at Vandevender Junior High School in Parkersburg, West Virginia.

A week after the announcement, the ten finalists were at NASA JSC undergoing medical and psychological evaluations and a flight on the KC-135 'vomit Comet'. On 19 July, Christa McAuliffe was named as the primary Teacher-in-Space candidate, with Barbara Morgan as her back-up. Training for their mission (manifested as STS 51-L) would begin on 9 September, when they were also assigned to the flight-crew in training for the mission. Though disappointed at not being selected, the other candidates congratulated McAuliffe and Morgan and would continue working for NASA for the next year to help prepare the lessons that McAuliffe planned to present from space in what was being called 'The Ultimate Field Trip'. As Kathy Beres explained: 'None of us felt we'd lost or were second best or were slighted. There has to be a bit of disappointment, but it's a new beginning. I look forward to going back to the classroom and telling my students all about it.'[5]

Sharon Christa McAuliffe (1948–1986) was born in Boston, Massachusetts. She began her teaching career in 1970, gained a Masters degree in education in 1978, and taught civics, American History and English in a number of schools in Maryland and New Hampshire. She joined Concord Senior High in 1982, teaching economics, law, American history and a course of her own devising called 'The American Woman',

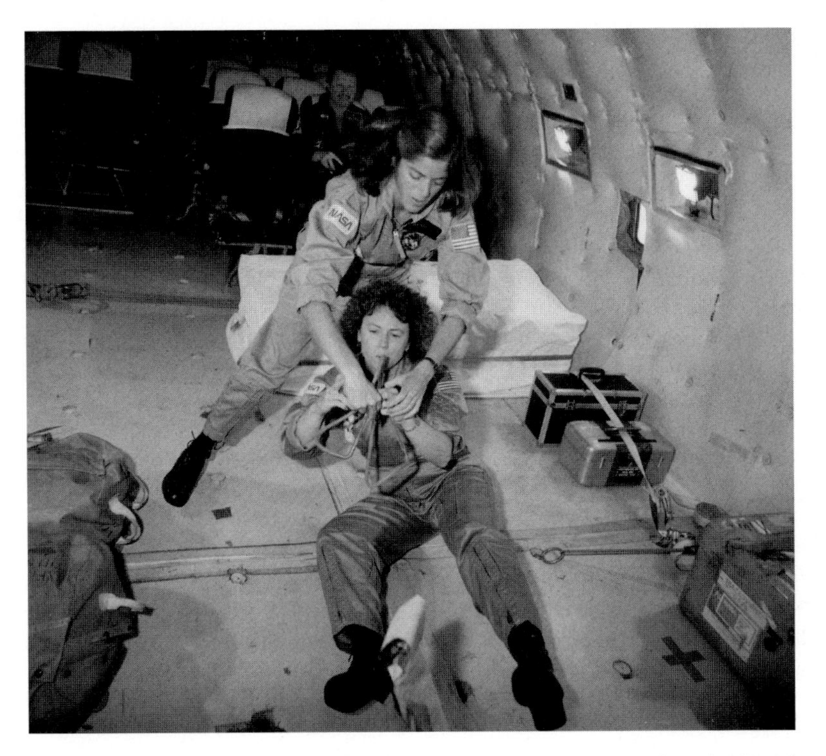

Barbara Morgan and Christa McAuliffe train with experiments for the Teacher-in-Space programme planned for STS-51L.

to grades ten to twelve. In applying for the position of Teacher-in-Space candidate, she realised that she could start her life over again as an astronaut: 'As a woman I have been envious of those men who could participate in the space programme and who were encouraged to excel in areas of maths and science. I felt that women had indeed been left outside of one of the most exciting careers available. When Sally Ride and the other women began training as astronauts, I could look among my students and see ahead of them an ever increasing list of opportunities. This opportunity to connect my abilities as an educator with my interests in history and space is a unique opportunity to fulfil my early fantasies. I watched the space programme being born, and would like to participate.'

Barbara Morgan (*b.*1951) earned a BA degree in human biology in 1973 and completed her teaching credentials the following year. She began as a remedial mathematics and reading teacher before becoming a secondary-school teacher in 1975. Her experience also included a year as a teacher in Quito, Ecuador, and when named as alternate to McAuliffe in 1985 she stated that she was not at all disappointed, as she believed that she and all the other TIS candidates would have the chance to fly in space in the future.

Journalist-in-Space programme

This programme was intended to be part of STS 61-I, originally planned for the end of 1986. After the announcement of the programme in October 1985, 1,703 applications were received. The selection process was handled by the Association of Schools of Journalism and Mass Communications, and was administered by the College of Journalism at the University of South Carolina, Columbia. The original plan was to select a hundred regional semi-finalists by March 1986, and then reduce this to forty national semi-finalists. A group of five finalists would have been named for a week of medical tests at JSC before the prime and back-up were selected and named on 17 April; but the loss of *Challenger* delayed this process, and the hundred regional semi-finalists, including more than twenty female applicants, were named in April. On 14 May 1986 the list was reduced to forty candidates, including eight female journalists:

Theresa 'Terry' Anzur, of NBS News, Chicago.
Marcia Bartusiak, a freelance journalist in Norfolk, Virginia.
Diane Eicher, of the *Denver Post.*
Joan Esposito of WLS-TV, Chicago.
Caroline 'Terry' Marotta, a freelance journalist in Winchester, Massachusetts.
Anne 'Kathy' Sawyer, of the *Washington Post.*
Lynn Sheer, of ABC News, New York.
Barbara Stanton, of the Detroit Free Press.

The five finalists were never named, as the *Challenger* tragedy and recovery led to the decision to suspend flights by non-career astronauts. In January 1989 NASA announced that future crew assignments would be limited to NASA career astronauts and scientists assigned to specific payloads or experiments critical to mission success. This decision effectively terminated the original Teacher-in-Space and Journalist-in-Space programmes.

New NASA selections

By the spring of 1991 NASA was about to begin a new round of astronaut candidate selections, but with an emphasis on a broader educational background rather than flying credentials. This reflected not only an increasing number of scientific Shuttle flights, but also the construction of the International Space Station that had been in the final planning stages since 1984. However, although five further selections were announced by NASA between 1992 and 2000, all of the new trainees would complete astronaut candidate training for Space Shuttle crew assignments in addition to any other specialist training required by other programmes in which they would participate.

Class of 1992

The process of selecting the fourteenth group of NASA astronauts began on 1 July 1991 and resulted in 2,054 applications. From these, eighty-seven were medically screened and interviewed early in 1992, resulting in nineteen new astronauts, announced on 31 March 1992. During the process, NASA announced that it would accept applicants from other agencies and foreign countries to train as Mission

Specialists, beginning with Group 14. The non-NASA crew-members would complete the Ascan training programme and be eligible for assignment as MS on further Shuttle missions, or would use the experience as part of their preparation for long-duration assignment on the ISS. Of the eighty-seven applicants short-listed by NASA for this group – which had extra emphasis in life science and space manufacturing requirements – there were nine female candidates. Of the nineteen named in March 1992, three were women, all MS candidates:

Catherine 'Cady' Coleman (Captain, USAF).
Wendy B. Lawrence (Lieut-Commander, USN).
Mary E. Weber (civilian PhD).

Class of 1994

Begun in 1993, this selection process resulted in 2,962 applications, which were then reduced to about 120 finalists (including thirty-two females). These were interviewed and medically screened during the summer of 1993. After a US government freeze on hiring federal employees, the nineteen new Ascans were named on 8 December 1994. In this group there were five females, two of whom were selected as Shuttle pilot candidates in response to the need to recruit more pilot candidates as a result of natural attrition from the earlier selections in the Astronaut Office pilot pool:

Kalpana Chawla (civilian PhD) MS candidate.
Kathryn P. Hire (civilian) MS candidate.
Janet L. Kavandi (civilian PhD) MS candidate.
Pamela E. Melroy (Major, USAF) Pilot candidate.
Susan L. Still (Lieut-Commander, USN) Pilot candidate.

Class of 1996

In announcing this selection NASA was fulfilling a need to increase the number of applicants to cover the approaching programme of construction of the ISS and to provide a long-duration crew-member training pool. This was the largest NASA astronaut group selected since 1978, and in the application period towards the end of 1995 more than 2,400 applications were received. In January 1996, 120 applicants, including twenty-three women, were interviewed and medically examined at NASA JSC, with the thirty-five Ascans being named on 1 May 1996. With nine international astronauts also joining the class at the start of training in August 1996, a two-year training programme was initiated for this group to alleviate the inevitable strain on the limited facilities available during Ascan training. All the women selected in this group were MS candidates:

Yvonne D. Cagle (civilian PhD)
Laurel B. Clark (Lieut-Commander, USN)
Joan E. Higginbotham (civilian)
Sandra H. Magnus (civilian PhD)
Lisa M. Nowak (Lieut-Commander, USN)
Heidemarie Stefanyshyn-Piper (Lieut-Commander, USN)

Peggy Whitson (civilian PhD)
Stephanie D. Wilson (civilian)

Class of 1998

Twenty candidates were expected to be named in the seventeenth selection. More than 2,500 applications were processed, with 120, including more than twenty females, being interviewed between October 1997 and January 1998. On 3 June 1998 the new group of twenty-five was named, and they were joined by six international candidates for their Ascan training programme. One candidate was announced in advance; former Teacher-in-Space back-up Barbara Morgan, as an Educator Mission Specialist. The four female MS candidates selected in this group were:

Tracy E. Caldwell (civilian PhD)
Patricia C. Hilliard (civilian MD)
Barbara R. Morgan (civilian schoolteacher)
Sunita L. Williams (Lieut-Commander, USN)

Class of 2000

The eighteenth selection programme produce more than 300 applicants in the summer of 1999, and from these, approximately 120 applicants, including about twenty females, were interviewed and medically screened at JSC between September 1999 and January 2000. This was also the last selection in which the Ascan training programme would be focused on Space Shuttle operations in addition to ISS operations, procedures and hardware. From the next selection (2004), candidates would prepare for missions in the new Crew Exploration Vehicle (CEV), to be developed for a return to the Moon and to support manned flights to Mars. The three female candidates chosen on 26 July 2000 were:

K. Megan McArthur (civilian PhD)
Karen L. Nyberg (civilian PhD)
Nicole P. Stott (civilian)

TRIUMPH TO TRAGEDY

By June 1985 all six women of the 1978 selection had flown in space – and Sally Ride had flown twice. They were all assigned to new flights, or were in the pool for assignment to a new mission, but the next American women into space would be the two selected in the 1980 Class: Bonnie Dunbar and Mary Cleave. Waiting in the wings were the members of the 1984 and 1985 selections, a number of female Payload Specialists, and Teacher-in-Space candidate, Christa McAuliffe. The launch schedule over the next couple of years was expected to be busy, but the triumph of the first flights by women on the Shuttle would be brought into perspective by the *Challenger* accident, which affected the whole Shuttle programme and vividly demonstrated the dangers of trying to enter space by rocket. (See Appendix 4 for flight details of Shuttle missions featuring female crew-members.)

A 'full-up' mission

STS 61-A, carrying the Spacelab D1 payload, was launched on 30 October 1985. It was the ninth flight of *Challenger* and the twenty-second of the programme. With eight crew-members, it was also the largest crew ever to be launched into space on one mission. and included representatives from NASA, West Germany and Holland. Flying as MS1 was Bonnie Dunbar, whose responsibilities on the mission included operation of the Spacelab science module and its subsystems and a number of experiments conducted during the week-long mission. She had been named to the mission on 14 February 198,. and in preparation for this international mission had completed a six-month experiment training programme in Europe. This mission featured round-the-clock crew operation in shifts, operating a suite of seventy-six experiments in basic and applied microgravity research, focusing on materials sciences, life sciences, and technology, communications and navigation. Dunbar, serving on the Blue Team, was termed a 'scientific mission specialist', focusing on the scientific aspects of the mission rather than the operation of the orbiter. She also served as the second contingency EVA astronaut (EV2). The mission concluded successfully after 7 days 44 min.

Bonnie Dunbar (*b.*1949) was raised on a farm in eastern Washington State, and due to this isolated location she became a keen reader at an early age. Her interest in science was encouraged by her physics teacher, and she majored in engineering while continuing to read extensively as a pastime. Dunbar had long dreamed of being an astronaut, and hoped that one day women would be selected for the programme; but after the military informed her that they did not accept women as pilot candidates, she followed an engineering career and entered university. Many years later, when she learned that women were being accepted for pilot training, she was told she was now 'too old'. She received a BS degree in ceramic engineering from the University of Washington in 1971, and while there she participated in pioneering research with ceramic tiles for a joint programme between the university and NASA, researching a tile-based thermal protection system for the new Space Shuttle. After completing her Masters degree and a term as a visiting scientist at the Harwell research facility near Oxford, England, in 1975–76, she accepted a position at Rockwell International Space Division in Downey, California, as a senior research engineer developing equipment and procedures for the manufacture of the Shuttle thermal protection systems. Upon hearing of NASA's call for women astronauts in the Class of 1978 she applied, but although she was a finalist she was not selected. Realising the need to broaden her experience and not become known as a 'tile expert' for the rest of her life, she accepted a position at NASA as a systems engineer, working as a payload officer and flight controller as well as serving as a guidance and navigation officer on the flight control team for the 1979 Skylab re-entry mission. She then worked on the integration of a number of Shuttle payloads. In 1983 she earned her PhD with a multi-discipline dissertation that encompassed both material sciences and physiology. She was selected to NASA in 1980, completed the Ascan training programme, worked on the support crews for STS-2 and STS-9 and in the SAIL facility, and was then assigned to the Spacelab D1 crew.

Cleave's comet

STS 61-B – the twenty-third Shuttle mission – was launched on 28 November 1985. Mary Cleave, serving as MS2, acted as Flight Engineer during the ascent to orbit. The mission was scheduled to deploy three communication satellites (one American, one Australian and one Mexican) and conduct two experiments in space construction techniques and a programme of mid-deck experiments. The crew also included the first representative from Mexico to fly in space: Payload Specialist Rodolfo Neri. After reaching orbit, Cleave activated a series of experiments to grow crystals, through different organic solutions, which revealed that crystals grown during the mission were larger and stronger than those grown on the ground in the control experiment. The experiments were a continuing evaluation of space processing procedures in lieu of long-duration processing facilities on space platforms or future space stations. Cleave was also primary RMS operator for the mission, supporting the two EVAs conducted on 29 November and 1 December in which Jerry Ross and Sherwood Spring constructed EASE and ACCESS in the payload bay of *Atlantis*. This provided baseline data for possible construction techniques and the mobility of large space structures in consideration of future space platforms or a space station. Halfway through the mission Cleave conducted a routine dump of waste water out of the Shuttle into the vacuum of space. The timing was perfect for Sun illumination of both *Atlantis* and the water dump, and observers on the ground recorded both the orbiter and the fifteen-mile long water dump. From Mission Control, Sally Ride informed the crew that the event was being christened 'Cleave's Comet'. STS 61-B landed on 3 December 1985 after a mission of 6 days 21 hrs.

Mary Cleave (*b.*1947) has always loved flying. She began at the age of fourteen, and gained a pilot's licence before she could drive. She first flew solo when she was sixteen, and received her private pilot's licence at seventeen. She paid for her flying with money earned from by baby-sitting and giving baton-twirling lessons. Following graduation from high school in 1965, Cleave had plans to become an airline hostess, but at 5 feet 1.5 inches (1.56 metres) she was too short. Instead she decided to become an animal veterinarian, specialising in large animals, as she had loved nature since she was a small child; but after two years she realised that her lack of height was a disadvantage, and instead followed in the footsteps of her mother and took up a career in teaching. She earned a BS degree in biological sciences in 1969, and afterwards worked as a biology teacher on a floating campus (a ship-borne college) while continuing her studies into ecology and pollution for two years. In 1971 she resumed her academic studies and earned her Masters degree in microbial engineering in 1975 and a PhD in civilian and environmental engineering in 1979. Throughout the 1970s she also held a number of research posts, by which she gained experience in both laboratory and field work. As she worked in the laboratory and research institutes, she noted that she was one of only a handful of women, and that none of them seemed to hold the certified engineering credentials of her male colleagues. She therefore decided to pursue her doctorate in civil and environmental engineering to improve her credentials and aim for a better position after her academic studies were completed. Although she applied for the astronaut 1978 selection she was not selected, but she succeeded in 1980. During the claustrophobia test (in which candidates were zipped up in the Shuttle rescue ball), she fell asleep! Following her Ascan training programme, Cleave was given technical assignments in SAIL, and helped to develop crew equipment, including the personal hygiene kit and flight documentation such as the Malfunction Procedures Data Book. One of her earliest assignments, based on her background of microbial ecology and water sewerage and pollutants, was in helping to develop the Shuttle Waste Management System. She was also chosen as the test subject (smallest size) for the Shuttle EMU development test programme. She also served a tour as CapCom on several missions prior to selection to her first Shuttle flight on 2 February 1984, including talking to Sally Ride during STS-7 on the first female-to-female (in space and at Mission Control) link. The two women acted so prosaically, with routine conversations, that one unnamed female reporter told her that the event was 'very disappointing' because nothing of historical value was said. Cleave was sorry to disappoint the reporter, but pointed out that both astronauts were merely doing their jobs.[6] She found the experience of serving as CapCom useful on her spaceflight – and the knowledge acquired of her first spaceflight was equally useful for her second assignment as CapCom.

A helping hand

As primary RMS operator for the space construction demonstrations, Cleave offered some suggestions for future RMS operations when constructing larger space structures.[7] Before she flew she sought the advice of another colleague who had been RMS operator to learn about extended use of the arm and how to stabilise herself; but at the time, the general consensus was that she did not really need to do so. Anna Fisher explained that she had hooked her feet around the top of the Commander's flight seat, but Cleave would not be able to do this, as her Commander, Brewster Shaw, would be occupying his seat during most of the time that she would be operating the RMS. Instead, she used a seat restraint tied low to keep out of the way of the other crew-members.

As with those before her, Cleave also experienced quite a difference between using the real arm in space and the training model on the ground or in the water tank. She was warned to be aware of how fast the real arm moved in space, and how it moved much more easily than in the simulations. One recommendation – to gain experience in moving the arm around before an astronaut rode it on EVA – was a very useful piece of advice; but during EVA she discovered that the back-drop of Earth distracted her when she was positioning the astronaut on the arm. She pointed this out during debriefing, bringing to mind a similar event during the Gemini programme (Gemini 9 in 1966) in which the astronauts learned to conduct visual rendezvous and acquisition against the blackness of space rather than against the colours and lighting conditions of Earth. Cleave suggested that small indicator lights should be added to the top of the RMS so that the end effector would still be visible through the windows in the dark – which was considered a valid point for a future space station RMS system. She also explained that the spotlights on the arm blurred the TV images and were not used, and revealed that she was constantly looking out of the windows when translating one of her colleagues about the bay: 'When you have a live body on the end of the arm, you tend to be really hawking them; you don't want to bump them into anything.'

Her evaluations of all aspects of RMS operations were valuable additions to the baseline data for future RMS operations and for refining training procedures and mission simulations. She also stressed the importance of other astronauts on the flight-deck in helping her operate the RMS and clearing other structures in the bay. RMS operation was certainly not a one-person task. A suggestion for additional audio-visual cues for the operator – such as a heads-up display facility – was also noted, but the procedures book and its repetitive use in training and simulations certainly proved very useful for the flight. Operating the system became second nature and allowed her to focus on real-time activities, trusting her long hours of training to complete moves and procedures without taking her eyes off what was happening outside. One significant problem that she noted was that the windows of the Shuttle had several 'dead spots' for a single operator, and did not offer perfect visibility. Having other crew-members looking out of other windows from different angles helped reduce these 'dead spots', but it was impossible to constantly complete camera operations by the book, and the training devices allowed only partial task training to be completed prior to flight. One highlighted problem during EVA was in keeping the video recorders supplied with tape and recording all aspects of the EVA; and the EVA was too fast and dynamic for the checklists to be repeatedly consulted.

McAuliffe's ultimate field trip

The next woman to enter space for the first time should have been schoolteacher Christa McAuliffe, on *Challenger*'s tenth mission STS 51-L. On the same mission, Judy Resnik was assigned as MS2, serving as launch and entry flight engineer. They were part of the seven-person crew who arrived at the Cape on 23 January for their final preparations for launch on 26 January. The mission had already slipped several weeks, and would actually not be launched until 28 January, due to constraints of the weather. The events of that fateful day have been recounted many times and are beyond the scope of this book (see Bibliography), but it is worth recording McAuliffe' intentions.

She planned to complete a personal journal of her experiences and activities on the mission, as a trilogy – the first part covering her selection and training for the mission, the second reviewing the flight, and the third dealing with her reactions and thoughts after her return. During the sixth day of the mission, McAuliffe would have broadcast two live lessons from space.[8] The first of these – 'The ultimate field trip' – would have been based on her own interpretation of being a teacher given the chance to fly in space. She would have conducted a video tour of the flight deck and mid-deck of *Challenger*, describing the controls, work stations, living quarters and facilities, as well as pointing the camera out of the window to the payload bay and describing the components of the orbiter in view. For the second lesson – 'Where we've been and where we're going. Why? – she would have used models of the 1903

Barbara Morgan and Christa McAuliffe monitor a Shuttle mission from Mission Control, Houston, during preparations for the STS-51L mission.

Wright Brothers Flyer and the proposed Freedom Space Station, and would have explained that only eighty-two years had separated that pioneering flight and her own flight. She would also have explained the advantages and disadvantages of manufacturing in zero-g, described the multitude of spin-offs that have been developed from the space programme to benefit all on Earth, and highlighted the advantages of the proposed Space Station, as well as discussing the reasons for continuing to live and work in space. She planned to cover several topics, including astronomy, observations of Earth, onboard experiments, *Challenger*'s payload (the Tracking Data Relay Satellite and Spartan–Halley free flyer), materials processing experiments, and technological advances. In addition to the lessons, McAuliffe would have conducted a number of demonstrations entitled 'Mission Watch' (manifested for the second and fourth days), which would have been filmed and recorded for use in educational packages distributed to schools after her mission, to which she would have added a narrative. NASA was also planning to arrange for schools to receive regular daily updates on the mission via satellite TV, under the Classroom Earth project specialising in direct satellite transmissions to elementary and secondary schools. A Mission Information Pack included educational materials, the TV schedule, an orbital map and backgrounds of the crew and mission. McAuliffe's back-up Teacher-in-Space, Barbara Morgan, would have served as moderator for these special daily broadcasts.

'The vehicle has exploded'

After several delays, the crew boarded *Challenger* on 28 January 1986. McAuliffe was presented with a large polished apple by one of the support team, but she passed it back, asking if it could be saved until she returned. Judy Resnik took her seat on the flight deck, and McAuliffe was seated on the mid-deck. As *Challenger* lifted off from the pad, a small puff of black smoke emerged from the bottom seal of the right-hand booster; but it was not noticed at the time, although automated cameras were recording the launch from several angles and caught the event on film. The flight crew, unaware of events behind them, went about their well-practised procedures on the flight deck, while those on the mid-deck enjoyed the thrill of acceleration and the beginning of their trip into space.

On arrival at the Cape, McAuliffe had reported that 'No teacher has ever been better prepared to teach a lesson.' She was pleased that a teacher had been selected as the first spaceflight participant because of the daily contact that her profession has with everyone at some point in their lives. In response to a reporter's question during an earlier interview she had commented on her thoughts about the moment of launch: 'I'm not sure how I'm going to feel sitting there waiting for take-off when those solid rocket boosters ignite underneath me and everything starts to shake. It's kind of like the first time you go on a carnival ride. You've said 'I've got enough courage', and you're really excited about doing this and conquering your fears.'[9]

Seventy-three seconds into the flight, those solid rocket boosters initiated the massive explosion that ripped the vehicle apart and claimed the lives of McAuliffe, Resnik and their five fellow crew-members. Later evidence revealed that the intense cold at the Cape had caused the right-hand lower o-ring seal to fail (resulting in the tell-tale puff of black smoke), causing hot gases to leak from the booster, and sealing the fate of *Challenger* and her crew. The white-hot gases, impacting on the external tank, burned through the insulation, the tank's skin and a support strut holding the booster to the ET as the SRB broke away. A flash of flame and a huge explosion signalled the abrupt end to the mission.

After *Challenger*

At the time of the STS 51-L launch there was a crowded manifest of eleven missions through to December 1986. Mission 26 (61-E) was to carry the first Astro astronomical payload in March; and in May, two launches just days apart were to deploy the Ulysses solar probe (Mission 27 – STS 61-F) and the Galileo probe to Jupiter (Mission 28 – STS 61-G). Mission 29 (61-H) would have been a comsat deployment mission in June, followed by Mission 30, the first launch from the Vandenberg AFB in California in July, the classified STS 62-A mission. Mission 31 was STS 61-M, deploying a TDRS satellite, and Mission 32 was to carry the Earth Observation Mission payload, utilising the first Spacelab Short Module configuration. In early September, Mission 33 would fly the classified DoD payload STS 61-N, and later that same month Mission 34 (STS 61-I) would include a satellite deployment and retrieval objective. In early November, Mission 35 (STS 61-K), the Hubble Space Telescope deployment mission, would be followed later that the month by Mission 36 (STS 61-L), a mixed science and commerce mission with satellite deployments and onboard materials science investigations. The year would have ended with mission 37 (STS 71-B), another DoD mission, in December.

In a total of sixty available flight-seats for these missions, only four were assigned to female crew-members. The STS 61-H crew included Anna Fisher for her second mission and Indonesian PS Pratiwi Sudarmono. The NASA astronauts assigned to this mission had been named to STS 61-C on 29 January 1985, but were reassigned to 61-H in March. Planned to deploy three communications satellites belonging to the Western Union (Westar), Indonesia (Palapa) and the UK's MoD (Skynet), representatives from the UK (Nigel Wood) and Indonesia (Sudarmono) would have observed the deployments, but not actively participate in them, and conducted experiment on the mid-deck.

The STS 61-I crew included Bonnie Dunbar on her second mission, as well as a PS from India and the finalist for the Journalist-in-Space position. An Indian Insat satellite was to be deployed, and the Long Duration Exposure Facility (LDEF – deployed on STS 41-C) would be retrieved by RMS.

The STS 61-K crew, including Kathy Sullivan, would deploy the Hubble Space Telescope. Sullivan was one of the two EVA-trained crew-members (the other was Bruce McCandless) prepared to perform contingency EVAs to assist in the deployment, abandonment or re-berthing of the HST during the mission.

Also in training was the crew assigned to 61-M (originally 61-I), commanded by Loren Shriver and including Sally Ride as MS2 on her third mission. All of these missions were suspended pending the inquiry into the *Challenger* accident, and the crews were assigned to generic training while the cause of the accident was determined and the recovery from the event completed. All assigned Payload Specialists were also stood down pending reassignment of their payloads or missions.

Return-to-Flight

Between the loss of *Challenger* in January 1986 and the return to flight in September 1988, several veteran astronauts decided to leave the programme to pursue new ventures and careers. Most of the astronauts were involved in supporting the official inquiry and activities to prepare the vehicle to return to flight.

With the tragic loss of the 51-L crew and with all assigned crews grounded, the opportunity was taken to review the question of flying Payload Specialists and Space Flight Participants on future missions. In April 1986 the Payload Specialist Liaison Office (PSLO) reported that NASA and the US Government had made commitments or signed contracts to fly eighty-five Payload Specialists. Six months later, a revised Shuttle manifest detailed the resumption of flights from STS-26 in 1988. No crews included Payload Specialists, and it was also revealed by NASA Administrator James Fletcher that the first five Shuttle missions would each be flown by a five-person NASA astronaut crew. Media reports at the time suggested that Fletcher wanted no Payload Specialists on any Shuttle flights for the next twenty missions, 'if not forever.'[10] It was also becoming clear that the Astronaut Office was in favour of eliminating all Payload Specialist assignments from Shuttle flights, with the exception of Spacelab-type missions, and wherever possible only one Payload Specialist would fly on any given mission. This was relaxed slightly from 1992, but the frequent flights by Payload Specialists as demonstrated in 1984–86 were not part of the resumption of Shuttle flights from 1988.

Several astronaut 'crews' participated in milestone ground training and simulation exercises as part of the return-to-flight programme. Though not officially 'flight crews', they mostly consisted of generic crews already in the training schedule for post-*Challenger* missions.

On 29 October 1986 a fully integrated simulation (linked to Mission Control) was completed by the NASA astronauts assigned to the former 61-H mission, including Anna Fisher. In April 1987 this same crew completed a long-duration Shuttle simulation at JSC using the Shuttle Mission Simulator, and also conducted a simulated EVA in the WETF. The crew was now identified as the STS 61-M (Training) simulation crew, prior to being assigned to a return-to-flight mission.

Kathy Thornton (one of the 1984 astronauts) was assigned to an Emergency Egress Test Crew under the command of Frank Culbertson. This crew included seven astronauts, none of whom had flight experience. It was therefore not a future flight crew, but rather was formed to develop their experience of pad operations and emergency training. Thornton participated as a 'Mission Specialist' during simulations carried out at KSC. A further simulation was completed by the same 'crew' on 4 May 1987.

In August 1987 Sally Ride – a member of the 1986 Presidential Commission investigating the loss of *Challenger* – resigned from NASA to join the Stanford University Center for International Security and Arms Control.

An increasing number of scientific flights began to be manifested to further Shuttle missions as commercial payloads declined because of the long delay. Many commercial payloads were reassigned to expendable launch vehicles. Coupled with the long lead-planning for the proposed Space Station Freedom, this led to members of the Astronaut Office forming an Astronaut Science Support Group on 29 February 1988.[11] This group was formed to provide direct interaction with prospective experimenters on Shuttle and space station missions. Using their experience of flying in space, group members would provide guidance for experiments proposed for spaceflight by increasing scientific and engineering flexibility while maintaining Shuttle-operating guidelines. As well as providing a point of contact between customers, NASA and the Astronaut Office, the astronauts believed that their increased participation in the design, development and operation of experiments would help to improve data returns and simplify the repair of hardware in space, thus maximising the return from each experiment. What was not mentioned in the release, but which was becoming clearer with fewer trained Payload Specialists flying on missions, was that the role of NASA Mission Specialists as experiment operators would increase when the Return-to-Flight programme had been completed. Members of the six-person group included Mary Cleave (specialising in biological materials processing), Bonnie Dunbar (materials processing) and Rhea Seddon (life sciences).

On 17 March 1988 NASA named three Shuttle crews, including the first confirmation of flight assignments for women astronauts. The former STS 61-H crew that included Anna Fisher was named to STS-29, planned to deploy another TDRS. Fisher was removed from the crew because she was pregnant with her second child. (The STS-26 Return-to-Flight crew was named in January 1987, the DoD STS-27 crew in September 1987, and the STS-28 DoD crew in February 1988.) The STS-30 crew was the former STS 61-G crew, with Mary Cleave added to the crew to produce the required five. STS-30 was manifested to deploy the Magellan probe to

Venus. In the same announcement, Kathy Sullivan was confirmed to the HST deployment crew, formerly known as STS 61-J and now designated STS-31.

When the crew for STS-29 was named in 1988, Anna Fisher was medically disqualified due to her pregnancy. Seventeen years later, and with no second flight to add to her experiences on STS 51-A in 1984, she remains an active NASA astronaut who has filled a wide range of administrative and management roles while still maintaining her training proficiency for future flight assignment. She served on the selection board for the 1987 class of astronauts, and was working in the Space Station Support Office until her second child was born in January 1989. From 1989 until 1996 she took an extended leave of absence to raise her children, and after returning to the Astronaut Office in 1996 she was assigned to the Operations Planning Branch, working on the Operational Flight Data File (the procedures book carried on the Shuttle) and ISS training issues, as well as serving a term (June 1997 to June 1998) as Branch Chief of the Operations Planning Branch. Following a reorganisation of the Astronaut Office she became the Deputy for Operations and Training for the ISS for the next year, and after this assignment she was appointed Chief of the Space Station Branch of the Astronat Office, coordinating all astronaut input into the development and evaluation of hardware, operations, procedures and training while awaiting a new flight assignment.

On 30 November 1988, five more crews were named by NASA, reflecting the growing confidence in the Shuttle system after the highly successful STS-26 Return-to-Flight mission in September. Kathy Thornton and Ellen Baker, from the 1984 class, were named, respectively, to STS-33 (DoD) and STS-34 (deployment of the Galileo probe to Jupiter). Marsha Ivins – a colleague from the 1984 selection – was named to the long-delayed retrieval ofthe LDEF on STS-32, with Bonnie Dunbar on her second mission. This was the first major departure from the crews assigned before 51-L and although some generic training continued (such as Seddon on Spacelab Life Sciences), most pre-*Challenger* crewing plans were abandoned. By now, assigned PS positions were suspended, and most of the PSs were stood down. Those assigned to Astro, Spacelab Life Sciences, Earth Observation Mission (later renamed Atlas), Spacelab J, and the German D2 and Italian Tethered Satellite projects, continued their independent and national training programmes prior to selection to specific crews by NASA.

On 12 January 1989, new NASA Administrator and former NASA astronaut Richard Truly announced that crew assignments for future missions would be limited to NASA astronauts and scientists considered essential for mission success. Consequently, the original Space Flight Participant programme, the Teacher-in-Space programme, the Journalist-in-Space programme and the guest Payload Specialists assignments were effectively discontinued in their original format.

LAUNCH AFTER LAUNCH

After the Shuttle had completed its Return-to-Flight mission in 1988, the manifest could be more clearly defined. With the flight schedule increasing, payloads and experiments were being allocated to the three available orbiters (*Columbia*, *Discovery* and *Atlantis*), while planning continued for the replacement orbiter (*Endeavour*). With an increasing number of missions being planned, the astronauts training for them were presented with further opportunities. Unlike the early days of the US manned space programme – in which a team of astronauts was assigned to support or back-up a particular mission, skip the next two flights, and fly the third – Shuttle assignments usuallu depended on availability, experience, skills, and an element of luck. Long lead-training for science crews led, in 1990, to the new role of Payload Commander, for the development, planning, integration and on-orbit coordination of payload/Shuttle activities during the mission. With a decline in commercial payloads on the Shuttle and a desire to create a large, long-duration space station, science operations on the Shuttle became more prominent during the 1990s.[12] The following summarises the missions and responsibilities of female crew-members.

Probes and observatories

During the period 1988–92 the Shuttle programme tried to catch up with the backlog of payloads and deployments caused by the recovery from the *Challenger* accident, including those not transferred to ELV or deleted from the manifest.

STS-30: Mary Cleave

After the *Challenger* tragedy, the first flight to include a female crew-member was STS-30, launched in April 1989. It was the twenty-ninth mission in the series, the fourth for *Atlantis*, and the tenth to include a female crew-member. On her first mission, as MS2/Flight Engineer on STS 61-B, Mary Cleave took Seat 4 on the flight deck, but because she was totally focused on what she was doing she came to realise, on her second flight, how much she had missed during the ascent. On STS-30 she flew with space rookie Mark Lee, and both were responsible for the Magellan payload, not the orbiter. Lee wanted to fly on the flight deck in Seat 3; and as there was only a five-person crew, Cleave was left alone on the mid-deck for the ascent, with the responsibility of deploying the bail-out pole should the need arise. She therefore had little to do during nominal ascent. However, on the mid-deck there was only the smaller side hatch window: 'I thought it was a real lousy deal – 'I'm going to be all by myself down there, and I can't see a thing.' But it was great, because I could hoot, I could holler, I could have a marvellous time, and really appreciate the ride. You know, when you're working so hard on the flight deck, you don't even appreciate it. It was really a great ride, so my bad deal ended up being fabulous.'[13] Cleave had worked on spacecraft deployment procedures for some time, but had not expected to be assigned to such a flight so quickly after the *Challenger* accident. However, the urgency in deploying the spacecraft in the best launch window available, and the need to bring the four-person crew up to five, resulted in her quick assignment. 'I was thrilled, because it was the first time we had deployed a spacecraft that was going to another planet from the Shuttle, so it was a lot of fun. It was great work.' She also enjoyed working with the staff of the Jet Propulsion Laboratory. To Cleave, this was a 'meaty assignment' for a Mission Specialist, as she had much to do on the very first day on orbit, followed by three days during which she could take many more pictures than she had been able to on her first mission. She did not sleep much during the rest of the flight, and instead made the most of 'just having a good time'. A certain amount of undue attention was focused on her being the first woman to fly in space after the loss of *Challenger*, but she made it quite clear that this was not important.

In addition to supporting Lee with the deployment of Magellan, Cleave was responsible for the crew equipment onboard the vehicle (drawing on one of her earlier Astronaut Office technical assignments), Earth observation tasks, habitability issues (another early Astronaut Office assignment), in-flight maintenance (including changing a computer in-flight), photographic and TV equipment operation, fluids experiment apparatus, the Mesoscale Lightning Experiment (MLE), and the eight Detailed Supplementary Objectives (DSO).[14]

STS-34: Shannon Lucid and Ellen Baker

The next mission to include female crew-members was also a planetary deployment mission delayed from 1986. The jovian probe Galileo was deployed from *Atlantis* on 18 October 1989, a few hours after launch. Onboard STS-34 were Shannon Lucid on her second mission, and Ellen Baker flying her first mission. For this mission, Lucid was the lead specialist on the IUS and the primary crew-member for the deployment of the IUS/Galileo combination. She also had crew responsibility for the Polymer Morphology (PM) experiment, the IMAX camera, the Shuttle Solar Backscatter Ultraviolet (SSBUV) instrument operations and MLE operations. Baker was lead specialist on the crew for Galileo systems, and was also assigned responsibility for MLE operation and the Growth Hormone Concentration and Distribution (GHCD) in Plants experiment. In addition, as EV2 she was responsible for contingency EVA operations and, with colleague Franklin Chang-Diaz, was suited up in the Shuttle airlock during the deployment of Galileo, ready to exit the spacecraft should EVA be necessary to re-stow, separate or deploy the payload. This was not required, however, and she was therefore not presented with the opportunity to perform an EVA. During ascent, Lucid rode in Seat 3 on the flight deck, while Baker rode alone in Seat 5 on the mid-deck. For descent, they exchanged places.[15]

Ellen Baker (*b.*1953) After earning a BA in geology in 1974 she pursued a medical career, gained her MD in 1978, and trained for three years in internal medicine before joining NASA in 1981 as a medical officer at JSC in Houston. She graduated from the USAF Aerospace Medical Course at Brooks AFB, San Antonio, and was working as a physician in the flight medicine clinic at JSC when she was selected for astronaut training in 1984. After Ascan training, her technical assignments included working on flight software, engineering and operational support roles, and also assignments in developing operations within the space station programme

STS-33: Kathy Thornton

This mission was the fifth in a series of classified DoD Shuttle missions, and very little detail about the flight has been released. The crew were thought to have deployed an electronic intelligence satellite called Mentor for the National Security

Agency, using the IUS upper stage. It is also thought that Thornton participated in a programme of medical experiments during the mission, and that her previous experience as a civilian working in military facilities involved in nuclear physics was an indication that the mission involved work on the Strategic Defense Initiative ('Star Wars').

Kathy Thornton (*b*.1952) earned a BS in physics in 1974, an MS in 1977, and a PhD in 1979. As a graduate student she took part in various nuclear research programmes. From 1979 she studied nuclear physics in West Germany, under a NATO post-doctoral fellowship, and in 1980 returned to the United States to work for the Army Foreign Science and Technology Center in Virginia, She was selected for the astronaut programme in 1984. Her various Astronaut Office technical assignments included serving on the escape test crew during the return-to-flight period, working in SAIL, and a tour as CapCom.

STS-32: Bonnie Dunbar and Marsha Ivins

On this mission (her second) Bonnie Dunbar was MS1, and had primary responsibility for RMS operations, including grappling and manoeuvring the LDEF while it was photodocumented and then berthing it in the payload bay for return to Earth. She also had crew responsibility for the American Flight Echocardiograph (AFE), the Characteristics of Neurospora Circadian Rhythms (CNCR) experiment, the primary Protein Crystal Growth (PCG) experiment, and the Fluids Experiment Apparatus (FEA). On her first flight, Marsha Ivins was primarily responsible for LDEF retrieval support, photographing the condition of the facility prior to berthing in the payload bay, and was jointly responible for the CNCR, both PCG experiment units, FEA, AFE, and the IMAX camera system. Ivins also served as the ascent and entry Flight Engineer seated on the flight deck, while Dunbar flew the ascent on the flight deck and moved to the mid-deck for the descent. Although no EVA was planned, Dunbar had trained for contingency EVA procedures.[16]

Marsha Ivins (*b*.1951) Following her studies at the University of Colorado, where she earned a BS degree in aerospace engineering in 1973, Ivins accepted a position at NASA JSC in 1974, working initially as an engineer on the Shuttle orbiter cockpit displays, including work on the heads-up display for the orbiter during 1978. She subsequently worked as a flight and test engineer onboard the specially modified Gulfstream I aircraft (the Shuttle Training Aircraft (STA)), designed to train astronauts in the approach and landing techniques required to bring home the Shuttle orbiter. After selection to the astronaut programme in 1984, Ivins completed the Ascan training and evaluation programme and took Astronaut Office technical assignments in checking out the orbiters at the Cape, in SAIL, and as CapCom in Mission Control. She has also worked on Shuttle safety and reliability issues and avionics upgrades in the cockpit of the orbiter, and represented the Astronaut Office in all areas of crew photographic equipment, procedures and techniques and for orbiter flight crew equipment.

STS-31: Kathy Sullivan

Kathy Sullivan – flying as MS3 on her second mission – had several primary and back-up assignments in support of the mission and the deployment of the Hubble Space Telescope. As lead on secondary payloads, she was primary crew-member responsible for Medical DSOs and the medical kit, crew equipment and habitation issues, the Flight Data File, Earth observations, and IPMP. She had back-up responsibilities on CCTV and photographic equipment carried on the flight, the student experiments, and the Protein Crystal Growth and Radiation Monitoring experiments. She was also back-up for the HST systems and as contingency EV2 crew-member, remaining in the airlock during deployment in readiness for a contingency EVA. During the mission, difficulties with deployment almost resulted in Sullivan and EV1 Bruce McCandless exiting the airlock to conduct manual deployment of the high-gain antenna masts, the solar array support poles, and the twin solar arrays; but controllers successfully worked around the problems by removing the need for a contingency EVA – to the pleasure and disappointment of the EVA crew.[17]

STS-37: Linda Godwin

This mission, in April 1991, featured the deployment of the Compton Gamma Ray Observatory – one of NASA's Great Observatories. At 17.5 tons it was the most massive object carried by the Shuttle to that date. It was deployed by Godwin, using the RMS, but an antenna failed to unfurl and so she had to position the observatory to allow her EVA colleagues (Jerry Ross and Jay Apt) to perform a contingency EVA to unfurl it manually. The following day she was again primary RMS operator while the two EVA astronauts evaluated space station construction techniques. In addition to her role as primary RMS operator, Godwin supported and participated in the operation of secondary payloads and experiments.

Linda Godwin (*b.*1952) graduated from high school in 1970 and then pursued an academic career, earning a BS degree in mathematics and physics from Southeast Missouri State University in 1974, followed by an MS in physics in 1976 and a PhD in physics in 1980, both from the University of Missouri in Columbia. At the same time, she taught physics, conducted research in low-temperature solid-state physics, and published several papers in this field. In 1980 she joined NASA at JSC in Houston, where she was a flight controller and payload operations officer, working on a variety of Shuttle missions. Following her selection to the astronaut programme in 1985 and completion of the Ascan training and evaluation programme, Godwin received Astronaut Office technical assignments in SAIL, and supported the development of the IUS upper stage and the development of several Spacelab missions.

STS-40: Rhea Seddon, Tammy Jernigan and Millie Hughes-Fulford

This flight was the first Spacelab laboratory mission since October 1985 and the first totally dedicated to life sciences research. STS-40 – with a crew which included four doctors – carried the Spacelab Life Science 1 payload to gather the most detailed and interrelated physiological measurements during a spaceflight since the Skylab missions of 1973 and 1974. Eighteen primary experiments were flown on SLS-1 in the Spacelab module and mid-deck area of the Shuttle, twelve other experiments were flown under the Get Away Special programme, and seven were in the NASA Orbiter Experiment programme. This was also the first spaceflight that included three women in the same crew.

Jernigan flew as MS2/FE, and held primary crew responsibilities for SPOC, SMIDEX, the Flight Data File, MODE and all photographic and TV equipment and objectives. She was also back-up for DPS, MPS, Orbiter Manoeuvring System/ Reaction Control System, APUs and hydraulics, Guidance Navigation and Control, EPS, Environmental Control and Life Support Systems, communications and instrumentation, caution and warning, payload bay doors and radiators, one DSO, and Spacelab Systems (computers, electrical and environment). In addition, she was a back-up on SLS-1 payload issues and was also designated contingency EV2 astronaut. During the mission a potential problem with the payload bay door seals indicated that a contingency EVA might be required to repair them, but tests revealed that the doors would close securely. Had EVA been required, Jernigan would have accompanied EV1 Jim Bagian outside to undertake the repair. One of her specialist areas on the mission was the payload bay doors and associated systems, including closing the payload bay doors manually – one of the contingency EVA scenarios for which she trained.

As MS3, Seddon was primarily responsible for the SLS-1 payload, and was back-up on SMIDEX and photographic and TV issues. She was also assigned crew responsibilities in Spacelab systems (primary for computers and environmental, and back-up for electrical). For the launch she flew on the mid-deck, but exchanged with Bagian for the entry phase of the mission. Hughes-Fulford's primary role was in working on the SLS science programme, and as such she assumed back-up crew responsibility for the SLS-1 payload. She was also back-up responsible for communications and instrumentation.[18]

Tammy Jernigan (*b*.1959) graduated from high school in 1977 and attended Stanford University, where she earned a BS in physics in 1981 and an MS in engineering in 1985. In the same year she earned a second MS degree in astronomy from the University of California at Berkeley, and in 1988, after joining the astronaut programme in 1985 and completing the Ascan programme, she received her PhD in space physics and astronomy from Rice University in Houston. Her initial Astronaut Office technical assignments included a tour in SAIL, working on the development of secondary payloads to fly on the Shuttle, and a tour as CapCom in Mission Control. In April 1989 she was assigned to STS-40/SLS-1, and flew as a replacement for John Fabian, who had resigned from NASA in 1985 and was not replaced until after the Return-to-Flight mission following the *Challenger* tragedy. She became the first female professional astronomer to fly in space.

STS-43: Shannon Lucid

This mission featured the fifth TDRS deployment and experiments related to the development of the Space Station programme. Four payload-bay experiments, eight mid-deck payloads, thirteen DTOs and nine DSOs were flown, and several of these were intended exclusively for investigations into extended-duration spaceflight. Lucid set two new records for a female astronaut: the most spaceflights (her third), and the greatest number of logged hours in orbit (more than five hundred). As MS1, Lucid rode in Seat 3 on the flight deck of *Atlantis*, and for entry occupied Seat 5 on the mid-deck. Her crew responsibilities for this mission were in ECLSS, FES, and the SSCE, BIMDA, and PCG secondary experiments, as well as the payload bay doors, radiator cycling (opening and closing), and communications and instrumentation. She was also responsibile for the DTOs and DSOs carried on the flight, served as crew medic, and was responsible for all crew equipment carried on the mission, the flight rules, and as primary for the IUS and the TDRS satellite.

STS-42: Roberta Bondar

The first mission of 1992 was STS-42, carrying the Spacelab International Microgravity Laboratory. The sole female crew-member on the flight was Canadian physician Roberta Bondar, on her first and only spaceflight. In the division of crew responsibilities,[19] she is listed with none, other than in working the experiment programme throughout the mission. IML-1 operations were focused around a two-shift system for continuous collection of data from the science payload of materials and life science investigations. Bondar was the PS assigned to the twelve-hour Blue Shift, and as a PS she rode into and out of orbit on the mid-deck of the Shuttle. The science payload was assembled by more than sixteen countries and 220 scientists, resulting in more than fifty investigations supplied through six international space science research organisations – NASA, the European Space Agency (ESA), the Canadian Space Agency (CSA), the French space agency (CNES), the German space agency (DARA), and the Japanese space agency (NASDA). Bondar had previously

Roberta Bondar inside the IML laboratory during mission STS-42 in January 1992.

served as PI of taste experiments flown on the Shuttle, and was pioneering the investigation of cerebral blood flow during weightlessness. Onboard STS-42 she focused her attention on the six packages of experiments provided by the CSA, and provided support for the suite of international experiments and research. Following the mission she resigned from the Canadian astronaut programme in August 1992 to resume her full-time medical career and lecture on her experiences as an astronaut.

STS-45: Kathy Sullivan

On 29 September 1989, Sullivan was named as MS1 to the first of a series of Atlas (formerly EOM) Earth-atmosphere observation missions, and became the designated science Payload Commander for that mission on 25 January 1990 (the first woman astronaut to be so assigned). For the launch she was seated on the mid-deck of the Shuttle, and for entry took Seat 3 on the flight deck. While on orbit, Sullivan worked on the Blue Shift, operating the Atlas payload from the afterdeck of the Shuttle, as there was no laboratory module carried on the mission. In addition to the Atlas 1 payload, her primary crew responsibilities were in crew equipment, Earth observations, EVA and EMU (also serving as contingency EVA astronaut EV1), TV and STL. She also held responsibilities for two DTOs and four DSOs.[20] In October 1992 she resigned from the astronaut programme to accept a position as Chief Scientist with the National Oceanographic and Atmosphere Administration (NOAA) in Washington. In 1996 she became a Director and Chief Executive Officer (CEO) of the Ohio Center for Space Research, at the Center of Science and Industry.

STS-49: Kathy Thornton

On 19 December 1990, Kathy Thornton was named as MS4 on the Intelsat retrieval/reboost mission. This was also the maiden flight of *Endeavour* – the vehicle which replaced *Challenger*. Thornton's crew assignments[21] included Payload Deployment and Retrieval System issues, in-flight maintenance, medical issues, Assembly of Station by EVA Methods (ASEM), EV3 (for the ASEM assembly experiment), Crew Self Rescue issues and CPCG Block II. In addition she had responsibilities for eight DSOs during the mission. For both launch and entry she rode on Seat 7 on the mid-deck. During the mission, as well as supporting the attempts to retrieve and redeploy Intelsat, she completed a record-breaking EVA of 7 hrs 45 min to test space construction techniques aimed at the development of procedures and hardware for the planned Freedom Space Station. She became the third woman to perform an EVA, and logged more time than Savitskaya and Sullivan together.

STS-50: Bonnie Dunbar and Ellen Baker

1992 was a busy year for science and international cooperation on Shuttle missions, with *Columbia* flying the first of a series of extended-duration missions of 10–18 days, using additional supplies of cryogenics and a series of modifications to the vehicle. STS-50 carried the US Materials Laboratory (USML-1), a programme of thirty-one experiments that featured research in fluid dynamics, crystal growth, combustion science, biological science, and technological demonstrations. Dunbar was named to the flight as MS1/Payload Commander on 3 September 1990, and worked on the Red Shift during the mission. During ascent she rode on the flight deck, and during entry occupied Seat 5 on the mid-deck. Baker was assigned to the mission as a replacement for a reassigned astronaut (Ken Bowersox) as MS2/FE, and was assigned to the Blue Shift.

Dunbar's responsibilities on the mission included in-flight maintenance of Spacelab and primary point of contact for the USML payload.[22] In addition, she was responsible for one DTO and six DSOs. Baker's crew responsibilities – in addition to being ascent and entry FE and contingency EVA crew-member EV1 – were as flight medic (due to her experience as a physician), in-flight maintenance for the Shuttle orbiter, Earth observations, photographic and TV equipment procedures and objectives, and SAREX (Amateur Radio), as well as participation in five DTOs and eight DSOs. The highly successful USML mission qualified the Shuttle EDO kit for longer science missions in preparation for Space Station research that was under development. It also established a new record of 13 days 19 hrs for a single Shuttle flight – just surpassing the Gemini 7 record set in December 1965. STS-50 became the fourth longest US mission, after the three Skylab missions of 1973 and 1974.

STS-46: Marsha Ivins

This was the mission manifested to deploy the ESA EURECA-1 free flyer, which would be retrieved on a later mission. Also performed were the first experiments with an Italian Tethered Satellite, which unfortunately failed to deploy past 256 metres instead of its planned twelve miles, due to faulty hardware. Useful data were retrieved from the abbreviated experiment operation, and it was planned for a re-flight under TSS-2 (STS-75 in January 1996). Ivins was assigned to the mission as MS2/FE on 23 August 1991, replacing a reassigned astronaut (Andy Allen). As a member of the Red Shift, her crew responsibilities on this mission included primary and support work as contingency IV crew-member, PDRS, rendezvous and proximity operations, three DTOs and one DSO, and three medical DSOs.[23] She was primary point of contact for the secondary payload and ICBC, in-flight maintenance and crew equipment, photographic/TV equipment, Earth observations and orbiter systems for the Red Shift (APU/HYD, Comm/Inst/Tags/Audio, caution and warning, EPS, mechanical and PDRS, and eight procedures lists). Additional duties on this flight included work on the post-flight crew movie.

STS-47: Jan Davis and Mae Jemison

STS-47 flew the Spacelab J (Japanese) long module configuration of forty-three experiments in materials sciences and life sciences, as well as nine Getaway Special canisters and three mid-deck experiments, including one from Israel. Thirty-four experiments were provided by Japanese investigators, seven were provided by the US, and there were two joint experiments. The crew of seven was divided into two twelve-hour shifts, with Jan Davis and Mae Jemison both working on the Blue Shift. Both astronauts had been named to the flight on 28 September 1989, with Davis as MS2/FE and Jemison as MS4 and also designated Science Mission Specialist – a role that was being evaluated by NASA to provide Mission Specialists instead of Payload Specialists, by which the astronaut would focus more on the scientific operations

The crew of spacelab J, flown on STS-47, included Mae Jemison – the first female African-American astronaut – and Jan Davis and her husband Mark Lee – the first married couple in space.

than flight operations (but not in the same way that a Payload Commander would assume a leadership role in organising the whole payload into the mission). Jemison sat on the mid-deck for both launch and entry. Jan Davis was also assigned as contingency EVA crew-member EV2, and sat on the mid-deck for launch and the flight deck for entry. Both astronauts worked on the LBNP, FTS, MRI, AFTE, and BCR experiments, and Jemison also was assigned to the FEE experiment, serving as a test subject for research into human responses to adjusting to spaceflight.[24] In addition to the mission operations, the media also emphasised that Jan Davis was flying with her husband Mark Lee on the same mission, that they were the first husband and wife to fly in space at the same time, and that Mae Jemison was the first African-American woman to orbit the Earth. STS-47 was also the fiftieth flight of the Space Shuttle system.

Jan Davis (*b.*1953) grew up in Huntsville, Alabama, and remembers the Saturn rocket engine tests, which shook the windows in the family home. With almost the whole neighbourhood involved in the space programme, she naturally developed a growing interest in the programme, taking an interest in science and mathematics; but it was not until the first women astronauts were selected in 1978 that she considered applying herself. She had earned a BS degree in applied biology from the Georgia Institute of Technology in 1975, and then a second BS in mechanical engineering from Auburn University in 1977. Continuing her studies at the University of Alabama at Huntsville, she gained an MS in 1983 and a PhD in 1985, both in mechanical engineering. From 1977 to 1979 she worked for Texaco as a petroleum engineer located in Bellaire, Texas, and afterwards joined NASA's Marshall Space Flight Center in Huntsville, Alabama, as an aerospace engineer working on the Hubble Space Telescope and the Chandra X-ray Observatory, and, in 1987, the redesign of the Shuttle's solid rocket boosters. She was selected for the astronaut programme in 1987. Her Astronaut Office technical assignments prior to her first spaceflight included technical support for a variety of Shuttle payloads in the Mission Development Branch (including the Tethered Satellite System) and a turn of duty as CapCom at Mission Control.

Mae Jemison (*b*.1956) was born in Alabama and raised in Chicago. She graduated from high school in 1973, and earned a BS degree in chemical engineering and a BA in African and Afro-American studies from Stanford University in 1977. Pursuing a medical career, she gained her MD from Cornell University in 1981 and took up a medical career in the US and with the Peace Corps in Sierra Leone, Liberia and West Africa, after which she returned to her career in the US. After selection to the astronaut programme in 1987, she worked as a Cape Crusader at KSC and in SAIL. Following the mission she reportedly became frustrated with her experiences in the space programme and declined a second flight. In March 1993 she left the programme and set up her own consultancy firm, lecturing on environmental and health issues. She had wanted to be an astronaut since watching the Mercury and Gemini missions on TV in the 1960s, and decided to try for the astronaut programme when she had acquired the appropriate educational and practical experience. A determined, strong-willed and independent woman, she was sometimes at odds with the 'team player' ethos of NASA, but she has tried to use her experiences as an astronaut to inspire younger generations. A fan of the 1960s TV series *Star Trek* as a child, she looked to communications officer Lieut Uhura as a role model, seeing her as the first female (and African-American) to have a 'technical' acting role on American television. As a tribute to Nichelle Nichols (Uhura), Jemison opened her shifts onboard the Space Shuttle *Endeavour* with the words: 'Hailing frequencies open.' In 1993 she appeared in an episode of *Star Trek: The Next Generation* ('Second Chances') as Lieut Palmer, and is therefore the only member of the *Enterprise* 'crew' to have actually flown in space. Jemison developed keen interests in dance, art and photography, and can speak several languages, including Russian, Japanese and Swahili.

STS-52: Tammy Jernigan

Jernigan flew her second mission as MS3, having been assigned to the flight on 23 August 1991. STS-52 deployed the LAGEOS-II (Laser Geodynamics satellite) and carried the US Microgravity Payload of three experiments in the payload bay and ten separate experiments under the CANEX (Canadian Experiments) CANEX-2 payload. The mission demonstrated the Shuttle's flexibility in being a satellite launcher, science platform and technology test-bed during one mission. Jernigan's crew responsibilities on the mission included primary roles for LAGEOS and IRIS and back-up roles for USMP-1, CANEX-2 and RMS operations. In addition, she was assigned a primary role on the CPCG and HPP experiments, back-up for the MAC (computer) in space demonstration and as a back-up crew medic, worked on four DSOs, and was contingency EV2 astronaut for the mission. She flew the ascent on the mid-deck and the entry on the flight deck.[25]

STS-54: Susan Helms

This flight carried a suite of mid-deck and payload bay experiments, and deployed the sixth Tracking and Data Relay Satellite (TDRS-F). In addition, two of the crew performed the first of a series of EVAs to develop techniques and procedures for further space station (Freedom) EVA operations. Helms – named to the crew on 23 August 1991 – was flying her first mission as MS3. She flew the ascent in Seat 5 on the mid-deck, and the entry in Seat 3 on the flight deck. Her crew responsibilities for this mission included back-up for the TDRS payload, primary for the DXS, and contingency EVA IV astronaut. She also had support duties as crew medic and for photographic and TV issues. She was assigned to work on four DTOs, and also worked on the secondary payloads, CGBA, PARE and SSCE.[26]

Susan Helms (*b*.1958) graduated from the USAF Academy in 1980 with a BS degree in aeronautical engineering, and in 1985 was awarded an MS in aeronautics and astronautics at Stanford University. She was commissioned into the USAF in 1980, and became a weapons engineer with the Armaments Laboratory at Eglin AFB, Florida, working on the F-16 and F-125 weapons systems. After gaining her MS she became an Assistant Professor of Aeronautical Engineering at the USAF academy, and in 1987 she attended the Flight Test Engineer course at the USAF Test Pilot School at Edwards AFB in California. At the time of her selection to the astronaut programme she was serving as a manager for CF-18 flight control system simulations at Canada's Engineering Test Establishment in Alberta. She was selected in 1990, and completed Ascan training the following year. Her Astronaut Office technical assignments included detachment to the mission development branch, working on the RMS and robotics issues.

During mission STS-54, Susan Helms found time to play an electronic keyboard, despite a busy flight programme.

STS-56: Ellen Ochoa

This mission flew the second ATLAS facility and the Shuttle Backscatter Ultraviolet payload, to record the relationship between the Sun's energy output and the chemical constitution of Earth's middle atmosphere, and how they affect the Earth's ozone layer. In addition, the Spartan-201 free-flyer was deployed and retrieved by RMS to study the velocity and acceleration of the solar wind and for observing aspects of the Sun's corona. Two educational experiments and several smaller mid-deck experiments completed the science investigations. Ochoa was named as MS3 to STS-56 on 16 March 1992, and was part of the Blue Shift, operating the payload in twelve-hour shifts throughout the mission. Although assigned to the 'orbiter' crew, she shared the operation of the ATLAS payload and served as primary RMS operator, deploying and retrieving the Spartan free-flyer. During both ascent and entry, she occupied a seat on the mid-deck.

Ellen Ochoa (*b*.1958) became the first Hispanic woman in space by flying STS-56. After graduating from high school in 1975 she was awarded a BS degree in physics from San Diego University in 1980, and an MS in 1980 and a PhD in electrical engineering in 1985, from Stanford University. For three years after receiving her doctorate, she worked as an optical recognition researcher at the Sandia National Laboratories in Livermore, California, after which she was a staff member at NASA's Ames Research Center, Moffett Field, California, working on optical recognition systems for space automation. She subsequently led a team of scientists and engineers developing improved aerospace computer systems, as Chief of the Intelligence Systems Technology Branch. She was selected to NASA in 1990, and completed Ascan training the following year before being assigned to SAIL.

STS-57: Nancy Sherlock (formerly Currie) and Janice Voss

This was the first flight of the SpaceHab mid-deck augmentation module, offering additional locker and logistics volume on Shuttle missions. The flight featured a rendezvous with the EURECA free-flyer that was deployed on STS-46 the year before. The first SpaceHab mission included thirteen commercial development experiments, six NASA experiments, three smaller payloads in the payload bay, and two experiments on the mid-deck. An EVA was also included in the flight plan as part of the space station EVA preparations. Janice Voss was named to SpaceHab 1 as MS3 on 21 February 1992, and Sherlock was named as MS2 on 16 March. Both astronauts were flying their first mission, with Sherlock flying on the flight deck for both launch and entry, and Voss occupying Seat 6 on the mid-deck. Sherlock had joint responsibility for several Shuttle systems (DPS, MPS, OMS/RCS, APU/Hyd, EPS, ECLSS Comm/Inst, PLBD/Rads, SPOC, RMS back-up operator, GAS Bridge Assembly, and supporting SpaceHab experiments and secondary investigations), and was also responsible for designing the mission emblem and producing the crew photograph – some of the smaller assignments that astronauts must address during the course of their mission preparations. Janice Voss worked mainly on SpaceHab issues, but also had responsibility for medical issues, RMS, Shoot, SpaceHab systems and experiments, educational experiments, and the development of the crew patch and crew photograph. Sherlock's primary payload roles included EURECA systems, RMS support during EVA, and the GBE. She was also responsible for the SpaceHab experiments BPL, HFA, Trans, ASC-2. Her back-up task and payload roles included EURECA RMS operations and the SpaceHab experiments Aspecs, HFA, EPROC and EFE. Voss's primary tasks lay in Shoot, with back-up tasks/payload responsibilities for EVA RMS, SpaceHab systems and SAREX. Her SpaceHab experiment responsibilities were extensive. She was prime on HFA, EPROC, SCG, TES-COS, APCF, CPDS, 3DMA, ECLISE-HAB, GPPM, LEMZ-1, ORSEP, SAMS, ZCG, and her back-up experiment roles included CR/IM-VDA, PSE, and CGBA.[27]

Nancy Sherlock (*b*.1958) was a qualified US Army helicopter pilot at the time that she was selected for astronaut training in 1990, and had been a flight simulation engineer on the STA in a secondment to NASA JSC since 1987. She was awarded a BA in biological sciences at Ohio State University in 1980, and was a research assistant in neuropathology at Ohio State University College of Medicine prior to entering the US Army in July 1981. She earned an MS in safety engineering from the University of Southern California in 1985, and progressed as a senior Army aviator and helicopter instructor, logging more than 2,700 hours flying time in helicopters and multi-engine fixed-wing aircraft. After joining the astronaut programme and completing Ascan training, her early Astronaut Office technical assignments were in crew equipment issues for the Mission Development Branch.

Janice E. Voss (*b*.1957) tried four times to enter the astronaut programme before finally succeeding in 1990. She had first worked at JSC in Houston as a co-op, working on computer simulations from 1973 to 1975. She was awarded a BS in engineering science at Purdue University in 1975, and an MS in aeronautics and astronautics at MIT in 1977. She then returned to NASA JSC, working as a crew trainer in Shuttle entry guidance and navigation issues. She received her PhD in aeronautics and astronautics from MIT in 1987, and joined the Orbital Sciences Corporation to work on their Transfer Orbit Stage. Her early Astronaut Office technical assignments focused on Spacelab and SpaceHab issues in the Mission Development Branch.

STS-58: Rhea Seddon and Shannon Lucid

The second Spacelab Life Sciences mission included two experienced female crewmembers. Originally, SLS-1 and SLS-2 were to fly as Spacelab 4, but they were divided into two separate missions in 1984, and expanded with new investigations. On 23 October 1991, Rhea Seddon was named MS1/Payload Commander for the mission and was joined by Shannon Lucid, who was named as MS4 on 6 December 1991. Seddon occupied Seat 3 on the flight deck during ascent, and Seat 5 on the mid-deck for entry, while Lucid remained in Seat 6 on the mid-deck for both phases of the mission. The SLS-2 flight featured fourteen life science experiments, a programme of six DTOs and thirteen DSOs, and the SAREX amateur radio experiment. Seddon had primary responsibilities for the SLS-2 payload and Spacelab activation and experiments. In addition, she served as back-up crew medic and supported the deactivation of the Spacelab module prior to entry. Lucid's roles were as primary for human factors DTOs/DSOs, and as contingency EVA crew-member EV1. She was also responsible for back-up roles in the SLS-2 payload and experiment programme.[28]

On 3 May 1993, Seddon broke four metatarsal bones in her left foot during egress training exercises at JSC. She was practising 1-g emergency egress procedures from the mock-up of the orbiter in the Building 9 training facility, when her left foot became pinned under her, causing the break as she used the inflatable slide chute. At the time, NASA announced that most of her crew training at that point consisted of refresher courses, and that because of this, plus her experience in past assignments, her assignment to STS-58 would not be affected. She returned to full training in June.[29] However, she was unable to carry out the fully suited egress testing, but she had done this for SLS-1, and was able to undertake the final training sessions at the Cape. Both she and NASA were satisfied that she could handle emergency egress.[30] Following SLS-2, assignments she supported preparations for the science payloads for the first residency by an American (Norm Thagard) on Mir in 1995. However, it was while working on this assignment that she declined a long mission on Mir and the ISS. She noted the different approach to science on Russian missions, and surmised that it would be about ten years before serious science was performed on the ISS. She firmly believes that scientific results have not been improved since Spacelab, especially regarding the life sciences experiments flown on SLS-2 or Neurolab in 1998. She was unwilling to go to Russia for two years to train for a mission, and instead opted for a position in an academic institute. In 1996 she took a leave of absence to work at Vanderbilt University in Nashville, Tennessee, to coordinate medical experiments for the STS-90 Neurolab Spacelab mission flown in April 1998. She left NASA in November 1997, and took a position as Chief Medical Officer of the Vanderbilt Medical Group.

STS-61: Kathy Thornton

This was the first Hubble Space Telescope service mission after the deployment of the telescope from STS-31 in April 1990. In December 1993, STS-61 flew to upgrade the instruments, install a corrective optics package, and exchange the solar arrays and other components to improve and increase the operational lifetime of the facility. On 25 August 1992, Kathy Thornton was assigned as MS1/EV3 for the mission. Her crew responsibilities included supporting the first, third and fifth EVAs as an IV crew-member, and completing the second and fourth EVAs herself with Tom Akers. She was jointly responsible for HST operations, payload communications systems, and photographic and TV issues, focusing on CCTV and the camcorder carried on board, as well as the Electronic Still Camera under evaluation. She served as back-up medical officer for the crew, and was point of contact for launch stowage, crew equipment and DTO 700-8 (Global Position System flight test) and DTO 312 (ET separation photography), and back-up for in-flight maintenance. Her first EVA, on 5 December, lasted 6 hrs 36 min, during which time which she and Akers installed the new solar arrays on the telescope. Two days later, during an EVA of 6 hrs 50 min, the pair of astronauts installed the new COSTAR corrective optics

Kathy Thornton carries out EVA operations on the first Hubble Space Telescope service mission in December 1993.

package and a new computer, and separated an old solar panel for replacement. In completing these EVAs, Thornton set a new female endurance record of 21 hrs 11 min.[31]

STS-60: Jan Davis

The second SpaceHab mission also featured the first flight by a Russian cosmonaut (Sergei Krikalev) on a US spacecraft, as part of the Shuttle–Mir programme. Jan Davis was assigned to the mission as MS1 on 28 October 1992. She was primarily responsible for the RMS, four GAS canisters and three SpaceHab experiments (CPCG, CGBA and PSB), and was back-up for SpaceHab systems, the ORSEP experiment, SEF experiments, one DSO, one DTO, photography and TV, and as crew medic. In addition she assumed the role of contingency EV2 crew-member. She was also responsible for several orbiter systems and tasks, including Comm/Instr, payload-bay doors/Rads, EVA/EMU, crew equipment and habitation. During ascent she occupied Seat 3 on the flight deck, and during entry she was on the mid-deck.

STS-62: Marsha Ivins

The second US Microgravity payload was carried on STS-62, continuing a programme of long-duration medical experiments and space technology demonstrations. Marsha Ivins was named to the crew as MS3 on 5 March 1993, and had primary responsibility for the Dexterous End Effector (DEE) – a new and improved end effector on the RMS – and the CGBA mid-deck payload. She was also assigned to three DTOs and six DSOs, and assumed back-up crew responsibilities for OAST-2, EIST/SKIRT, SAMPOIE and TES, and the PSE and CPCG mid-deck payloads. She was also responsible for all photographic and TV equipment procedures and objectives on the mission, and for orbiter systems (Comm/Inst, crew equipment, back-up medical issues, PDRS, PLBD (mechanical issues), and PGSC/PADM). During launch, Ivins rode in Seat 5 on the mid-deck, and for entry, in Seat 3 on the flight deck.[32]

STS-59: Linda Godwin

This was the first of two Space Radar Laboratory missions flown in 1994 to map the changing global environment by using a package of imaging radar hardware, located in the payload bay of the Shuttle, to distinguish human-induced environmental changes against natural changes. Godwin had been assigned as PC for the mission on 23 August 1991, and was assigned to the Red Shift. Her crew responsibilities included being primary for SRL-1 (lead for SIR-C and X-SAR issues for the Red Team), PGSC SW/HW, and for one DTO and six DSOs. She was also assigned back-up responsibilities for SAREX, Earth observations connected to SRL objectives, and as contingency EVA astronaut EV1. As with all Shuttle crews, each astronaut received supplementary assignments for peripheral areas of mission preparation, completion and conclusion. For Godwin on STS-59, this included organising the 'Touchdown Party' after the completion of the mission. During the mission she occupied Seat 5 on the mid-deck for both launch and entry.[33] While in training for this mission, Godwin was also named (19 March 1993) as Deputy Chief of the Astronaut Office – the first female astronaut to assume this role since the Office was established in the early 1960s. She remained in this role through January 2000.

STS-65: Chiaki Mukai

The first Japanese woman to fly in space was named to the STS-65 crew on 19 October 1992. As prime PS she would participate in the second International Microgravity Laboratory mission (IML-2), which included more than eighty investigations developed by more than two hundred scientists and an international team from NASA, ESA, CNES, DARA, CSA and NASDA, in a programme of materials and life research studies. As a member of the Red Shift, Mukai had a primary role in operating the IML science payload, including primary assignments for EDOMP, FFEU, LIF, RRMD, and TEI/CCK, and with back-up roles for almost all the other major investigations on the laboratory.[34]

Chiaki Mukai (*b.*1952) After attending Keio University she gained her MD in 1977 and completed a two-year residency in general surgery at the university's hospital.

She then spent six years as an instructor in the university's department of cardiovascular surgery until being selected to the Japanese astronaut team in 1985. After serving as back-up PS for Spacelab J (STS-47), she continued her medical career as a visiting scientist in Houston, Texas, and in Japan. She was the only the second foreign female astronaut to fly on the Space Shuttle. For both launch and entry phases of the flight, Mukai occupied Seat 7 on the mid-deck.

STS-64: Susan Helms

This mission featured a payload of atmospheric research, robotic processing of semiconductor materials, the deployment and retrieval of a free-flying astronomical satellite, and the testing of a new untethered manoeuvring backpack for use as an emergency rescue system (SAFER) during an EVA. Helms was named as MS2/FE for the mission on 17 November 1993. and in addition to her role as ascent/entry Flight Engineer she had responsibilities for rendezvous, PDRS, the Spartan-201 free-flyer, ROMPS, SPIFEX and BRIC. and participated in four DTOs and six DSOs. She was also the primary RMS operator.[35] Since her return from her first flight (STS-54), Helms had completed a tour as CapCom from June 1993 to February 1994.

STS-66: Ellen Ochoa

The third ATLAS Earth environment observation mission was also the last of the series. Ellen Ochoa flew as MS1/Payload Commander as a member of the Red Shift. Having flown on the previous ATLAS mission, her experience was useful in organising the new scientific and flight objectives. Mike Foale had flown on Atlas 1 and as PC for Atlas 2, and as he did not want to be come known as 'Mr Atlas' by flying the third mission he moved to a new assignment and nominated Ochoa as Payload Commander for the new mission.[36] Ochoa occupied Seat 5 on the mid-deck during ascent and entry. Her primary payload roles included Atlas 3, Christa-SPAS and SSBUV, and she was back-up for ESCAPOE, HPP and STL. She was also assigned to several orbiter/systems assignments, including primary on Comm/Inst (for orbit), KU-Band antenna, payload-bay doors and radiators, and RMS operation and systems, and was back-up for crew equipment and medical issues, and PGSC/PADM. She also worked on two DTOs.[37]

STS-63: Eileen Collins and Janice Voss

The crew for the STS-63 'Near-Mir' mission was named on 8 September 1993. This mission was manifested to fly the third SpaceHab mission and the second Russian cosmonaut on the Shuttle (Vladimir Titov), and to complete the first rendezvous with the Russian Mir space station. Eileen Collins was the first female astronaut to be assigned as a Pilot to a flight crew, while Voss was assigned as MS1. As Pilot, Collins was the first woman to occupy the right-hand Seat 2 on the flight deck, while Voss occupied Seat 5 on the mid-deck. The seating was the same for ascent and entry.

Collins' crew responsibilities included back-up for Mir rendezvous operations, primary for the MSX secondary payload. and back-up for GLO-2, ORERACS-2,

In 1995, Eileen Collins became the first female Pilot of a Shuttle.

AMOS and CONCAP-II secondary payloads. For the SpaceHab 3 experiments she was responsible for 3-DMA, and was back-up for BPL-3, F-GBA, and WINDEX-01. She was also assigned support work on six DSOs. As Pilot for the mission she had additional responsibilities for the orbiter vehicle, including MPS, OMS/RCS, APU/HYD, EPS and ECLSS, and the Flight Data File.[38] When named to the crew, she said: 'Being recognised as the first woman Shuttle Pilot gives me a chance to show young women that these kinds of jobs are available. It's good that public attention is being drawn to my assignment – not so much as it has taken so long, but because it may encourage other young women to seek a career in maths and science and the astronaut programme. People need to understand, Shuttle Pilots must first be military pilots with degrees in maths and science. It was only recently that the military even let women qualify as test-pilots. and not many women choose maths, science and engineering as a track to a career.'[39]

Voss's crew responsibilities included the payload in SpaceHab 3 and Spartan-204, as well as secondary experiments ICBC and CTOS. She was also assigned work on eight DSOs. For the SpaceHab experiment package she was assigned responsibilities for ASC-IV, BRIC-03, CHARLOTTE, CROMEX-06, CGBA-05, CPCG-VDA, GPPM-02, IMMUNE 02, PCG-STES-03 and PCF-LST-03. Her orbiter/systems responsibilities included PGSC/PADM, HP48, Lasersa, and primary on RMS operations.

Eileen Collins (*b.*1956) was always fascinated by aircraft, flight and the space programme, and grew up enthralled by the exploits of the Mercury, Gemini and

Apollo astronauts, often imagining herself as an astronaut in a 'far-off dream'. Realising the need for academic qualifications for any flying career, she focused on mathematics, and earned an AS degree in mathematics and science in 1976, a BA in mathematics and economics in 1978, an MS in operations research in 1986, and an MA in space systems in 1989. She joined the USAF in 1978. In the same year, NASA selected the first female astronauts, and this spurred her on towards her goal of being a test-pilot and, one day perhaps, a Shuttle Pilot. After pilot training she became a T-38 instructor pilot, then a C-141 pilot and instructor, followed by assignment as a T-41 instructor pilot and assistant professor of mathematics at the USAF Academy (1986–89). She logged combat missions during the invasion of Grenada in October 1983, and during 1989–90 she attended the USAF Test Pilot School at Edwards AFB, California, where she was studying at the time of her selection to the astronaut programme. She was selected in 1990, completed the Ascan programme the following year, and was assigned to the Mission Support Branch of the Astronaut Office, working on orbiter engineering systems support and with the Cape Crusaders in KSC launch/landing support. She also served as CapCom at Mission Control, Houston.

STS-67: Wendy Lawrence and Tammy Jernigan

This mission included the second flight of the Astro payload – a programme of experiments studying ultraviolet radiation from distant stars and galaxies – with the crew operating three telescopes in the payload bay from the aft flight-deck. Jernigan had been named Payload Commander on 3 August 1993, and flew as MS3 assigned to the Blue Shift, flying the ascent in Seat 5 on the mid-deck, and entry in Seat 3 on the flight deck. Since flying her previous mission (STS-52) she had served as Chief of the Astronaut Office Mission Development Branch. Her primary responsibilities on STS-67 were in payload-bay doors and radiators, medical issues, onboard duties in medical, stowage and crew equipment, cabin configuration, SPOC, the opening and closing of the payload-bay doors, TIPS and PADM, and the ET photographic DTO objectives. As PC she was primarily responsible for the Astro payload and science objectives, and also assumed pre-flight duties for medical DSOs, post-insertion planning, descent orbit planning, preparations, and flight rules. She was also responsible for four DSOs, and was trained as contingency EVA astronaut EV1.

Lawrence was named to the crew as MS2/FE on 10 January 1994, and this was her first flight assignment. Also assigned to the Blue Shift, she rode in the Seat 4 position for both launch and landing, assisting the CDR and PLT during the ascent and entry phases of the mission. She was responsible for several systems, including GNC, DPS, MPS/SSME, OMS/RCS, ECLSS, Comm/Inst, payload-bay doors and radiator (mechanical issues), WCS, and Galley. Her onboard duties included stowage and crew equipment, in-flight maintenance issues, DUMPS, photography and TV, Earth observations, the filming of the crew movie, the ET photographic DTO, and as contingency IVA crew-member IV1. She was also responsible for the GAS payloads, PGCG and CMIX. Her pre-flight duties included working on the crew mission emblem, the pre-flight crew photographs, and the crew shirts worn during the mission.[40]

Wendy Lawrence (*b*.1959) has an impressive family background. Her father is Vice Admiral William P. Lawrence, USN (Retired) – a recipient of the Congressional Medal of Honour during the Vietnam War, and a finalist for the 1959 Mercury astronaut selection. Following her father into the US Navy, Lawrence entered the US Academy at Annapolis in 1977 after graduating high school. She was awarded a BS in oceanographic engineering in 1981 and an MS in oceanographic engineering in 1988. After graduating from Annapolis she completed helicopter pilot training and became one of the first two female helicopter pilots to be assigned to carrier deployment with Helicopter Support Squadron 6. She has also served a two-year assignment as officer-in-charge of Detachment Alfa, Helicopter Anti-Submarine Squadron Light 30 (HSL-30), and at the time of selection to the astronaut programme was instructing physics classes at Annapolis. She was selected to NASA in 1992, and after Ascan training she was assigned Astronaut Office technical roles in the mission support branch, where she worked in SAIL and as an assistant training officer.

THE SHUTTLE–MIR YEARS

The series of American Shuttle flights to the Russian Mir space station between 1995 and 1998 offered the opportunity to develop rendezvous and docking experience (which NASA had not had since the end of the Apollo era in 1975) and techniques and procedures for the impending International Space Station. This cooperative programme with the Russians also provided the astronauts with long-duration mission experience, and enabled visiting crews to experience space-station operations and logistics transfers in preparation for the ISS. (In order to present a complete chronological sequence of female Shuttle flights, the Shuttle–Mir missions are included in this section.[41])

STS-71: Ellen Baker and Bonnie Dunbar

The main task for the first docking mission Shuttle–Mir Mission 1 (SMM-1) was to return the first resident NASA astronaut (Norman Thagard) from Mir and exchange the Mir 18 resident crew with the Mir 19 cosmonauts. The Shuttle crew, named on 3 June 1994, included Ellen Baker as MS1 and Bonnie Dunbar as MS3. Dunbar was Thagard's back-up for his Mir flight, and had completed Soyuz/Mir training at the Russian cosmonaut training centre prior to joining the STS-71 crew. On later flights, NASA resident crew-members would be exchanged via the Shuttle, but Dunbar would only be visiting Mir and would not take over from Thagard (who had launched to Mir onboard a Russian Soyuz). Riding to and from orbit on mid-deck Seat 5, Dunbar's only primary crew responsibilities were for Mir payloads and systems and for Spacelab payloads. Baker flew to and from orbit in flight-deck Seat 4, and her primary responsibilities on the flight were on the Spacelab payloads, the medical operations (including the examinations on the returning Mir resident crew), and IMAX camera operations and the electronic still camera. She was also contingency EVA astronaut EV2. Her back-up role was in SAREX experiment.[42]

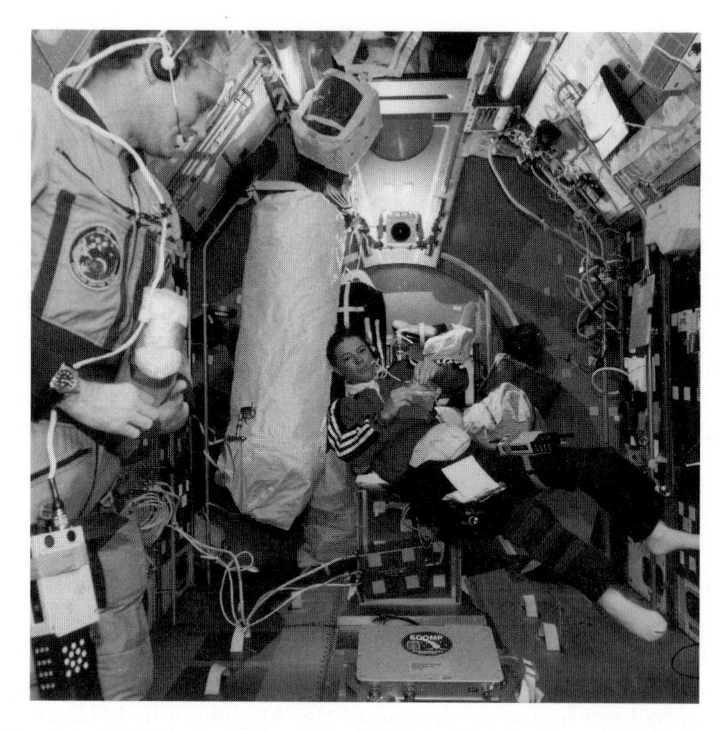

Bonnie Dunbar and Norman Thagard in the Spacelab science module during the STS-71 docking mission to Mir in July 1995.

STS-70: Nancy Sherlock and Mary Ellen Weber

This was a TDRS deployment mission that also carried out a number of secondary experiments. The TDRS-G was the seventh and final Shuttle-deployed satellite of the series. The crew was named to the flight on 25 August 1994, with Sherlock being named MS2/FE and Weber as MS3. Sherlock rode the ascent and entry in Seat 4 on the flight deck, while Weber flew the ascent in Seat 5 on the mid-deck and the entry in Seat 3 on the flight deck. Nancy Sherlock was also assigned as contingency IV crew-member, and held orbiter systems responsibilities for APU/HYD, EPS, payload-bay doors and radiators, in-flight maintenance, Flight Data File, HP48, Earth observations and lasers. She was also given responsibilities for the secondary payloads WINDEX, STL, RME, the NIH-R (Pare) Hercules camera, CPCG, BRIC, SAREX and the DSO 677 (microgravity) investigation, as well as participating in the human factors (904) and cardiovascular (624) DSOs. She also had additional duties in the preparation of the post-flight crew report. Weber trained as contingency EVA astronaut EV1 and had a back-up role with the TDRS-G payload. She held responsibilities for the orbiter systems Comm/Inst, photography and TV, medical issues, crew equipment and habitation, PGSC/PADM and work on the secondary payloads STL, MIS-B, CPCG, BDS, VFT, DTO 656 (PGSC) and all human factors and cardiovascular DSOs. In pre-flight preparations, Weber was also given the task of securing the distinctive and unique crew flight shirts used during the mission.[43]

Mary Ellen Weber (*b*.1962) earned a BS degree in chemical engineering from Purdue University in 1984 and a PhD in chemical engineering from the University of California at Berkeley in 1988. As an undergraduate she gained experience during internships at Ohio Edison, Delco Electronics and the 3M Corporation, and after receiving her doctorate she joined Texas Instruments. From 1988 to 1990 she worked on researching new techniques for microelectronic manufacturing and then transferred to TI's Sematech Division to work as a material engineer. Following selection to NASA in 1992 and the Ascan training programme she was given Astronaut Office technical assignments in the Cape Crusaders, a tour in SAIL, and work in the payload development branch.

STS-73: Cady Coleman and Kathy Thornton

This flight carried the second US Microgravity Laboratory payload in the Spacelab Long Module, conducting research in fluid mechanics, biotechnology and combustion sciences. Both Coleman and Thornton were named to the flight on 17 March 1994, with Coleman as MS1 and Thornton as MS3 and Payload Commander. The crew worked a two-shift system – Coleman on the Blue Shift and Thornton on the Red shift. Both women were also given primary responsibilities for the USML-2 payload and for the systems of the Spacelab module. Thornton was assigned to help activate the Spacelab Module, while Coleman's main task was to ensure correct deactivation at the end of her last shift. Coleman was also assigned to work on two DTOs and one DSO, and was trained as contingency EVA crew-member EV2, while Thornton was assigned as contingency IV crew-member IV1, one DTO and two DSOs, and medical issues on the crew. Both astronauts were assigned duties for in-flight maintenance issues.[44] Coleman rode the ascent in Seat 3 on the flight deck, and exchanged seats with Thornton in Seat 5 on the mid-deck for re-entry.

Cady Coleman (*b*.1960) earned a BS degree in chemistry from MIT in 1983 and a PhD in polymer science and engineering from the University of Massachusetts in 1991. She was commissioned in the USAF in 1983, began active duty in 1988, and was assigned to Wright-Patterson AFB as a research chemist. She carried out research on new polymer materials, and was a surface-analysis consultant for the LDEF facility. She was also a volunteer for the centrifuge programme at the Crew System Directorate at the Armstrong Aeromedical Laboratory at Wright-Patterson, setting several endurance and tolerance records, participating in physiological tests, and evaluating new equipment. She was selected to NASA in 1992, and her first technical assignments, after completing Ascan training, were in the Astronaut Office Mission Support Branch, working in SAIL.

STS-76: Linda Godwin and Shannon Lucid; STS-79: Shannon Lucid

This was the third mission to dock with the Russian Mir space station (the second – STS-74 – carried an all-male crew) and the first to deliver an NASA astronaut (Lucid) for a residency visit. Lucid was the first American female astronaut to complete a long-duration mission. Linda Godwin was named to the STS-76 crew as

MS3 on 14 April 1995. Her primary assignments on the mission included Kid Sat and SAREX, and serving as EVA crew-member EV1. On 27 March 1996 she completed an EVA of 6 hrs 2 min. during which she attached four Mir Environmental Effects Payloads (MEEP) to the station's docking module to record the environment around the station over the next eighteen months. Neither Godwin nor her colleague Richard Clifford crossed over from the US-supplied docking module to the Russian segment, as this still fell under the jurisdiction of the Russian cosmonauts for EVA purposes. It was therefore a Shuttle-based EVA; but it was the first American EVA at a space station since the end of the Skylab programme in early 1974. Godwin's back-up responsibilities included SpaceHab issues, rendezvous procedures, the orbiter docking systems, and the transfer of frozen samples.[45] Lucid had minimal assignments on the STS crew, and assumed the identification role of MS4 for ascent only. She became MS4 for the entry phase on STS-79 – the mission that brought her home and delivered the next NASA Mir resident (John Blaha). There were no other female crew-members on STS-79.

STS-78: Susan Helms

This mission carried the Life and Microgravity Spacelab (LMS) Long Module configuration and set a new Shuttle endurance record of 16 days 21 hrs, with the crew working a single-shift system. Susan Helms was assigned to the flight as Payload Commander and MS2/FE on 8 May 1995. She served as contingency EVA astronaut EV1, and had primary roles in the forty-one different LMS experiments in life sciences, microgravity and acceleration research fields, as well as PGF, other microgravity investigations, and PAWS. Her back-up roles were in Spacelab module systems, AGHF, BDPA, COIS and SAREX.[46]

STS-80: Tammy Jernigan

This flight of 17 days 15 hrs 54 min broke the previous Shuttle endurance record. It will probably remain the longest Shuttle flight in history, with future missions of the Shuttle planned for longer than twelve to fourteen days during the completion of ISS construction. The payload included the ORFEUS-SPAS astronomical observation free-flyer satellite and the Wake Shield Facility for growing a thin-film semiconductor material in a near-perfect vacuum. The programme of research on this mission also encompassed EVA development exercises and ISS-related techniques and investigations. On 17 January 1996, Jernigan was named MS1 for the mission and was given the primary responsibilities for ORFEUS-SPAS orbiter Space Vision Systems and as EV1 crew-member. Her back-up responsibilities included RMS operations, BRIC and VIEW-CPL. She flew the ascent phase of the mission in Seat 5 on the mid-deck, and the descent in Seat 3 on the flight deck. Two EVAs were planned to evaluate equipment and procedures for use during the construction and maintenance of the ISS. However, both were cancelled in-flight when a hatch door latch on the outer airlock exit jammed, preventing access to open space.[47]

STS-81: Marsha Ivins

This fifth Shuttle–Mir docking mission exchanged John Blaha for Jerry Linenger. Marsha Ivins was named to the crew as MS3 on 2 February 1996, and flew the ascent in Seat 5 on the mid-deck and the descent in Seat 3 on the Flight Deck. As with all Shuttle–Mir missions, logistics transfer was a feature of the flight, as was a programme of secondary scientific experiments and research conducted in the SpaceHab module. Ivins was assigned primary responsibility for logistics and science transfers (Loadmaster) and supervising cargo movement to and from the Shuttle, as well as primary stowage, ALIS (down), the transfer of water supplies from the Shuttle to Mir, and IELK. Ivins also had primary responsibilities for the hand-held laser used for distance measurement during rendezvous and docking operations with Mir, and was tasked with photographing the separated external tank. She was also assigned as contingency EVA crew-member IV1, and was responsible for the Kid Sat experiment DTO and three DSOs. She also had back-up responsibilities for SpaceHab science and systems, Earth observations, geography, oceanography, meteorology, and Risk Mitigation Experiments OPM, GN2 Dewar and the CGBA. She was also back-up for the SpaceHab module payload and its science operations and for its subsystems operation, including activation and deactivation, set-up and tear-down. In addition, she had minimal responsibility for various orbiter systems, including rendezvous and proximity operations.[48]

STS-83/94: Susan Kilrain (née Still) and Janice Voss

Microgravity Science Laboratory 1 was a Spacelab Long Module research mission designed to conduct experiments and evaluate facilities connected with microgravity research, bridging the gap between Skylab, the Shuttle, Spacelab, and what was planned for the ISS. MSL-1 tested some of the hardware, facilities and procedures being developed for use on the long-term ISS research programme. Designed as a long-duration Shuttle flight, the first attempt under STS-83 was terminated four days into the mission due to an errant fuel cell, but despite the shortened mission, much valuable information and experience was gathered. Plans for a re-flight were pressed as quickly as possible due to the importance of the payload, and the same vehicle, payload and crew were re-launched as STS-94, completing a highly successful and important mission in preparation for ISS science operations. The crew responsibilities remained the same for both missions, but their experience on the short STS-83 mission helped them to increase productivity on STS-94, which a few weeks later resulted in a wealth of information from both missions.

Janice Voss was named PC/MS1 on 17 January 1996, and Pilot Susan Kilrain was assigned to the crew on 31 May 1996. On 18 February 1997,[49] Cady Coleman was reportedly in training as a back-up MS to Don Thomas, who on 29 January had sustained a broken right ankle following a routine training exercise. He was expected to make a full recovery before the flight in April (and did so), but Coleman was assigned as a precautionary measure, as due to her previous experience that she required minimal training to bring her up to speed. For launch and landing, Kilrain occupied the right-hand pilot seat on the flight deck, but Voss rode in Seat 3 on the flight deck for ascent and changed to Seat 5 on the mid-deck for entry. Voss was assigned to the Blue Shift and had the primary roles of MSL-1 activation and deactivation, science objectives and Spacelab systems. Kilrain was assigned to the Red Shift, in a back-up role, and was responsible for some MSL 1 activities, monitoring Spacelab systems, and secondary experiments. Her primary responsibility was as contingency EVA IV crew-member, and she was also given charge of all Earth observations. To gain experience she was also assigned to the de-orbit programme at the end of the mission, and was given responsibilities in the orbiter systems and procedures, including MPS, OMS/RCS, APUY/HYD, EPS, Comm/Inst, photo/TV, crew medic, in-flight maintenance and HP48.[50]

Susan Kilrain (*b.*1961) graduated from high school in 1979, earned a BS in aeronautical engineering from Embry-Riddle University in 1982, and in 1985 earned an MS degree in aerospace engineering from Georgia Institute of Technology. While pursuing her Masters degree she worked as a wind-tunnel project officer for the Lockheed Corporation. She was commissioned in the US Navy in 1985, qualified as a naval aviator two years later, became an instructor pilot on the TA-4J Skyhawk, and later flew EA-6A Intruders with Tactical Electrical Warfare Squadron 33. She graduated from the US Navy test-pilot school in 1993. At the time of her selection to NASA in 1994 she was in training as an F-144 Tomcat pilot with Fighter Squadron 101. After completing Ascan training in 1996 she was assigned to the vehicle systems and operations branch of the Astronaut Office, and was afterwards assigned to STS-83. Following the reflight, she married Colin Kilrain and worked as a CapCom in Mission Control. She was subsequently reassigned to the Office of Legislative Affairs at NASA HQ in Washington before being recalled to active duty with the US Navy in December 2002. Reflecting on her experiences as a pilot and at NASA, she realised that she was lucky to have been born at the right time and to have been in the right place when the opportunity arose. She often tries to inspire others to develop an interest in science and engineering by explaining how it enabled her to enter the astronaut programme and to fly in space.

STS-84: Eileen Collins and Yelena Kondakova

This sixth Shuttle–Mir docking mission delivered Michael Foale to the station to replace Jerry Linenger. Eileen Collins was named as Pilot on 15 July 1996, and on 22 August 1996 veteran cosmonaut Yelena Kondakova was named as MS4. For the launch and landing, Collins occupied the Pilot's seat, while Kondakova remained in Seat 6 on the mid-deck. Collins assumed orbiter responsibilities, including (primary) MPS, EPS, OMS/RCS, APU/HYD, water dumps, IFM, WCS and RPOP/TCS, and (back-up) GNS, HUM Sep, Comm/Inst, aborts, photography and TV, PGSCs, rendezvous and ODS. She also had back-up responsibilities on SpaceHab Double Module science objectives (CVDA, RRMO, and SSD MOMO) and the in-flight maintenance issues for SpaceHab, plus a primary role in two mid-deck science payloads (CREAM and PCG-STES). In addition, she completed work on six DTOs, one DSO, and three RME objectives. During the transfer operations she was responsible for water transfers to the station, powered hardware, PCG-GN, the Beetle kit, and ambient items. She was also responsible for SIMPLEX, and was

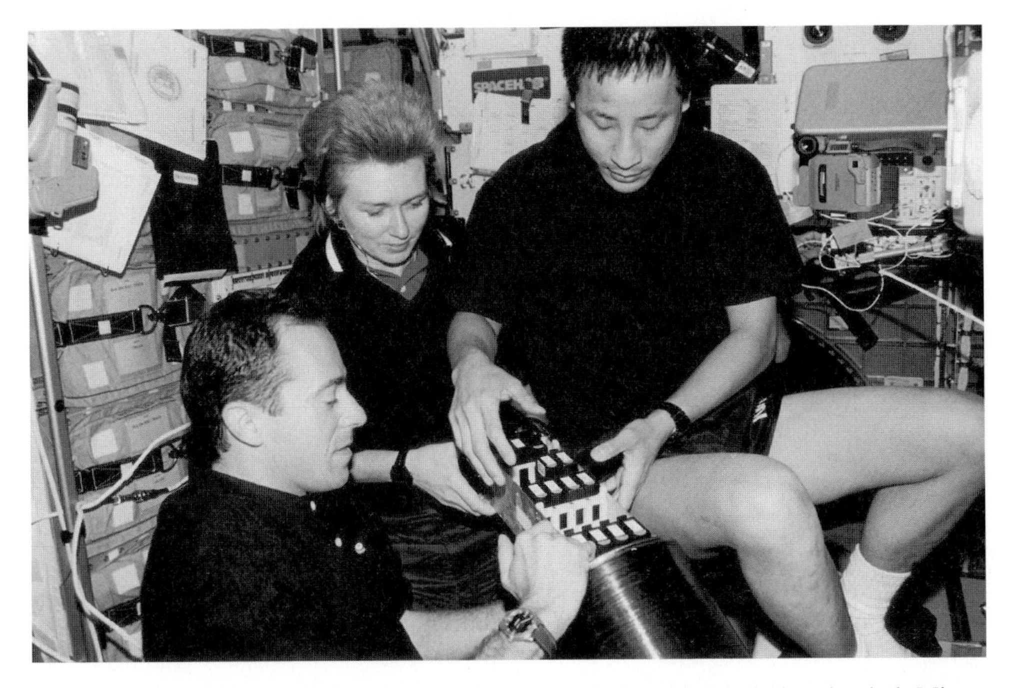

Yelena Kondakova works on the SpaceHab module experiments during the sixth Mir docking mission in 1997.

back-up for the Mir Structural Dynamics Test and Mir photographic surveys. Kondakova had orbiter responsibilities for payload bay and radiator opening and closing, photographic and TV objectives, contingency EVA crew-member IV2, OCA, IFM, TIPS, post-insertion activities and de-orbit preparation, radio communications (Russian) during rendezvous. She was also responsibile for thirteen experiments in the Bio rack on the SpaceHab module, and participated in the CREAM experiment on the mid-deck, and two DSOs. Her training and flight activities included supporting the activation and deactivation of the SpaceHab module and SpaceHab systems, and science investigations, as well as the refrigerators and freezers located in the module. During transfer operations, Kondakova supported ambient logistics and the transfer of Russian hardware into the Mir station. She was primary crew-member for Russian-language issues, and assisted Foale in joint US–Russian science programmes.[51]

STS-85: Jan Davis

This mission carried the CHRISTA-SPAS-2 pallet satellite designed to study the Earth's atmosphere, and featured tests and evaluation of a new Japanese robot manipulator system as part of its development prior to being installed on the ISS. There was also a package of technological applications, science investigations and astronomical research projects. Davis flew the ascent in Seat 3 on the flight deck and the entry in Seat 5 on the mid-deck. Named as MS1 on 12 September 1996, this was her final mission into space. Since her previous flight (STS-60) in 1994, she had served as Chairman of the NASA Education Working Group until 1995 and completed a tour as Chief of the Payload Branch of the Astronaut Office until selection to STS-85. Following the mission she was appointed Director of Human Exploration and Development of Space (HEADS) at NASA HQ in Washington, and in July 1999 was transferred to MSFC, in Huntsville, as Director of Flight Projects Directorate and then Director of Safety and Mission Assurance. Her crew responsibilities on this mission included primary payload assignments for MFD, PCG-STES, one DSO, payload safety issues and back-up roles on CHRISTA-SPAS, OSVS, SWUIS, BRIC, ACIS, one DTO, and supporting two DSOs. She was also assigned as contingency EVA crew-member IV1 and primary operator of the RMS, and participated in Earth observation objectives. Her orbiter responsibilities included Comm/Inst for ascent and orbital operations, opening and closing of the payload-bay doors and radiators, crew medic, crew equipment and stowage, de-orbit preparation, flight rules, and flight plan issues.[52]

STS-86: Wendy Lawrence

This seventh Shuttle–Mir docking mission delivered David Wolf to Mir, replacing Michael Foale for NASA's sixth long-duration residency on the station. Lawrence, flying as MS4, had been named to the crew on 6 December 1996, but originally as the astronaut who would replace Foale on Mir. She was replaced as resident by Wolf on 30 July, due to concerns about the suitability of the Russian Orlan EVA suit (she had never trained for EVA, and was too small to work inside the suit). However, NASA announced that Lawrence would remain on the Shuttle crew due to her training, flight experience and knowledge of Mir systems. She would assist in the crew and logistics transfers during the flight. Since her previous mission (STS-67 in 1995) she had completed Russian-language training and Mir-residence training for the NASA-6 mission on Mir, as well as a rotational tour as the sixth NASA Director of Operations in Russia. She flew ascent and entry on the mid-deck, but due to the late change to her role she had minimal responsibilities on orbiter issues. Her primary role was to assist the change-over of Foale and Wolf. Her crew roles included back-up for WCS, post-insertion and de-orbit preparations, transfer of soft stowage items, participation in the DTO (1118) photographic survey of Mir, and participation in Earth observation objectives.[53]

STS-87: Kalpana Chawla

This mission carried the fourth US Microgravity Payload (USMP-4) and the deployable pallet satellite CHRISTA-SPAS, as well as conducting ISS-related EVA demonstrations. Chawla was assigned as MS1, and was primary RMS operator during the CHRISTA-SPAS deployment operations. She reportedly failed to enter a command during the operations, preventing the satellite from responding to flight control commands and forcing the crew to retrieve it during one of the EVAs. The Spartan operation was delayed for one day, and a new flight plan was uplinked to the crew, but the final command required to prepare Spartan for deployed operations was a crew input via the computer, which was not received by the

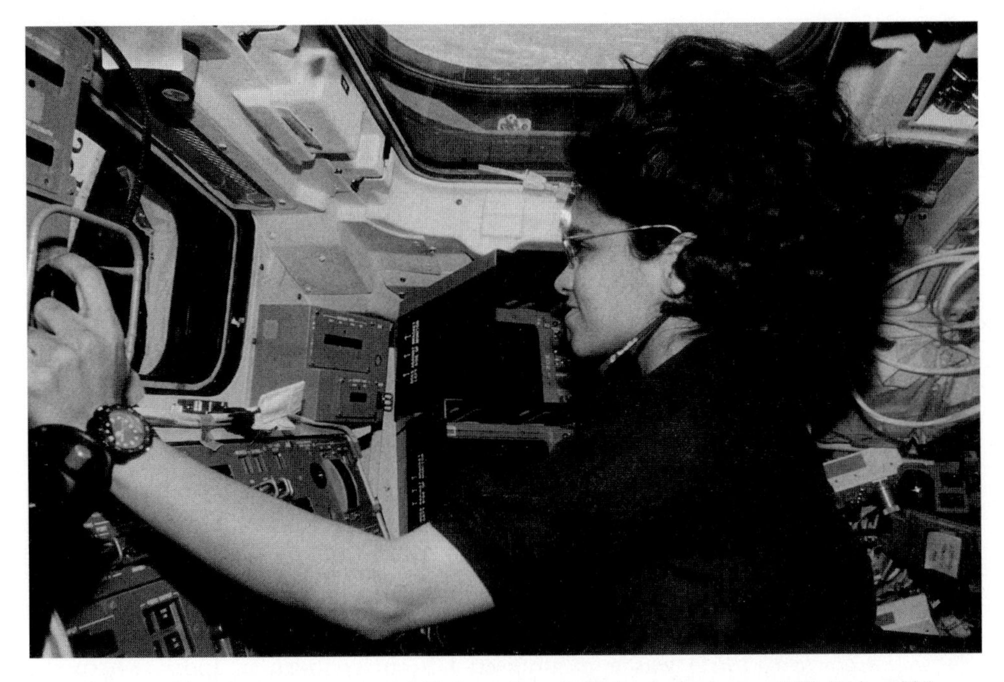

Kalpana Chawla operates the RMS from the aft flight deck during STS-87 in 1997.

spacecraft. Lack of telemetry and onboard verification processes resulted in this error being undetected by both the crew and Mission Control. The subsequent inquiry into the incident by an STS-87 Close Call Investigation Board determined that the crew had inadvertently omitted the Spartan stand-by step. The satellite was released by the RMS, but when the expected preprogrammed manoeuvre was not executed, in accordance with flight rules and with the agreement of Mission Control, the crew tried to re-grapple the satellite. During this process, Chawla input a premature capture followed by release in close proximity to Spartan while still driving the RMS forward towards the grapple fixture. The investigation determined that the RMS snare wires contacted the grapple pin on the satellite and imparted rotational and translational rates into the satellite. Rather than attempt to grapple, Chawla pulled the RMS away. The induced rotation of about 2° per second was judged too large to attempt another grapple, and matching the rotation by using the orbiter was deemed too expensive in fuel usage. The recapture by EVA was therefore planned, and was successfully implemented two days later. The Board's nine-point recommendations included improvements in the payload training accountability, the assessment and thoroughness of crew-member training, the retention of critical training objectives, and re-examination of RMS training, including increasing the failure-to-capture element. In addition, upgrades in display information, new 'ready to deploy' information and rate matching flying techniques and procedures were to be developed and trained for.[54] Chawla's lack of previous flight experience with contingency RMS operations certainly contributed to the error, and this – combined with the limited pre-flight training programmes for such incidents, and the lack of on-orbit information for the crew and the controllers on the ground – was included in the report. Although this was a setback, the manual recapture and the rest of the mission progressed smoothly.

Chawla was named to the flight on 14 November 1996. She rode the ascent in Seat 3 on the flight deck and the entry in Seat 5 on mid-deck. In addition to the RMS, her primary responsibilities were for Spartan, USMP and SOLSE, supporting the EVAs as IV crew-member, assisting in the EVA preparations and post-EVA activities, the EVA Development Flight Test objectives, and conducting Earth observations. During the ascent she was responsible for Comm/Inst, payload-bay doors and radiators, photography of the separated external tank, and one DTO. Her orbiter responsibilities during the mission included PGSC, cable configurations, de-orbit targets, the world map used for ground recognition references, photographic and TV objectives, stowage, and CCTV. During the rendezvous operations she was responsible for CCTV and photography, RSAD, grapple and deploy, and RMS systems. She also worked on Spartan deployment and retrieval issues. For the USMP payload she had crew responsibility for MGBX, ELF, WCI, PEP, AASDF, IDGE and MEPHISTO, one DSO, documentation, TGDF, and de-orbit preparations.[55]

Kalpana Chawla (*b*.1961) was a naturalised US citizen born in India, where she received her early education, including earning a BS in aeronautical engineering from Punjab Engineering College in 1982. Moving to the US for graduate work, she earned her MS in aerospace engineering from the University of Texas in 1984 and her PhD in aerospace engineering from the University of Colorado in 1988. From 1988 until her selection to the astronaut programme in 1994 she worked as a researcher in aeronautical studies of airflow for exotic aircraft such as VTOL aircraft, and hovercraft designs for MCAT Institute in California, supporting studies at NASA Ames. In 1993 she joined Overset Methods Inc as VP and research scientist involved in simulations of multiple-body aerodynamic problems. After completing Ascan training she was assigned to the EVA and robotics branch of the Astronaut Office, and following the mission was assigned to the development of Robotic Solution Awareness Displays (to improve crew awareness using robotic systems), and testing software in the SAIL.

STS-89: Bonnie Dunbar

This was the eighth Shuttle–Mir docking mission, during which Andy Thomas changed with David Wolf as the seventh and final American astronaut resident on the station. Bonnie Dunbar was named to the crew as MS3/PC on 4 March 1997, and flew the ascent in Seat 5 on the mid-deck and the descent in Seat 3 on the flight deck. She assumed crew responsibilities for Spacelab, as contingency IV crew-member and for Russian-language issues. She was also back-up for the logistics transfers to and from the Mir station, and assisted in the rendezvous with the station, with back-up responsibility for the Orbiter Docking Systems. Her orbiter roles included APU/HYD, cable configurations, CMO, Comm/Inst, EPS, photography of the separated ET, HHL, OCA, ODS ingress and egress, OMS/RCS, PGSCs, closing of the payload-bay doors and radiators, post-insertion activities on the mid-deck, RPOP/TCS, crew stowage and equipment, and TIPS. In SpaceHab operations, Dunbar was primarily responsibile for the Spacehab Module, its systems, deactivation and in-flight maintenance and operations within the module, with back-up responsibility for SpaceHab activation. In addition, she was assigned responsibilities in all nine SpaceHab science and technology investigations. During the transfer to Mir she was back-up coordinator, and participated in the transfer of water supplies, batteries, cold stowage items, the gyro dome, IELK and OPM. She took primary responsibility for a number of critical item transfers to and from the station, and was also assigned to fifteen mid-deck and science payloads or objectives, including Earth Kam, Earth observations, GAS canisters and photographic surveys.[56]

Following STS-50 in 1992, Dunbar served as Deputy Associate Administrator for Material Sciences at NASA HQ in Washington, and was later assigned to long-duration Mir flight training. From October 1995 until November 1996 she was reassigned at JSC as Assistant Director for ISS readiness and Russian/American cooperation, reviewing training strategies and procedures. Following STS-89 she was appointed Assistant Director for University Research and Affairs, and was subsequently assigned in an Astronaut Management role as Deputy Associate Director for Biological Sciences and Applications.

STS-90: Kathryn Hire

This flight carried the last Spacelab Long Module payload prior to the start of ISS construction and habitation. Neurolab consisted of twenty-six experiments in eight fields of neuroscience on humans and animals, in preparation for long-duration missions on the ISS. Kathryn Hire was assigned as MS2 on 18 April 1997, and flew the ascent and entry in Seat 4 on the flight deck, acting as launch and entry Flight Engineer. The crew worked a single-shift system, and Hire's responsibilities included serving as contingency EVA crew-member IV1. She was also jointly responsible for photographic and TV objectives, stowage and habitability, DPS, Comm, PGSC, MPS/EPS/OMS/APU/RCS, ECLSS, Spacelab CDMS and ECS, and back-up on Spacelab science issues, crew equipment (including crew shirts), and the Flight Data File.[57]

Kathryn Hire (*b*.1959) graduated from the US Navy Academy with a BS degree in engineering management in 1981, and subsequently earned an MS in space technology from the Florida Institute of Technology in 1991. She qualified as a naval flight officer in 1982, and participated in airborne oceanographic missions as a project coordinator, mission commander and detachment officer in charge. She then became a navigation instructor and course manager at the Naval Air Training Unit at Mather AFB in California, and left active duty in 1989 to join EG&G Company at NASA KSC as an orbiter processing engineer. She subsequently transferred to Lockheed Space Operations Company, as a Shuttle Test Project Engineer for Shuttle maintenance, checkout of EMU pressure suits, and the Orbiter Docking System. She had joined the US Navy Reserve in 1989, and became the first American woman assigned to a combat aircraft, as a patrol aircraft navigator and communicator with several overseas deployments. She was selected to NASA in 1994, and after completion of Ascan training she worked as a CapCom in Mission Control prior to assignment to STS-90. Following the mission she worked as an Astronat Office Lead Astronaut in SAIL and on Shuttle payload and crew equipment issues, after which she was recalled to active duty in 2003 and assigned as a member of the US Navy Central Command Staff during the second Gulf War. When she returned to JSC she was assigned to the Cape Crusaders at KSC.

STS-91: Wendy Lawrence and Janet Kavandi

This was the ninth and final Shuttle–Mir docking mission, returning Andy Thomas from his resident stay on the station. The crew was named to the flight on 23 October 1997, with Wendy Lawrence assigned as MS2/FE flying ascent and entry in Seat 4 on the flight deck. Janet Kavandi was named as MS3 and flew ascent in Seat 5 on the mid-deck and entry in Seat 3 on the flight deck. Lawrence had been assigned to the flight partly as compensation for losing her residency mission and to call upon her experience with Mir systems, rather than calling upon another astronaut when most were in training for ISS operations. Lawrence's responsibilities on the crew for orbiter issues included joint support for DPS, MPS, EPS, OMS/RCS, APU/Hyd, ECLSS, HUM Sep, water dumps, medical officer, Comm/Inst, aborts, photographic and TV issues, IV crew-member, PGSCs/OCA, in-flight maintenance, WCS, post-insertion activities, rendezvous, RPOP/TCS, ODS, hand-held laser/CCTV, ergometer, three DTOs and DSOs, one RME, primary RMS operator, and SHUCS under the Spacelab Science objectives. She also had support roles in four mid-deck science experiments. Her main orbital activities focused on transfer operations between the Shuttle and Mir, including cold stowage and water transfers, batteries, ambient payloads, Russian hardware, and powered hardware.

Janet Kavandi was assigned shared orbiter responsibilities for OMS/RCS, EPS, APUs/Hyd, payload-bay doors and radiator opening and closing, photographic and TV issues, post-insertion activities, rendezvous operations support, HHL/CCTV, orbiter docking system, ergometer, support work on three mid-deck science experiments, three DSOs/DTOs, Earth observations and two RME investigations, and was contingency EVA astronaut EV2. She was also back-up RMS operator responsible for the freezers and refrigerators in the SpaceHab, as well as Spacehab set-up, activation, deactivation and tear-down, backing-up system operations and IFM, and working on ambient cargo transfer to Mir and Russian hardware transfers.

Janet Kavandi (*b.*1959) graduated from Missouri Southern State College with a BS in chemistry in 1980. She later earned an MS in chemistry from the University of Missouri in 1982 and a PhD in chemistry from the University of Washington in 1990. After two years working as an engineer for Eagle-Pitcher Industries she was employed at Boeing Company in Seattle in 1984, working as an engineer with a speciality in power systems on various missile projects, as well as for the ISS, IUS, and the Advanced OTV. Selected to NASA's astronaut programme in 1994, she qualified as an MS in 1996 and was assigned to the payloads and habitation branch of the Astronaut Office, working on payload integration issues for the ISS.

STS-95: Chiaki Mukai

This mission featured the return to space, after thirty-six years, by former Mercury astronaut John Glenn, who in February 1962 had been the first American to orbit the Earth. At the age of 77 he was making a second flight as a Payload Specialist to perform a programme of medical experiments related to the ageing process. The flight also included a Spartan free-flying satellite to perform solar observations, Hubble Space Telescope Orbital Systems Test platform for evaluating hardware for forthcoming Hubble servicing missions, and a variety of science payloads in a SpaceHab module. Mukai was assigned as PS1 on 13 February 1998, and flew the mission ascent and entry in Seat 7 on the mid-deck. She was responsible for a number of SpaceHab science investigations – specifically, a suite of Japanese experiments that included a saltwater aquatic life experimental unit, plant growth experiment, sleep experiments and a protein turnover experiment. She was also responsible for the SpaceHab experiments AGHF, ADSEP, APCF, Advanced Separation payload, Aerogel, Astroculture, BioBox, BRIC, commercial BioDyn Payload, CGBA, CIBX, CPCC, EORF, FAST, MEPS, MGBX, NIH-C, NHK camera, OSRF, OCC, OSTEO, PCAM, PTO, VDA/STES, MOMO, SAMS and VFEU.[58] In addition, she utilised her medical training to support the medical research objectives of the mission, and served as back-up crew medic. Following this mission, Mukai became Deputy Mission Scientist for STS-107 in August 2000, coordinating science operations for the SpaceHab research mission flown in January 2003. In 2004 she accepted at appointment at the International Space University.

AN ALL-FEMALE SHUTTLE CREW

The success of the flight of 77-year old former Project Mercury astronaut and US Senator John Glenn on STS-95 in 1998 generated a positive outcome in gathering biomedical data from a wider variety of test subjects, such as flying some of the Skylab astronauts. However, the one topic that became more widely discussed was the possibility of flying an all-woman crew on the Space Shuttle, which at the time raised both positive and negative comments. However, the idea was not new.

An all-female Salyut visiting mission

As the sole crew-member of Vostok 6, Valentina Tereshkova had, of course, been the first 'all-woman crew' in space; but following the EVA performed by Svetlana Savitskaya on Soyuz T-12/Salyut 7 in July 1984 the Soviets were eager to make the international headlines by mounting a mission that could upstage the expanding Space Shuttle programme and also be attainable in the strained resources of the national space station programme.[59]

The mission of an all-female Soyuz T crew was planned as an eight-day visiting mission to Salyut 7 in spring 1985, independent of any crew exchange with the main resident crew, which therefore could be flown at any time that a resident crew was on board. Plans for the exchange of resident crews onboard Salyut 7 had been abandoned after a Soyuz launch pad abort in September 1983, and would be delayed by the difficulties encountered on the station over the ensuing two years.

At that time, the Soviet's ruled that any Soyuz crew assignment had to include an experienced cosmonaut in command – and since only one Soviet woman was currently flight-experienced it fell to Savitskaya to take command of the Soyuz T mission. With two spaceflights and an EVA to her credit she was the obvious and only choice for the seat. At the time there were at least another eight female candidates available for flight assignment to the mission. Three of the eight unflown women cosmonauts had been previously assigned to support the Soyuz T-7 flight and T-12, as well as the planned but unflown T-8 female flight. Irina Pronina had served as back-up to Savitskaya on T-7, and was to have flown to Salyut 7 as a crew-member on Soyuz T-8, but she was taken off the flight for political reasons. Yekaterinia Ivanova had served as back-up to Savitskaya on T12, whilst Natalyia Kuleshova was on the original back-up crew for Soyuz T-7 but was replaced by Pronina due to her being unready for specific mission training. Pronina was not selected for the mission, despite her experience in mission training. The Flight Engineer seat on the all-woman crew was therefore given to Ivanova, and the third seat (Cosmonaut Researcher) was given to physician Yelena Dobrokvashina.

There were too few flight-experienced female cosmonauts to form a back-up crew, and it was therefore comprised of male cosmonauts, as it was not thought sufficiently 'interesting' to assign a male Commander with two female rookies. In his article in *Spaceflight* published in 1999, Bert Vis pointed out that assigning male back-up crew-members would not raise any additional headlines should they have to fly in place of a female crew-member, and such a replacement would ruin the sole objective of flying an all-female crew, which was never a major contribution to the developing Salyut space station programme and had no real importance.

As the crew trained, events on the unmanned space station were to contribute to the fate of their mission. By April 1985 the space station Salyut 7 had been in orbit for over two years and had already undergone several systems failures (notably, a main engine failure which had been repaired during 1984) and was currently unmanned pending the launch of its next resident crew. When the Salyut suffered a serious power failure that affected its onboard environmental subsystem as well as its automatic rendezvous and docking and attitude control subsystems, leaving the Salyut 'dead in the water', a decision was made to mount a special rescue and repair mission, and then, if successful, resume resident operations, thus at first delaying the proposed female visiting mission into spring 1986 at the earliest, and then in May 1985, cancellation of the flight. Even after the success of the two-man Soyuz T-13 crew to restore the Salyut to operational use after more than a hundred days of difficult work and the partial crew exchange with the T-14 crew to resume a three-man residency, the subject of an all-female crew was apparently not raised again. An official reason was a lack of Soyuz T spacecraft (pending the introduction of the upgraded Soyuz TM version) on which the female cosmonauts had trained. It has also been suggested that the difficulties encountered on Salyut 7, and the illness of the T-14 commander Vladimir Vasyutin, forcing his early return, prompted officials to terminate the Salyut 7 programme and move on to the larger modular Mir space station. In addition, during an interview with Savitskaya in 1990 she suggested that Director of Cosmonaut Training Lieut-General Vladimir Shatalov was one of the officials not interested in flying an all-female crew and was not at all enthusiastic about flying females in space. This lack of support for rescheduling the flight on a Soyuz TMA to Mir essentially terminated the mission.

Considering an all-female Shuttle mission

Following Glenn's Shuttle mission, NASA's Associate Administrator Arnauld E. Nicogossian, of the Office of Life and Microgravity Sciences and Applicants at NASA HQ in Washington, wondered whether an all-woman crew might be subjected to different stresses than men are subjected to during a Shuttle mission, and was instructed by NASA Administrator Dan Goldin to investigate the question.[60]

In response to negative comments on such a mission from critics who thought it would be nothing more than a publicity stunt (an opinion which had surfaced during Glenn's flight), Goldin stated that he considered that such a flight might inspire young women to enter science and engineering careers. He also could not understand why such questions were being asked, as no-one ever considers why all-male crews fly. Since 1983 NASA had launched twenty-eight female astronauts, and Goldin initiated an investigation as to whether NASA was launching female astronauts into space in direct proportion to their representation in the astronaut corps. During 1999 more than 140 astronauts, including 33 females, were available for flight assignments. The Administrator revealed that only when any possible scientific benefits and 'personal issues' had been reconciled, would any consideration for an all-female Shuttle crew be considered for any additional inspirational benefits. The fact that NASA was considering such a flight was, according to some sources, a legitimate consideration following the public mood after Glenn's flight.

Although no official timetable was established for such a mission, the Executive Director of the Society of Women Engineers stated that a female Shuttle crew could indeed attract the attention of young girls who were thinking of a career in either science, engineering or technology: 'There has not been enough awareness-training for girls to understand that these professions are open to them, that it's just not little boys who grow up to be astronauts.'

The idea progressed to a formal review stage, but was not pursued to flight manifest – primarily due to the lack of strong scientific data expected to be returned from the flight, and hesitancy in the female astronaut corps to be singled out for flight assignment.

Arguments for its lack of scientific merit focused on the fact that nothing new could be gained by flying a crew of seven women on a short Shuttle mission or by flying one or two women on a short mission (which had already been done several times), and that no significant or additional data would be obtained – which was more important than the promotion of the flight as an 'inspirational mission'. However, NASA appointed former astronaut Rhea Seddon to head a committee designed to evaluate whether gender, age and other criteria should be taken into consideration when assigning females to long-duration flights on the the ISS, and whether an all-female 'payload crew' on the Shuttle could be useful in obtaining maximum data on a dedicated mission and in also providing baseline data for long-duration missions on the station.

At the time (1999), Seddon revealed that no studies of female hormone levels had been carried out, and that such data, combined with studies on deconditioning and degenerative issues associated with long-duration spaceflight, would be useful. In her report, presented to NASA in 2000, she stated: 'If NASA really does want to do the science of female physiology on a reputable science mission, for all the right reasons and with all the right equipment, then I think that would be fine ... I don't think they should pull together a female crew for public relations reasons.'

During an interview in 2004,[30] Seddon was again questioned about the validity of flying an all-female crew. She agreed that although the purpose of John Glenn's flight was to obtain good scientific data, the real reason was for inspiration; and though it was possible to fly an all-woman crew, it would not be appropriate to do so solely for the same reason: 'You can't sell it based on inspiration. You have to say, 'well, we could get a lot of good data on women.'' When Seddon talked with female members of the Astronaut Office, they told her that they did not want to be assigned to a flight to be considered as guinea pigs, nor to be chosen simply because they were female. They wanted assignments based on their competence as astronauts, not as women astronauts. Seddon's committee had to take this into consideration along with scientific, health-care issues and research issues on the mission. If there were to be little support from the Astronaut Office for such a high-profile mission, then it would be difficult to support the funding and organise the resources.

There was also the question of being criticised in the media if anything went wrong during the mission. The proposal for flying an all-female crew came at a time when Eileen Collins became the first female Shuttle Commander (STS-93). As the mission approached, the media headlines announced that the mission was originally scheduled to be launched on the thirtieth anniversary of the first landing on the Moon (20 July), and the historic fact that the first female astronaut was in charge of the mission. Unfortunately, several delays forced postponement until 23 July. During ascent, there were two major anomalies. Five seconds after launch, a drop in voltage caused two primary controllers on engines 1 and 3 to shut down, although both redundant controllers on these engines performed normally, allowing *Columbia* to reach orbit successfully. In addition, a low level of liquid oxygen resulted in a premature shut-down 0.15 second before Main Engine Cut Off, triggering early shut-down. This did not seriously affected the performance of the mission, and the spacecraft entered orbit 16 f/s under speed and at 168 miles rather than the planned 176 miles.[61] The crew was not responsible for either incident, but the media produced headlines about 'women drivers running out of fuel'.

The two anomalies were traced to an arc in the wiring, and a small item of debris in the engine which had penetrated the coolant tube, causing a leak. These findings

Eileen Collins, first female Commander of a Shuttle, during STS-93.

were never to receive such prominent headlines as those initially reporting the incident, and there was no mention of the abort scenario that the crew would had to have undertaken – calling on all the experience of Collins and her colleagues – had the failures been more serious. The suggestion that such negative headlines also affected the decision to fly an all-woman crew was put to Seddon in the interview in 2004. She agreed that probably such a concern for media coverage was in the minds of those who were considering the all-woman Shuttle crew flight. There were no concerns that if assigned, all the women would do an extremely good job; but as Seddon explained: 'You always have to take into consideration if something happened and you know then you get a bum wrap, you know maybe it was something that nobody could have done anything about, or no one could have handled better, but I think that increased scrutiny is always very stressful.'

Several male crew-members had encountered such stress, and the experiences during STS-93 did nothing to enhance the support for flying an all-woman crew other than for publicity reasons. The idea was therefore quietly dropped, and there seems to be no talk of reviving it in the wake of the *Columbia* accident. With retirement of the Shuttle a few years away, and the introduction of a new Crew Exploration Vehicle for ISS supply flights and exploration of the Moon, the assignment of an all-woman crew remains many years in the future.

THE SHUTTLE–ISS YEARS

Construction of the International Space Station began in November 1988, and over the ensuing four years or so, ten of the fifteen Shuttle missions were connected with the construction and supply of the new space station.[62]

STS-88: Nancy Currie

The crew for this first ISS assembly mission was named on 16 August 1996, with Nancy Currie named as MS2/FE flying the ascent and entry in Seat 4 on the mid-deck. Her primary responsibility was the operation of the RMS during deployment of the Node-2 facility and its attachment to the Zarya.[63]

STS-96: Tammy Jernigan, Ellen Ochoa and Julie Payette

The crew for this second Shuttle mission to the new station was named on 4 August 1998. Tammy Jernigan was assigned as MS1/Payload Commander and flew ascent and entry on Seat 5 on the mid-deck; Ellen Ochoa was named MS2/FE, flying ascent and entry in Seat 4 on the flight; and Canadian astronaut Julie Payette was named MS4 and flew ascent in Seat 6 on the mid-deck and entry in Seat 3 on the flight deck.

Tammy Jernigan during EVA on STS-96 at the ISS in May 1999.

Jernigan had served as Astronaut Office Deputy for the ISS since 1997, and this was to be her final spaceflight. Following the mission she worked as Astronat Office lead for ISS external maintenance, and point of contact for the Italian logistics modules, until she resigned from NASA in October 2001 to take a position with the Lawrence Livermore National Laboratories. For the STS-96 mission she was assigned crew responsibility for the APAS docking system, EVA/EMU (EV1), payload-bay doors and radiators and post-insertion operations, and was back-up for photography of the ET, flight rules, RMS operations, and de-orbit preparations. She was also responsible for ingress into the vacant ISS station. During the mission she also participated (with Dan Barry) in an EVA of 7 hour 55 min on 28 May.

Ellen Ochoa's responsibilities included primary on rendezvous RMS operations and the SpaceHab facility, and back-up on APAS, APU/HYD, DPS, ECLSS, EPS, FDF, Flight Plan, GNC, IFM for SpaceHab, MPS and OMS/RCS. She was assigned to support VRA, RSAD, PILOT, and FRED investigations, was also assigned a primary role for the activation, deactivation and operational aspects of the SpaceHab module, and had additional duties in securing the crew shirts and providing taped messages for NASA PAO. She was also assigned as primary Loadmaster for the mission. Julie Payette was given crew assignments for Earth Observations, IV crew-member, photographic and TV objectives, payload-bay and radiators, and back-up responsibilities as medical officer, APU/hyd, crew equipment, EVA and EMU hardware and procedures, cable routing and plug-in, PGSCs, post-insertion activities, and SpaceHab module systems and procedures. She also served as back-up Loadmaster and primary stowage master on the ISS, and took

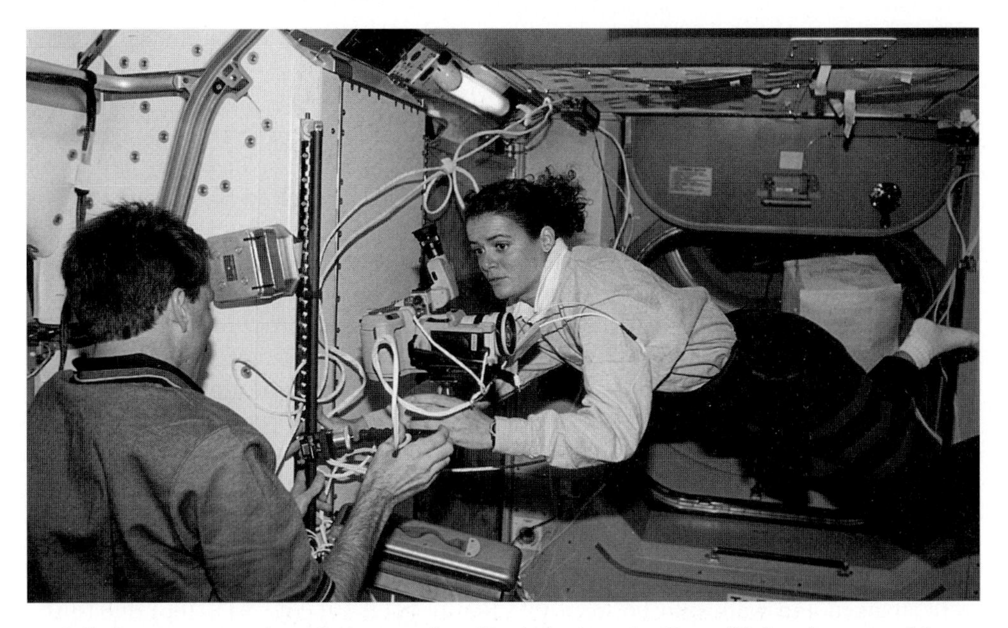

Julie Payette proceeds with the transfer of logistics into the Zarya/Unity elements of the ISS during mission STS-96 in May 1999.

responsibility for ISS hardware in Node 2 and FGB/Zarya, primary for ISS IFM, and back-up for RSAD. She was assigned two DTOs, acted as back-up for the activation, operation and deactivation of the SpaceHab module, and was primary on ISS systems CandDH, CORE, ECLSS, EPS, GNC and PROP, medial operations, multi-segment, OOM and TCS, and back-up for MIRTS and the radiation monitors. Finally, in preparation for the flight she also provided taped messages for PAO, and was responsible for both the official crew photography and for Space Flight Awareness/PAO issues.[64]

STS-93: Eileen Collins and Cady Coleman

This was the mission to deploy the Chandra X-Ray Observatory (initially called the Advanced X-Ray Telescope Facility). The flight also marked the first time that a woman (Collins) had commanded an American space mission (occupying Seat 1 on the flight deck). The crew was named in February 1998, and in addition to Eileen Collins, Cady Coleman was named as MS1, flying the ascent in Seat 5 on the mid-deck and the entry in Seat 3 on the flight deck. As well as taking overall command of the mission and assuming the responsibility for mission success and crew safety, Collins took crew assignments for DPS, ECLSS, rendezvous operations and the CCM experiment, with back-up assignments in MSX, SIMPLEX, SAREX-II and BRIC. Coleman was responsible for Comm/Inst, closing of payload-bay doors and radiators, FDF and post-insertion operations, and was back-up for Earth observations. She also took on the secondary payloads of GOSAMR, LFSAH, CCM, PGIM, MEMSD and the Earth Kam. In addition, she was assigned one DTO and two DSOs, was responsible for organising the official crew portrait, and was primary point of contact for Chandra and contingency EVA astronaut EV2.[65]

STS-99: Janet Kavandi and Janice Voss

This flight carried the Shuttle Topographical Radar Mission (STRM) payload to provide detailed topographical maps of the Earth's surface. The crew was named on 26 October 1998, with Kavandi being assigned as MS2/FE (Red Shift) and Voss as MS3/PC (Blue Shift). Kavandi took the additional responsibilities of contingency EVA astronaut EV1, and shared orbiter responsibilities for GNS, DPS, MPS, EPS, OMS/RCS, APU/HYD, ECLSS, Comm/Inst, opening and closing of the payload-bay doors and radiators, EVA and EMU equipment and procedures, crew medic, and IFM issues, as well as participating in gathering science data. Voss was responsible for payload science operations, and shared responsibilities in Comm/Inst, crew medic, post-insertion operations, and de-orbit preparation, FDF and flight rules.[66]

STS-101: Mary Ellen Weber and Susan Helms

The original five-person NASA crew (including MS1 Weber) for the third mission to the ISS was named on 16 November 1998. On 18 February 2000 the flight objectives were changed and were split between two flights (STS-101 and STS-106) due to the delays in launching the Russian Service Module, Zvezda. Weber was assigned as RMS operator supporting the EVA, and was responsible for transfer of logistics to the station (Loadmaster) as well as for SpaceHab systems. She occupied Seat 3 on the flight deck during both launch and entry. Her shared orbiter responsibilities included GNC, DPS, MPS, EPS, OMS/RCS, APU/HYD, ECLSS, HUM SEP, water dumps, Comm/Inst, opening and closing payload-bay doors and radiators, abort scenarios, primary on photographic and TV objectives, CCTV and camcorder crew equipment and stowage computers, WCS, TIPS, Flight Data File, orbiter stowage, one DTO, the orbiter docking system, Earth observations/geography/oceanography/meteorology, and the mission-specific crew shirts.

On 18 February the ISS-2 crew then in training (Jim Voss, Susan Helms and Yuri Usachev) was assigned to the crew to provide additional support in preparing the station for the first resident crew and also to provide them with ISS experience prior to the exchange between them and the ISS-1 crew in 2001. For STS-101, Helms rode ascent and entry in Seat 7 on the mid-deck, and during the mission she assisted in the replacement of batteries and associated equipment onboard Zarya and replaced and inspected other equipment onboard Zarya and Unity. She assumed additional shared responsibilities for this flight in photographic and TV objective, CCTV and camcorder operations, hardware, crew medic, computers, WCS, TIPS, FDF, rendezvous and prox ops, RPOP, TCS, RMS camera operations only (no arm operations), American ISS equipment installation, point of contact (shared) for Russian segment systems, US segment systems, nominal ingress and egress activities, cargo transfer, and photography of the separated ET. In addition, she shared responsibility for Russian-language issues.[67]

STS-92: Pam Melroy

This was the planned third mission for ISS assembly. Melroy – named to the crew as Pilot on 3 February 1998 – was only the third woman to fly in the right-hand seat on the flight deck during a Shuttle mission. She assisted Commander Brian Duffy in the rendezvous and docking to the ISS, and performed the undocking from the station at the end of docked operations, prior to conducting a fly-around of the complex to gain experience in future docking operations on a subsequent mission. She helped prepare the berthing port mechanisms for the mating of the Z1 truss to the US Unity connecting node, and then participated in the internal outfitting of the newly-attached Truss structure with grounding straps and logistical hardware. She was also assigned the operation of the 3D IMAX camera for collecting new movie footage of station construction. As Pilot, her Shuttle shared responsibilities included APU/HYD, GNC, MPS, EPS, OMS/RCS, ET connection door closure, payload-bay doors and radiators, TIPS, APCU/OIU, serving as crew medic, IFM, post-insertion and de-orbit preparations on the flight deck, RPOP/TCS. and the transfer of water.[68]

Pam Melroy (*b.*1961) earned a BS degree in physics and astronomy from Wellesley College in 1983, and later earned an MS degree in Earth and planetary sciences from MIT in 1984. She entered active duty in the USAF in 1984 and completed pilot training, qualifying as a KC-10 pilot. Over the next six years she served as an aircraft co-pilot, commander and instructor pilot, and logged more than two hundred combat and combat support missions in the first Gulf War. A graduate of the USAF

Test Pilot School in 1991, she was working as a test-pilot with the C-17 Combined Test Force when selected for astronaut training in 1994. She completed Ascan training in 1996.

STS-98: Marsha Ivins

STS-98 was also known as ISS assembly mission 5A, which would deliver the US Destiny laboratory to the complex. The crew for this flight, named on 4 August 1998, included Marsha Ivins as MS2/FE. In addition to supporting and backing-up the Commander and Pilot from the centre seat (Seat 4) on the flight deck during launch, landing and rendezvous to the station, Ivins assumed a number of primary and secondary tasks in monitoring orbiter systems and mission objectives. She was the primary RMS operator for the mission, and moved the Destiny laboratory across to its permanent location on the Unity module. In addition, she had to relocate a station docking port to the aft berthing mechanism on Destiny, and support and transport the two EVA crew-members to a variety of workstations across the ISS and Shuttle cargo bay during three EVAs. She also assisted Commander Ken Cockrell in the initial activation of the laboratory systems, and later joined the rest of the crew to assist with the continued outfitting of the laboratory in preparation for the commencement of science operations from inside the lab. She was also the Loadmaster for the mission, supervising the transfer of logistics from the orbiter into the station and directing the flow of used items from the station back into the orbiter for return to Earth. She was also responsible for many of the TV and photographic documentation tasks on the mission.[69]

Marsha Ivins peers out of the overhead flight deck windows during the STS-98 docking mission to the ISS in 2001.

STS-102/105: Susan Helms

Helms was identified as MS4 for ascent only on this mission, as she was a member of the ISS-2 residency crew. She served in the same capacity for the return on STS-105. Prior to transfer to the ISS station as a crew-member, she participated (as EV2, with ISS-2 crew-member Jim Voss) in an EVA of 8 hrs 56 min on 11 March 2001, and was EVA coordinator for the second EVA by the STS-102 orbiter crew.

STS-104: Janet Kavandi

This was another ISS assembly mission, delivering the US Quest airlock to the station. Janet Kavandi was named as MS2/FE on 28 September 2000, and flew ascent and entry in Seat 4 on the flight deck, assisting and backing-up the Commander and Pilot during ascent and entry as well as during rendezvous with the ISS. In addition to a number of responsibilities for orbiter systems and mission objectives, Kavandi was assigned as primary RMS operator, supporting and assisting the EVA astronauts during their excursions outside. From the aft flight deck, she operated the Shuttle RMS to position the arm's cameras to gain the best visual advantage for ISS-2 crew-member Susan Helms, who was operating the station's own arm from the Destiny laboratory. As Helms removed the Quest airlock from the payload bay of the Shuttle, Kavandi used the Shuttle RMS cameras to assist her in berthing the airlock to the station. Once Quest was placed onboard the station, Kavandi was responsible for leading the outfitting of the facility, and as Loadmaster was responsible for the relocation of logistics from the Shuttle to the station and for used items transferred the Shuttle for return to Earth.[70]

STS-108: Linda Godwin This was the first ISS Utilisation Flight, the crew of which was named on 29 January 2001. Godwin was assigned as Loadmaster for the mission, responsible for the transfer of supplies and equipment to the station and unwanted material back in the orbiter. In addition, she was responsible for the Androgynous Peripheral Docking System for docking and undocking with ISS, and completed Shuttle pressure and leak checks on docking with the station. As the primary RMS operator she relocated the Multi-Purpose Logistics Module Raffaello from the payload bay and attached it to the station for unloading, and then relocated it in the payload bay for return to Earth once operations with the MPLM had been completed. In addition to being responsible for a number of scientific payloads and orbiter systems and procedures, she was assigned to opening and closing of the payload-bay doors, and deployment and stowage of the orbiter's radiators. On 10 December, she and Dan Tani completed an EVA of 4 hrs 12 min, during which they installed several insulation blankets and performed a range of 'get-ahead' tasks in support of later EVAs during 2002.[71]

STS-109: Nancy Currie

This was Hubble Space Telescope Service Mission 3B. The orbiter crew was named to the flight on 26 March 2001. Currie was assigned as MS2/FE, with responsibilities for assisting and backing-up the Commander and Pilot during ascent, rendezvous with the HST, and entry. In addition, as primary RMS operator she captured the

Nancy Currie operates the RMS during the STS-109 Hubble Space Telescope service mission 3B in early 2002.

HST and brought it into the bay for servicing, then redeployed it using the RMS at the end of the servicing phase of the mission. She also supported and assisted the EVA teams as they worked on the telescope in the payload bay, and assumed a number of responsibilities for orbiter systems, procedures and mission objectives.[72]

STS-110: Ellen Ochoa

This was another ISS assembly flight, including delivery and installation of the S-0 Truss and elements of the Mobile Transporters. Ellen Ochoa was named MS2 on 11 April 2001, and served as launch and entry Flight Engineer in Seat 4 on the flight deck, supporting the Commander and Pilot during ascent and descent, and docking with the ISS. In addition to a number of orbiter system responsibilities and assignments with other payloads, experiments, and objectives, Ochoa was operator of the Canadarm2 from the Destiny laboratory on the ISS during the first, second and fourth EVAs. During the installation of the S-0 truss, she worked with ISS-4 crew-member Dan Bursch from the Destiny laboratory, operating the Canada2 arm to lift the truss out of the payload bay and install it on top of the Destiny module.[73]

STS-111/113: Peggy Whitson

Whitson was assigned to the STS-111 mission as MS5 for ascent only, as she was a member of the ISS-5 residency crew. She served in the same capacity during her return on STS-113.

STS-112: Pam Melroy and Sandra Magnus

The crew for this ISS assembly mission, to install the S-1 Truss, was named on 17 August 2001. Melroy was assigned as Pilot for the second time, and in addition to supporting and assisting Commander Jeff Ashby during the ascent, docking with ISS and landing phases of the mission, she had primary responsibility for a number of orbiter systems and procedures. She was also in charge of undocking the Shuttle from ISS and conducting a fly-around inspection for photographic and TV documentation objectives. In addition, she served as lead EVA choreographer IV1 and as back-up RMS operator. Sandra Magnus was assigned as MS2/FE, and occupied Seat 4 during launch and entry, assisting Ashby and Melroy during ascent, rendezvous and docking, undocking, ISS operations, and entry and landing. As well as being assigned to a number of orbiter systems and procedures, she was charged with opening and closing the payload-bay doors and radiators, and assisted in the operation of Canadarm2 onboard ISS, supporting the installation of the S-1 Truss during the three EVAs. As Loadmaster she was also in charge of logistics transfers to and from the orbiter.[74]

Sandra Magnus (*b*.1964) attended the University of Missouri-Rolla, earning a BS degree in physics in 1986 and an MS degree in electrical engineering in 1990. Her doctorate in materials science and engineering was earned from Georgia Institute of Technology in 1996. She worked as a stealth engineer at McDonnell Douglas Aircraft Corporation from 1986 to 1991, working on the effectiveness of radar signature reduction techniques, on internal research and development studies, and on the A-12 Attack Aircraft propulsion system for the USN until its cancellation. Between 1991 and 1996 she completed work for her doctorate, and supported this with graduate work at the NASA Lewis Research Center. Selected to NASA in 1996, she completed a two-year Ascan training and evaluation programme and a variety of Astronaut Office technical assignments prior to her first flight assignment on STS-112. During 1997 and 1998 she was assigned to the Astronaut Office Payload Habitability Branch, working on ISS facility payloads (freezers, glove boxes, and so on) with international partners (ESA, NASDA and Brazil). In 1998 she was assigned as a 'Russian Crusader', which involved working in Russia supporting hardware testing and operational project development roles.

THE END OF AN ERA

At the beginning of 2003, the long-delayed STS-107 SpaceHab research mission finally left the ground to conduct research and test procedures aimed at supporting science research on the ISS and for long-duration spaceflight. This two-week mission was one of the last non-ISS missions planned, and during the flight it was deemed an outstanding success in gathering and returning data. This success prompted mission planners to evaluate a re-flight of the Shuttle with a SpaceHab module independent of the ISS while STS-107 was still in orbit. However, the plans were changed after the tragic events of 1 February 2003. Sixteen minutes from home, *Columbia* was destroyed as the vehicle broke up at altitude, resulting in the loss of its seven astronauts. After the subsequent enquiry, plans were put in place to finally ground the Shuttle after completion of ISS construction around 2010. Non-ISS missions (such as a planned revisit to the Hubble Space Telescope) were cancelled, and in early 2004 proposals were announced for a new programme to support resupply of the ISS, exploration of the Moon, and pioneering human flights to Mars.

STS-107: Kalpana Chawla and Laurel Clark

The science crew for this SpaceHab research mission was named on 28 September 2000. Chawla was assigned as MS2/FE and occupied Seat 4 on the flight deck, and was responsible for supporting and assisting Commander Rick Husband and Pilot William McCool in the ascent and entry phases of the mission. She worked on the Red Shift, assisting Husband in manoeuvring *Columbia* to support a number of

STS-107 Mission Specialist Kalpana Chawla and Laurel Clark take a break during training.

The crew of STS-107 pose for the traditional in-flight crew photograph during the research mission in January 2003.

science objectives. In addition to responsibilities for various orbiter systems and procedures she was assigned to several experiments, including Astroculture, APCF, CPCG, BDS, eight experiments in the ESA Biopack, CM-2 (which included three separate experiments in combustion science), MGM, VCD-FE and ZCG. Clark was assigned as MS4, and also worked with the Red Shift on the mission. In addition to a number of orbiter systems and procedures responsibilities, she also worked several experiments during the two-week mission, including ESA ARMS hardware, AST 1 and 2, BDS, ESA Biopack (eight experiments), APandBT plant studies, CEBAS, CIBX, MPFE, OSTEO, PhAB4, SLEEP and VCD-FE.[75]

Laurel Clark (*b.*1961) earned a BS in zoology from the University of Wisconsin in 1983 and a doctorate in medicine from the same university in 1987. After completing an internship in paediatrics at Bethesda Naval Hospital in Washington, she graduated from the US Navy undersea medical officer course in 1989, and served for two years as a submarine medical officer and diving medical officer at Holy Loch, Scotland. After qualifying as a US Navy flight surgeon in 1992 she served with Marine Night Attack Squadron 211 and then Naval Training Squadron VT-86 as a flight surgeon until selection to NASA in 1996. Following Ascan training she worked in the Astronaut Office Payloads/Habitability Branch from 1997 until her assignment to STS-107.

Return-to-Flight

On 17 August 2001, Eileen Collins was named as Commander of the crew of STS-114 (LF-1) – the mission scheduled to follow STS-107 in 2003 and continue assembly and outfitting of the ISS. The loss of *Columbia* and its crew grounded the programme and delayed this mission into 2005. As part of the Return-to-Flight programme, NASA expanded the crew, and on 7 November 2003 three additional astronauts were named, including Wendy Lawrence as MS4 in charge of the robotic arm operations and logistics transfer to the station. The crew will also complete new objectives that have changed the mission from a purely logistics flight and crew rotation mission to one of testing and evaluating new procedures for flight safety. This includes orbiter inspection, repair demonstration objectives, and a reduced number of ISS tasks. The selection of the crew and their skills reflected this change of emphasis in the mission. Crews had also been named for the next five missions (STS-121, 115, 116, 117, 118), but these are all dependent upon the success of STS-114. Five women astronauts have been assigned to four of these missions as Mission Specialists: Lisa Nowak and Stephanie Wilson on STS-121; Heidemarie Stefanyshyn-Piper on STS-115; Joan Higginbotham on STS-117; and, on STS-118, former Teacher-in-Space back-up Payload Specialist and now NASA Educator Mission Specialist Barbara Morgan. Two more flights will be flown to complete the assembly of major US elements of the ISS. After this there remain fifteen or sixteen assigned

The STS-114 crew-members review the constantly changing flight plan in preparation for their Return-to-Flight Mission in 2005.

Shuttle missions designed to orbit major elements from international partners and to complete the outfitting and logistics supply, concluding the major assembly of the station. With this successfully completed, the Shuttle fleet will be retired. There are therefore only a limited number of available flight seats on the remaining Shuttle missions; and with the decision to retire the Shuttle by 2010 and the selection of new astronauts in 2004, a new training programme has been developed – a programme which, for the first time in thirty years, does not include substantial courses on Space Shuttle systems.

REFERENCES

1 David M. Harland, *The Story of the Space Shuttle*, Springer–Praxis, 2004.
2 *Space Facts* web site, http://www.spacefacts.de.
3 Michael Cassutt, 'The Manned Spacecraft Engineer Program', *Spaceflight*, **31**, 1989, 26–33.
4 Michael Cassutt, *Who's Who in Space: International Space Station Edition*, Macmillan, 1999.
5 Colin Burgess, *Teacher in Space: Christa McAuliffe and the Challenger Legacy*, Bison Books, 2000.
6 Mary L. Cleave, JSC Oral History Project, 5 March 2002.
7 STS 61-B Post-flight EVA Debriefing, 11 December 1985, NASA JBJ Space Center (no reference number).
8 Teacher-in-Space Project, NASA Public Affairs Release, 1985.
9 Ref. 4, pp. 67–68.
10 *USA Today*, 29 January 1987.
11 NASA News Release (JSC) 88-004, 29 February 1988.
12 The Shuttle missions are detailed in Harland, *The Story of the Space Shuttle* (ref. 1).
13 Mary L. Cleave, JSC Oral History Project, 5 March 2002.
14 STS-30 Mission Report, NSTS-50702, July 1989, NASA, STS-30 Flight Data File, Flight Plan, Final, 10 March 1989, NASA, JSC-23007-30.
15 'The mission of STS-34, pre-flight information', *World Spaceflight News*, **4**, No. 12, 1989.
16 STS-32 Press Information, December 1989, Rockwell International, PUB 3546-V REV 12-89.
17 'The mission of STS-31, pre-flight information', *World Spaceflight News*, **5**, No. 2, 1990; also STS-31 Crew Assignment Tasks, STS-31 Query Book, Public Affairs Office, NASA JSC.
18 STS-40 Crew Responsibilities, in the STS-40 Query Book, NASA JSC Public Affairs Office.
19 STS-42 Crew Assignments, in the STS-42 Query Book, NASA JSC Public Affairs Office.
20 STS-45 Query Book, NASA JSC Public Affairs Office.
21 STS-49 Query Book, NASA JSC Public Affairs Office.
22 STS-50 Query Book, NASA JSC Public Affairs Office.

23 STS-46 Query Book, NASA JSC Public Affairs Office.
24 STS-47 Spacelab J Payload Information Briefing Notes, NASA JSC, undated, *c*.1992.
25 STS-52 Crew Assignments, Query Book, NASA JSC Public Affairs Office.
26 STS-54 Crew Assignments, Query Book, NASA JSC Public Affairs Office.
27 STS-57 Press Kit, Release 93-78, June 1993, NASA.
28 SLS-2 Science briefing notes, NASA JSC, undated.
29 NASA News Release (JSC) 93-032 4 May 1993
30 AIS interview with Rhea Seddon, 21 April 2004.
31 STS-61 Crew Responsibilities, Query Book, NASA JSC Public Affairs Office.
32 STS-62 Crew Task Assignments, Query Book, NASA JSC Public Affairs Office.
33 STS-59 Crew Assignments, Query Book, NASA JSC Public Affairs Office.
34 STS-65 Press Kit, July 1994.
35 STS-64 Crew Assignments, Query Book, NASA JSC Public Affairs Office.
36 AIS Interview with Michael Foale, June 1993.
37 STS-66 Crew Assignments, Query Book, NASA JSC Public Affairs Office.
38 STS-63 Crew Assignments, Query Book, NASA JSC Public Affairs Office.
39 Trigg Gardner, 'First woman Shuttle pilot to fly, *The Citizen*, Houston, 15 September 1993.
40 STS-67 Query Book, NASA JSC Public Affairs Office.
41 David J. Shayler, 'American Flights to Mir (Space Shuttle), in *The History of Mir, 1986–2000*, British Interplanetary Society, 2000, pp. 71–85.
42 STS-71 Press Kit, June 1995.
43 STS-70 Query Book, NASA JSC Public Affairs Office.
44 STS-73 Query Book, NASA JSC Public Affairs Office.
45 STS-76 Press Kit, March 1996.
46 STS-78 Press Kit June 1996.
47 STS-73 Press Kit, November 1996.
48 STS-81 Query Book, NASA JSC Public Affairs Office.
49 NASA News Release 97-27, 18 February 1997.
50 STS-84/94 Query Book, NASA JSC Public Affairs Office.
51 STS-84 Query Book, NASA JSC Public Affairs Office.
52 STS-85 Query Book, NASA JSC Public Affairs Office.
53 STS-86 Query Book, NASA JSC Public Affairs Office.
54 Executive Summary of Investigation Report for the STS-87 Spartan Close Call, undated extract, NASA JSC.
55 STS-87 Query Book, NASA JSC Public Affairs Office.
56 STS-89 Query Book, NASA JSC Public Affairs Office.
57 STS-90 Query Book, NASA JSC Public Affairs Office.
58 STS-95 Query Book, NASA JSC Public Affairs Office, and STS-95 Press Kit, October 1998.
59 Bert Vis, 'Soviet Woman Cosmonaut Flight Assignments, 1963–1989', *Spaceflight*, **41**, 11, November 1999, 474–480.
60 Mike Snyder, 'All-woman crew defended', *Houston Chronicle*, 14 April 1999; and Irene Brown, 'NASA Consider all-female Shuttle crew', Space.com, 28 August 1999.

61 Dennis R. Jenkins, *Space Shuttle: The History of the National Space Transportation System – The First 100 Missions*, Midland Publishing, 2001, pp. 316–320.
62 David J. Shayler, 'NASA Shuttle Missions to the ISS', in *The International Space Station: From Imagination to Reality (1988–2001)*, British Interplanetary Society, 2002.
63 STS-88 Press Kit, December 1998.
64 STS-96 Query Book, NASA JSC Public Affairs Office.
65 STS-93 Query Book, NASA JSC Public Affairs Office.
66 STS-99 Query Book, NASA JSC Public Affairs Office.
67 STS-101 Query Book, NASA JSC Public Affairs Office; and STS-101 Press Kit, May 2000.
68 STS-92 Query Book, NASA JSC Public Affairs Office; and STS-92 Press Kit, October 2000.
69 STS-98 Press Kit, February 2001.
70 STS-104 Press Kit, July 2001,
71 STS-108 Press Kit December 2001.
72 STS-109 Press Kit, March 2002.
73 STS-110 Press Kit, April 2002.
74 STS-112 Press Kit October 2002.
75 STS-107 Press Kit, January 2003.

Stations in space

If we are to live and work in space for prolonged periods, the creation of an orbital space platform or space station affords opportunities for research, observations and activities not possible on short-duration missions.

Despite a number of women having visited a space station for several days, only four have performed residency missions: Yelena Kondakova (Russian), who spent 169 days in space, mostly onboard the Mir space complex during 1984 and 1985; Shannon Lucid (American), who spent 188 days in space (also mostly onboard Mir) in 1996; Susan Helms (American), who was a member of the second ISS resident crew during 2001, spending 167 days in space; and Peggy Whitson (American), a member of the fifth ISS resident crew who logged 184 days onboard the station during 2002. In several research programmes, scientists are investigating the medical aspects of females on extended-duration spaceflight. Future female crew-member assignments onboard the ISS are an important element in this long-term planning prior to the proposed return to the Moon to establish a research base, and expeditions to Mars. A French woman could have claimed to have been the first woman considered for a space station mission. In April 1979 the French officially accepted the offer of a spaceflight onboard a Soviet spacecraft (Soyuz T/Salyut 7) in 1982. The criteria issued in September 1979, for primary and back-up positions, focused on the aerospace industry and the military with restrictions to 82 kg, 95 cm in height and an engineering background. Included in the initial 400 candidates were A.C. Lavasseur-Regourd from the 1977 French Spacelab 1 selection, and Françoise Varnier (*b.* 1944), a lecturer at the University of Marseilles and a skilled private pilot. After a scrutiny of applications, a series of aviation-orientated medicals, a centrifuge test and interviews, a short-list of five – including Varnier – were put forward in October 1979 for further tests in Moscow. Embarking on an intensive Russian language course and parachute training (during which Varnier broke her leg) the May 1980 announcement of four to go to Moscow did not include her. She was tragically killed on 27 October 1996 whilst piloting her CAP10 aerobatics aircraft at Bourges, France. She was 52.

SAVITSKAYA SPACEWALKS

The first woman to visit a space station was Svetlana Savitskaya, who in 1982 flew on Soyuz T-7 and spent a week onboard the Salyut 7 station. This was, on the whole, for the purposes of propaganda, as the Soviets wanted to upstage the pending launch of the first American woman in space, Sally Ride, in 1983. It was decided that with the successful completion of Savitskaya's mission, another female cosmonaut should be sent to the Salyut on a long-duration mission, to be in orbit during the time that Sally Ride was flying.

Soyuz T-8

Irina Pronina was assigned to fly the mission as Cosmonaut Researcher after completing her assignments backing up Savitskaya on Soyuz T-7. Savitskaya would therefore serve as back-up. Joining Pronina as prime crew would be cosmonauts Vladimir Titov as Commander and Gennady Strekalov as Flight Engineer. The back-up crew would be Vladimir Lyakhov as Commander, Alexander Aleksandrov as Flight Engineer, and Savitskaya. The mission was scheduled for launch in April 1983, and was planned to last for more than 100 days, landing in July or early

Irina Pronina with Titov and Strekalov – the original T-8 crew. (Courtesy Tony Quine.)

August – well after the planned launch of STS-7, carrying Sally Ride, in June. However, shortly after the crews were assigned the plans were altered when Savitskaya was removed from the back-up crew, to be replaced by Alexandr Serebrov, who began training as back-up to Pronina. It has never been clearly explained why Savitskaya was not replaced with another female cosmonaut (possible physician Yelena Dobrokvashina).

For six months the crew prepared for their residency flight until, shortly before launch, Pronina was removed from the crew and replaced by Serebrov. She thus lost her chance to fly in space.[1] Almost a decade later, two of the male cosmonauts were interviewed by Bert Vis, who questioned them about this change. During an interview in 1991, Strekalov explained that it was due to a heavy EVA programme to erect and deploy additional solar-array panels during the mission, although Pronina would not have participated in them and would only monitor them from the inside of the station. A year later, Aleksandrov added that the most probable reason was political. Subsequent reports indicated that Marshall Ustinov, the Soviet Defence Minister, had not wanted Pronina to fly, and that the State Commission which confirmed crews prior to flight was also opposed to her participation. Could the opposition have arisen due to the prospect of a female being in sole residency onboard the Salyut? It appears so, as later reports indicated that the Central Commission was uncomfortable about a possible untoward occurrence during EVA which could leave Pronina alone on the station and during the return in the Soyuz – which would indicate the limited amount of Soyuz system training that she had undergone. Other reports indicated that the members of the Military Industrial Commission were opposed to flying women in space *at all*, even though the Americans were training female astronauts for their Shuttle programme. Despite Valentin Glushko's involvement in placing Soviet women in space, the idea was not wholly supported.

This discussion certainly indicated that Pronina would not fly. Soyuz T-8 was launched on 20 April 1983 on a planned 110-day mission to Salyut 7; but the flight lasted for only two days, due to a docking system failure and a replanning of crew assignments. Titov and Strekalov had trained extensively for the EVAs, and were assigned to return to the station as soon as possible. In the meantime, the planned T-8 mission was split into two, with the first part flown by Lyakhov and Aleksandrov (who would be replaced by Titov and Strekalov in September, and would conduct the EVAs) who were launched on the T-9 mission to Salyut 7; but on 26 September Titov and Strekalov were again thwarted in their attempt to board Salyut 7 when their launch vehicle exploded on the pad seconds before lift-off, resulting in a pad abort. As a result, Lyakhov and Aleksandrov had to perform EVAs for which they had hardly trained, and the whole crewing sequence for future Salyut 7 missions was revised, due to the launch abort and a leakage of fuel from the Salyut engine system on 9 September.[2]

Pronina had not felt that being a woman had markedly affected her career until she reported for cosmonaut training; and though sex discrimination was evident during the formation of crews, she reportedly stated that though 'unpleasant, I always believed in a better future.'[3] During 1985 she became involved with the plans for an all-woman crew to Salyut 7, but fell pregnant and remained inactive until

1986. Although still officially an active cosmonaut during the late 1980s, and working towards her Candidate of Technical Sciences advanced degree, she was stood down from the Energiya Cosmonaut team on 23 July 1992, after which she returned to her work at the design bureau. In 1991, former cosmonaut Konstantin Feoktistov indicated that she was a candidate for a planned mission to the Mir space station in 1991, but this was not pursued.

Soyuz T-12

The mission of Soyuz T-12 was planned as the second visiting mission to the 1984 expedition crew. For some time, several cosmonauts in training for Buran shuttle missions were also completing qualification for flights onboard Soyuz spacecraft to Salyut (or Mir) space stations in order to gain spaceflight experience prior to flying a Buran shuttle in space. This also provided for the Soviet government requirement (since the 1977 docking failure of Soyuz 25 to Salyut 6) that at least one cosmonaut should be flight experienced. This availability of a third seat allowed for the rookie cosmonaut to fly without disturbing assigned two-person crews.

The leader of the Buran group and the highly experienced test-pilot Igor Volk took the first assigned mission of T-12 research-cosmonaut. The Commander would be Vladimir Dzhanibekov (one of the most of the most experienced Soviet cosmonauts available) and Savitskaya would be Flight Engineer flying on her second mission, which would also involve her completing a spacewalk with Dzhanibekov. In December 1983, training began for a planned flight to Salyut 7 in July 1984, three months before Sally Ride flew her second mission on STS 41-G, during which Kathryn Sullivan would complete the first US female EVA with colleague Dave Leestma. If Savitskaya was successful, then she would not only become the first woman to fly in space twice, but also the first to perform an EVA. In February 1984 the back-up crew was assigned, with Vasyutin as Commander and Savinykh serving as Flight Engineer – both of whom were already in training as a future core crew to the Salyut. Pronina was the most experienced candidate and the most logical choice as back-up to Savitskaya; but according to an interview with Viktor Savinykh in 1992, the Chief Designer of NPO Energiya, Valentin Glushko, personally ordered the EVA to upstage the Americans, and also assigned Ivanova as the back-up to Savitskaya – primarily for the experiments, as she was not trained for EVA. She was not, however, assigned as Flight Engineer (Savinykh had that assignment), but as Research Engineer, so that technically she was Volk's back-up. It is known that had Volk not been able to make the flight he would have been replaced by another civilian test-pilot from the Buran group. Pronina was fully capable of making the flight as Flight Engineer had she been called upon to do so, and she could probably have completed the EVA. However, Savitskaya was considered the primary candidate for the EVA – not only due to her previous experience, but also her physical strength

No details of the intended EVA were announced when the flight was launched on 17 July, and even some official Soviet news reports again focused on the positive influence of having a woman on board a space station, rather than the skills she was about to demonstrate. Moscow World Service reported that Savitskaya's five male colleagues were shaving every day, and were being chivalrous and helping her with her experiments. Savitskaya, however, had taken on more of the cooking chores – and Vladimir Solovyov had begun eating potatoes with renewed pleasure now that Savitskaya was cooking them![1]

The first female steps into space

The more serious reason for her being on the flight was demonstrated on 25 July during an EVA of 3 hrs 35 min, during which Savitskaya, assisted by Dzhanibekov, tested a 'new general purpose hand tool' that had the capacity to cut, weld and solder.

The news report highlighted this milestone in human spaceflight: 'It is for the first time now that a woman is heard speaking in outer space. The woman Svetlana Savitskaya is offering comment on her walk in outer space: She is just now cutting steel plates, welding samples of different metals, and coating aluminium with silver with the help of an electronic device.'[4] The tool was a URI (a Soviet acronym) – an improved version of a more bulky instrument tested inside the Orbital Module of Soyuz 6 in 1969. Automated welding had been conducted in space before, but this was the first manual demonstration of the technique, and so much care was needed during the test programme. The programme featured six welding, six cutting and six soldering demonstrations using tin, stainless steel, aluminium and titanium samples. During the EVA, which was supposed to be completed during daylight passes,

Svetlana Savitskaya onboard Salyut 7.

Savitskaya complained of excessive sunlight in her eyes, preventing her from seeing the work area clearly. At the end of her activities she exchanged places with Dzhanibekov, who had filmed and photographed her activities, and then completed a series of tests with the URI.[5]

In TASS reports following the end of the EVA, Savitskaya's achievements were overemphasised as a major contribution to future participation of women on space stations: 'Her successful performance of unique experiments in space conditions has vividly shown the possibility of women's effective activity in the performance of complex research not only onboard manned orbital complexes, but in open space as well.'

Savitskaya, Dzhanibekov and Volk returned to Earth on 29 July, to the normal reports of a further successful step in cosmonautics and the usual round of awards and honours. From these reports it seemed that female cosmonauts would soon become regular members of space station crews – perhaps on long-duration missions, or perhaps as Commanders of all-woman crews. But it was not to be.

The 1980 selection disbanded

Savitskaya was the first and only Soviet female cosmonaut in the 1980 selection who flew in space. Although an all-female Soyuz crew was planned to visit Salyut 7 to celebrate International Women's Day, problems with the station and a limit of Soyuz T spacecraft contributed to the cancellation of such a flight, and as a result, each of the members of the 1980 female selection retired from active flight status.

Savitskaya lost the chance of a third flight as Commander of the all-woman crew due to constant delay and her pregnancy, although she continued as a civilian

Yelena Dobrokvashina wears a Soyuz Sokol suit. (Courtesy Tony Quine.)

Energiya cosmonaut until leaving the cosmonaut team in October 1993. She had risen to become Deputy to the Chief Designer at Energiya in 1987 and became a member of parliament in 1989. When she left Energiya in 1993 she chose a political career. In 2001 the Voice of Russia reported on her comments on extending her cosmonaut career: 'What I could do as a cosmonaut is done. Although I could have done more than I did, I could and was eager to fly the Buran shuttles, but the shuttle programme [had] folded. Cosmonautics demands an all-out effort. You devote yourself to cosmonautics or you choose to so something else. I've opted for politics.'

Of the other members of the 1980 selection, Klyushnikova does not seem to have finished her basic training for the cosmonaut certificate, whilst Almelkina was medically disqualified in May 1983. Ivanova retired in 1987, and Kuleshova retired in April 1992, after training as a potential FE crew-member to the Mir station. Pronina also retired in April 1992 and remained at Energiya, working in the flight control centre. The following March, both IMBP cosmonauts – Dobrokvashina and Pozharskaya – retired to return to their position at the medical institute, and they were followed by Savitskaya seven months later. Latysheva retired in 1994 to continue work as a senior scientific member of staff at the Institute of Space Research. Zakharova retired in October 1995. She had been eligible for assignment to a Soyuz/space station crew for more than fifteen years, but she never flew.

COMMERCIAL CUSTOMERS

In February 1986, Salyut 7 was replaced by a new space station: Mir. The first element was the base block, to be expanded over the next ten years, with the fifth and final module being added in April 1996. These ten years saw a dramatic change of politics and social and economic fortunes in the Soviet Union, resulting in the demise of the Communist state, the collapse of the USSR, and the re-establishment of Russia and a number of independent states. This was also a period of redirection in the US space programme, which was recovering from the loss of *Challenger*. The proposed space Station Freedom was far too large and expensive, and in 1992 it was replaced by the more economic International Space Station, with Russia as a full partner. In addition to participating in the ISS, the Russians agreed to host a series of American Shuttle flights to the Mir space complex, fly cosmonauts on the US Space Shuttle, and incorporate American astronauts on Mir-resident primary crews in Phase I of the International Space Station programme. During the late 1980s the Russians initiated a series of 'commercial spaceflights', offering (for a fee) the opportunity for representatives of other countries to fly to the Mir space station.

Journalist cosmonauts

In March 1989, a reported $12-million agreement between the Soviet Union space agency Glavkosmos and the Tokyo Broadcasting System (TBS) was drawn up, featuring a purely commercial flight of a TBS employee to the Mir space station for one week. Applications for the one flight-seat and a back-up position were sought from members of the TBS staff and its associated companies. A total of 163 applications was received; but as none of the candidates fell within the selection criteria, rules were released allowing seven finalists to be nominated for more detailed medical examination in Moscow. On 18 September 1989 the two finalists were named as male journalist Toyohiro Akiyama and female TV camera operator Ryoko Kikuchli. They began training in October. Shortly after being selected, Kikuchli was asked about possibly being chosen as the first woman to fly to Mir. She said that she was looking forward to it. 'When I first started work for TNBS, my goal was to be assigned to Beijing ... I never imagined I would be doing anything like this.'[6]

Unfortunately, however, Kikuchli did not fly to Mir – nor even into space. In August 1990 she was assigned as back-up to Akiyama, who would fly as Cosmonaut Researcher on Soyuz TM-11 on 2 December 1990. Kikuchli (*b*.1964) had studied Chinese at the Tokyo University of Foreign Studies, and began working at TBS after graduation. Assigned to the News Service, she became the only female camera operator employed by the company, and was assigned to China as well as covering the 1988 Summer Olympic Games in Seoul, Korea. A week before the launch of TM-12 on 26 November 1990 she was in hospital undergoing surgery to remove her appendix; but she recovered in time to attend the launch of her colleague from the Baikonur cosmodrome. After the flight she returned to work at TBS.

The agreement to fly a Japanese journalist into space did not settle well with their Soviet counterparts, many of whom felt that a Soviet citizen should be given the honour of being the first professional journalist in space. As plans for the flight of the Japanese journalist continued, NPO Energiya finally implemented and decided upon a flight of a Soviet journalist during the Internal Space Year, 1992. After a considerable number of disagreements between the Soviet Writers' Union and the leadership of the cosmonaut training centre, six candidates were nominated for medical examination in February 1990, and on 11 May the Joint State Commission approved them as cosmonaut trainees. One of the successful candidates was female journalist Svetlana Omelchenko, and three of the identified but unsuccessful candidates (of which nothing else is known) were Yelmina Akmedova, Lyudmilla Chernyak and Olga Sasoreova.

Svetlana Omelchenko (*b*.1951) was born into a military family. When her father retired from military service the family moved to Stavropol district of southern Kavkas, and after Svetlana left secondary school in 1968 she worked for a local newspaper, *Primanycheski Stepi*. In 1969 she moved to Moscow to work in the building trade as an electrician. She also studied at a technical school, and later undertook a journalism course at Moscow State University, from where she graduated in 1975. Her first job in journalism involved a small publication called *Air Racing*. She had at last entered the aviation world – if only as a passenger. She then worked for three years for the newspaper *Vozdushny Reys* (*Air Travel*), focuusing on articles on air safety for *Vozdushny Transport* (*Air Transport*) and *Delovoy Mir* (*Business World*). As a child she had dreamed of flying to the Moon, and she recalled the flight of Yuri Gagarin. Although not in good health and not strong, she had wanted to fly. When she was a teenager she applied for membership of a flying club, but was told that girls were not accepted for classes. When she applied for the Soviet Journalist-in-Space programme she realised that she was not

Svetlana Omelchenko undergoes training.

a probable candidate, as she wore glasses and was afraid of heights! Nevertheless, she succeeded.

The training programme – for qualification as a Cosmonaut Researcher on a Soyuz flight to Mir for a one-week residency – began in October 1990 and ended in February 1992. It included wilderness training, parachute training, flying in jets, zero gravity simulation in parabolic flights, working in simulations, scuba diving, altitude-chamber sessions to 10,000 metres, and centrifuge runs to up to 8 g. The state examinations were conducted at the end of January 1992, and the official designation 'Cosmonaut Researcher' was conferred on each of them on 7 February. Omelchenko was the first female cosmonaut to undergo the 'Neptune' graduation ceremony – a tradition begun at the time of Gagarin in the 1960s, in which the trainees are pushed by 'Neptune' into the water of the training centre swimming pool, to officially enrol them as cosmonauts. Apparently, Svetlana Savitskaya declined the initiation ceremony, allowing Omelchenko the chance to make a little space history, though her chance of a spaceflight was already dwindling.[7]

With the prgramme short of funds, cut-backs were required in various areas, including the Journalist-in-Space programme. Indeed, the priority was to feed families, as times were hard. The journalists were disbanded, and returned to their former positions. Omelchenko resumed work at *Deliovoy Mir* and *Vozdushny Reys*, writing on problems encountered with space exploration. Although she did not fly in space, she is proud of her achievement in passing the full cosmonaut research training programme, and of her framed certificate: Cosmonaut No. 151.

An actor's life . . .

Five years after the journalists qualified as Cosmonaut Researchers, a new suggestion was proposed. In December 1997, Russian film director Yuri Kara suggested making a film on Mir early in 2000, for release in April 2001 in time for the fortieth anniversary of the flight of Yuri Gagarin. This was but one of a number of confusing and speculative ideas that emerged at the end of the Mir programme in an attempt to acquire funding for the struggling Russian space programme, in order to maintain a presence on Mir at a time when the ISS was being constructed. Ultimately, however, it was impossible to raise the cash to sustain Mir in orbit. In August 1999 the station was vacated for the first time in ten years. Except for a brief occupation by an inspection crew during 2000, it was not crewed again, and was deorbited in March 2001 after fiftten years of operations.

According to contemporary reports, several Russian actors and pop stars, including three women, applied for the leading role in Kara's proposed film. In addition, two actresses (Emma Thompson and Demi Moore) expressed interest, but they soon declined any further participation. The script required two males and two females (prime and back-ups), but a search for Russian actors resulted in only three suitable candidates: the actor Vladimir Steklov and the actresses Olga Kabo (*b.*1968) and Natalya Gromushkina (*b.*1975). However, it soon became clear that if funding and a Soyuz seat could be found, only one candidate would fly – most probably Steklov. By April 1998 a fourth actor still could not be found, and Kabo withdrew from the project, as she had recently married and wished to start a family. Over the next two years, the project stumbled. Funds were not forthcoming, and the fate of Mir seemed to be sealed. In addition, a mission of only seven days was becoming less feasible, and Steklov was not required for a long-duration flight. During this time, Gromushkina's six-month training course was delayed due to lack of finance, but it was reported that she would fly with Steklov and not serve as back-up. Then the decision was made to fly only Steklov and not Gromushkina, who left the project and was replaced by another actress who would film her scenes in an aircraft flying parabolic profiles. But by March 2000, insuffucient funds had been raised, and the film project was cancelled.[8]

'Astronaut wanted – no experience necessary'

This was the advertisement that appeared in British newspapers and on the radio in June 1989, in response to an agreement between the Soviet space organisation Glavkosmos and the British company Antequara Ltd, to fly a British citizen in space for one week onboard Mir, under the project name Juno. The response to the advertisement resulted in more than 13,000 applications for the four final places and just one flight seat with a back-up, for a planned week-long stay on Mir in 1991. One of the applicants was a young chemist named Helen Sharman, who had heard the announcement over her car radio. She ticked all the requirements: a UK citizen aged between 21 and 40; a formal scientific training background; proven ability to learn a foreign language; and in a good state of health and fitness. After applying by telephone, and a short interview, she received an application form (more than 3,000 were issued). She completed it, posted it, and then forgot about it – but not for long.

After scrutiny of about 3,000 application forms, 150 suitable candidates (including about fifty women) were short-listed for a series of medical tests. The top thirty-two candidates (including ten women) were introduced to the press in August 1989. In her autobiography,[9] Sharman recalled that once the press learned of her success she was called 'the girl from Mars'. When the candidates and finalists were announced, it was observed that in each case about a third of them were women. As all the candidates had to have a scientific or technical degree, it was suggested that the women faired much better than the men. In figures released at the time (the late 1980s), the number of female university graduates with engineering or technical degrees was fewer than a third of the number of male graduates in the same fields.

A series of physiological tests and medical tests followed, after which only twenty-two candidates continued through the selection process. Only three of these were women – a dramatic change in the proportion of candidates. Exhaustive medical tests by civilian doctors reduced this number to sixteen, including two women. The next round of medical tests was held at RAF Farnborough, after which a further medical review by British and Russian specialists produced the final four. The names of the quartet were announced on 5 November 1989: Lieut-Commander Gordon Brooks, Major Tim Mace, Mr Clive Smith and Miss Helen Sharman. Then, after

The four Juno finalists: Gordon Brooks, Helen Sharman, Clive Smith and Tim Mace.

further medical tests in Moscow, the prime and back-up – Helen Sharman and Tim Mace – were announced on 25 November. Several months passed before it was decided which of them would fly – and then Helen Sharman was told that she had won the coveted seat into space in 1991.

Helen Sharman (*b*.1963) After leaving secondary school in 1981 she gained a BSc degree in chemistry from the University of Sheffield in 1984. She then worked as an electronics engineer for the MOV division of General Electric in London, and at the same worked on her PhD. In 1987 she joined Mars (the confectionary company) near London, conducting research into the chemical and physical properties of chocolate. Following her flight in 1991 she worked as a television commentator and public speaker on spaceflight, and became a lecturer, broadcaster and author, enthusiastically promoting science to a variety of audiences (mainly school children) across the UK.

New Russian female cosmonauts, 1989–2004

Meanwhile, female Russian cosmonauts were few and far between. With the introduction of the Mir space station it was assumed that there would be more flight opportunities for female cosmonauts in the 1980 selection; but it was not the case. Since 1980 only a few females have progressed to the medical examinations for cosmonaut candidate selection, and only two have been assigned as 'career' cosmonauts.

On 20 July 1988, Yelena Kondakova was passed by the GMK Chief Medical Commission as a prospective cosmonaut for the NPO Energiya group. She was formerly accepted for cosmonaut training on 9 February 1989, but was not accepted for cosmonaut training at TsPK until 1990. An unsuccessful candidate for this section has been identified as Irina Khokhlachova (*b*.1954), an engineer at NPO Energiya.

Yelena Kondakova (*b*.1957) graduated as a mechanical engineer from the Moscow Higher Technical School in 1980. In the same year she joined NPO Energiya, at the time of the selection of the first Soviet female cosmonaut candidates since the Tereshkova group. Over the next eight years she worked in a variety of assignments at the design bureau, including assignment as a flight controller for the Mir space station programme. She officially joined the cosmonaut detachment on 25 January 1989, and commenced her basic training programme in August 1990. She completed this by passing her examinations in March, and received her Cosmonaut Researcher certificate on 11 March 1992. Almost immediately she joined the Mir cosmonaut training group, making her first flight the following year and a flight on the US space Shuttle in 1997.

On 15 April 1994, three years after applying to join the Energiya cosmonaut team, another Energiya female engineer, Nadezhda Kuzhelnaya (*b*.1962). was selected for cosmonaut training, which she completed in April 1996. After leaving secondary school she initially studied at an institute for the construction industry, but decided to work in aviation after joining an aeroplane and parachuting club. In 1984 she went to the Moscow Aviation Institute, and graduated in 1988. For the next six years she worked at NPO Energiya as a design engineer specialising in flight control systems. In September 1996, shortly after completing basic training, she was assigned to the cosmonaut group preparing for flights to the International Space Station, and commenced training as a member of a Russian Soyuz taxi crew from August 1997. In November 1998 she took maternity leave, and resumed training in May 1999.

Two other women have been identified as possible but unsuccessful cosmonaut candidates.[10] Captain Svetlana Protasova (*b.*1966) is one of only a few Russian female fighter pilots in the Air Force. Asked if she had any desire to become a cosmonaut, she said that she had applied, but that her application was not successful. It was presumably examined by the Joint State Commission (MKV) selection board which met on 25 July 1997. However, it is not clear whether she was a formal candidate. Protasova was apparently a graduate of the Moscow Aviation School and the DOSAAF Zaporozhe Aviation School, logging more than 400 hours in advanced jet trainers and the Mig-29, and in a variety of aerobatic aircraft as a member of the Strizhi (Swift) aerobatic team.[11]

In the most recent cosmonaut selection of 2003, another Energiya engineer, Anna Zavylova (*b.*1975), was a strong candidate, but was dropped late in the selection process – possibly as a reaction to the *Columbia* accident in February of that year, and the resultant hesitation in flying female cosmonauts.

WOMEN ON MIR, 1991–98

During the fifteen years of Mir operations (February 1986 to March 2001), 104 astronauts and cosmonauts visited and worked onboard the complex. Nine American Shuttle crews and representatives from a dozen different countries (mostly from Russia) flew both short- and long- duration missions. Of these, only ten were women, and from this select group only two took part in long-duration missions.

The first female on Mir

Although Russian female cosmonauts hoped to visit Mir in its early years, and Japanese cosmonaut Akiyama hoped to fly as the first woman to Mir, it was the British Juno project's Helen Sharman who holds that distinction, spending one week on the station during May 1991.

Sharman's training began in November 1989 and lasted 18 months – a period she later described as 'the most significant period of my life; in its own way even more influential on my outlook and ideas than the time I spent in space.'[12] The training included mastering the Russian language and adapting to life at the Cosmonaut Training Centre near Moscow.[13] During most of 1990 the future of the mission was in some doubt, as various sponsorship deals to fund it were not immediately forthcoming. As preparations continued, much of the estimated $11 million fee, which the Russians expected, was not produced, and it was therefore significantly reduced to around $2 million to allow the project to proceed. Meanwhile, the training continued with a programme of physical and medical fitness tests and more Russian language classes. In addition there were the academic courses (all in Russian), the survival and emergency drills, weightlessness training in parabolic flights, and parachute training, plus the hours spent in the Soyuz simulator and Mir mock-up in becoming familiar with hardware and procedures. Specific crew training commenced in December 1990, and on 21 February 1991 the two cosmonauts were told which one of them would fly the mission.

Soyuz TM-12 was launched on 18 May 1991, and docked with Mir two days later. Sharman carried out a small programme of observations and experiments whilst on board the station – the medical experiments being mostly on herself. A series of biological experiments included using some wheat seedlings, potato roots and a miniature lemon tree to investigate the effects of weightlessness on plants to determine the benefits, or otherwise, of growing food for consumption during extended missions. An experiment developed by school-children carried 125,000 seeds into space to be later dispersed to schools across the UK. Two larger onboard experiments were the growth of protein crystals, and the electrotopographical experiment which exposed thin films of ceramic oxide to the vacuum of space for the evaluation as future coatings for spacecraft. There were also several photographic opportunities, linked to the mission sponsors, including ordering a bunch of flowers from space (for her mother) through Interflora (a Juno sponsor), and talking with amateur radio hams.

After the week in space, Sharman retuned to Earth – not with the two cosmonauts with whom she launched (who were to carry on the permanent occupancy of the station) – but with the pair of cosmonauts who had been in orbit since December 1990, and who had flown to the station with the Japanese cosmonaut. Following the completion of her activities under the Juno programme, Sharman fulfilled a number of post-flight obligations, but moved into a more private role and appeared less and less as a public figure. In Bettyann Holtzmann Kelvles' book *Almost Heaven*, the author wrote: 'The world of science and technology was not enormously enriched by [Sharman's] week's work, but the British public, and perhaps the British national ego, was. Sharman's good nature and flawless performance compensated for her awkward position ... From being a paying visitor to an expensive guest, she won the affection and admiration of her Soviet colleagues and the British public ... In 1993 she published her co-authored memoir ... She had 'seized the moment' [the title of her autobiography], but it was only an episode in her life and she closed that chapter quickly, no longer a cosmonaut and public figure.'[14]

Enamouring the British public, maybe – but for only for a short time, and certainly not influencing the Government, which has avoided any support or participation in human spaceflight. More than a decade after Sharman flew, the UK has no commitment to human spaceflight, and the interest in science and technology which could have inspired successive generations of school-children seems to have passed by. Apart from a few British citizens now serving as American astronauts for NASA, the UK does not have any plans to supply its own members for the ESA astronaut corps ... male or female.

A Russian female long-duration mission – at last

The next female visitor to the Mir space station – Russian cosmonaut Yelena Kondakova – was launched from Baikonur onboard Soyuz TM-20 on 4 October 1994, the thirty-seventh anniversary of the launch of Sputnik. Kondakova was to stay onboard Mir with her Commander Alexandr Viktorenko and Dr Valeri Polyakov, who had been onboard Mir since January 1994, forming the seventeenth resident crew on the station. Polyakov would return with his colleagues in March 1995, after setting a new endurance record of fourteen months in space. Kondakova and Viktorenko would return with Polyakov after some five months in space. Launched to Mir with Viktorenko and Kondakova was ESA astronaut and twice Shuttle veteran Ulf Merbold, who would return with the sixthteenth resident crew Yuri Malenchenko and Talgat Musabayev.

Kondakova had been assigned to long-duration training in August 1993, and was assigned to the back-up crew of what became the sixteenth resident crew, whom she would replace on her own mission. She was named to the prime crew of Euro Mir 94/TM-20 on 30 May 1995. Her career also followed a different path from that of either Tereshkova or Savitskaya. Unlike Tereshkova, she was not a parachutist, nor was she a pilot like Savitskaya. She did not have a father in an influential position in the Soviet military; but she *did* have an influential husband. In 1985 she had married former cosmonaut and flight director Valeri Ryumin, who had become a Director of the Shuttle–Mir programme and Deputy Chief Designer at Energiya, where Kondakova worked. Ryumin was no stranger to long missions, as on two of his first three flights he had spent a year on Salyut 6, which he helped design. He would also select himself to fly to Mir on the final Shuttle docking mission STS-91 in 1998 – his fourth spaceflight.

When selected for long-duration mission training, Kondakova was a little worried about her husband's reaction, as he opposed the idea of assigning women to such arduous missions. However, he allowed her to continue because he did not expect her to pass the medical examinations; but when she *did* pass, he was pleased that he had such a healthy wife.[15] When asked about her husband's objection to women flying in space, she said that he had to become adjusted to *her* plans; and at the pre-launch press conference she announced that one of her primary objectives on such a long mission was to demonstrate that 'a woman can do as much as a man'.[16] Following the two-day chase to the station, as Soyuz approached the docking port the Kurs automated docking system malfunctioned, forcing Viktorenko to perform a manual docking, with Kondakova assisting him from the observation port in the Orbital Module of the small ferry spacecraft. Over the next month, Kondakova assisted her Commander in performing hand-over duties from the previous main crew to ensure a smooth continuation of the occupation and activities on the station. In addition, she assisted Merbold with his programme of experiments, and participated in Polyakov's detailed medical experiments on long-duration residence and adaptation to spaceflight by new crew-members.

When the previous main crew returned to Earth with Merbold on 4 November, Viktorenko and Kondakova worked with Polyakov on their research programme, including the unloading of the Progress resupply craft. In January 1995 they tested the Kurs docking system automatically by all three cosmonauts entering Soyuz, undocking, flying out to 160 metres, and then approaching again to dock successfully. This 'spin in a Soyuz' took about 26 minutes.

The research work carried out during the mission included physiological, medical and astrophysical investigations, measuring the noise intensity inside the modules, using French equipment to measure radiation levels in low Earth orbit, and retrieving data on micrometeoroid fluxes from data points across the exterior of the station to determine the effects of orbital flight on the station structure and exterior-mounted equipment.

From the beginning of the new year they also continued astrophysical research with the Roentgen Observatory and other equipment, worked with the onboard furnace on technological experiments, examined the effects of spaceflight on various biological cultures in the Maksat experiment, provided information on the habitability of Mir, and performed routine maintenance and repairs on the heating system, the water supply, the pressure control system and a water leak. During February 1995 the Shuttle *Discovery*, flying the STS-63 'Near Mir' rendezvous mission, approached the station to as close as 11 metres, and remained stable for ten minutes in a demonstration of proximity operations that would be required during the forthcoming Shuttle docking missions. Onboard *Discovery* were American astronauts Eileen Collins (Pilot) and Janice Voss (Mission Specialist).

On 16 March, two days after launch, Soyuz TM-21 arrived at Mir carrying the Mir resident replacement crew: Commander Vladimir Dezhurov, Flight Engineer Gennady Strekalov, and American astronaut Norman Thagard – the first of seven astronauts to live on Mir over the next three years. Five days later, after handing over to their replacement crew, Kondakova joined Viktorenko and Polyakov in Soyuz TM-20 for the return to Earth. The third Russian female cosmonaut had set a new endurance record for a woman in both single and cumulative spaceflights of 169 days 5 hrs 22 min.

Following her first spaceflight, Kondakova followed the well-established Russian post-flight recovery programme for all long-duration crews, and had shown that she had stood up well to her five-month mission. Moreover, she was so successful that the following year she was assigned to return to Mir in 1997 – not on a Russian spacecraft, but on the American Shuttle. In August 1996 she was assigned to train as a Mission Specialist on the crew of STS-84, thus becoming the first (and probably only) Russian female to fly onboard a Space Shuttle (May 1998), Following this mission she returned to her post at Energiya, retired from the cosmonaut team in December 1999, and entered politics the following year.

Back-up to Norm

In 1993, following STS-50 assignments, Bonnie Dunbar took a management position as Deputy Associate Administrator in the Office of Life and Microgravity Sciences at NASA HQ in Washington. At this time, NASA was in working with the Russians in preparing to train two cosmonauts to fly on a Shuttle mission and two Americans to go to Moscow and train for an extended-duration flight on Mir. It was known for some time that Norman Thagard would be the primary candidate, and so his confirmation in February 1994 was not a surprise. However, it had been a problem persuading other astronauts to serve as back-up without the firm commitment that there would be a spaceflight at the end of a gruelling year-long training programme in Moscow. In general, American astronauts had not served in regular back-up roles

Norman Thagard and Bonnie Dunbar undergo Soyuz training at the Gagarin Cosmonaut Training Centre in preparation for Thagard's mission to Mir in 1995.

since the beginning of the Shuttle programme in 1982, and no American had trained for a mission longer than a couple of weeks since Skylab, twenty years earlier. Apparently, even the promise of a flight as a Mission Specialist on the first Shuttle docking mission, which would return Thagard, did not arouse much interest in the Astronaut Office. When Bonnie Dunbar was approached for the position in December 1993, she initially declined. It required some persuasion to change her mind, but after only a few days she relented, and agreed to the assignment.[17]

Although adjusting to life in Russia and training so far from home would be difficult for all of the Americans, Dunbar adapted well to the task. However, she was not particularly happy with some of the aspects of working with the Russians, They were evidently dismissive and condescending towards not just their women colleagues but to women in general; it was difficult to laugh at their jokes; and she did not like to be complimented on her looks, or hugged in the traditional Russian greeting. To Dunbar, her experience and her PhD degree were more important. In addition, the intense Russian training in close proximity to all crew-members and in the confines of the Cosmonaut Training Centre (effectively a military base) also revealed difficulties not noticed in Houston. She and Thagard had been selected two years apart in the first two Shuttle astronaut groups. They had therefore known each other for more than fifeen years; but they had not flown together, and though their relationship was not close, the close proximity to each other during their preparations for the flight to Mir actually underlined differences that resulted in both stopping talking to each other. The rift revealed human responses and differences that need to be overcome on really long spaceflights – especially to Mars.

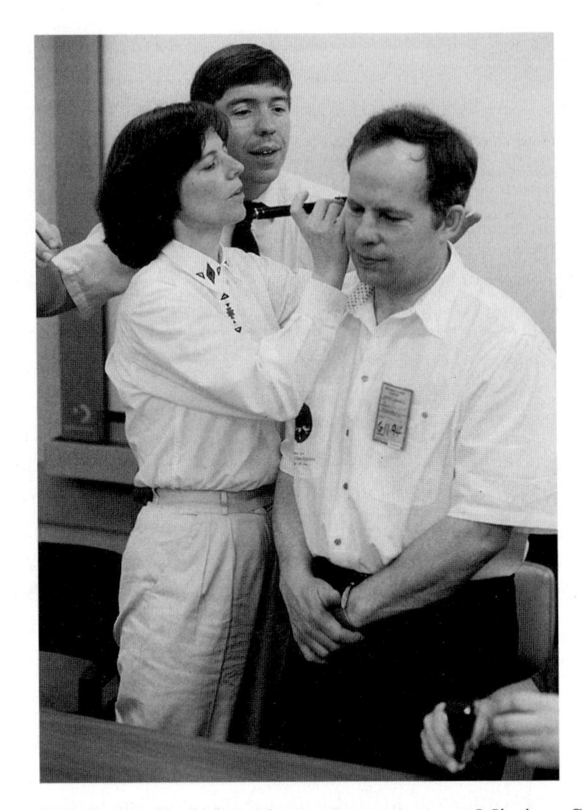

As part of the STS-71 medical experiment programme, Mission Specialist Bonnie Dunbar inspects the ear of Russian cosmonaut Anatoli Solovyov, with whom she could have flown a long-duration mission on Mir, together with Nikolai Budarin, as a member of the Mir-19 resident crew. Instead, she accompanied them into orbit onboard STS-71 and returned with the Mir-18 crew onboard the Shuttle.

In early planning, Dunbar could have taken over from Thagard to maintain an American presence on Mir, arriving on STS-71 in July 1995 and returning home on STS-74 in November; but this was not pursued. Although she hoped for such a mission, delays in negotiations and in starting a committed long-duration training programme with the Mir main crew, and having worked with the STS-71 crew too long, prevented her from rejoining the Russian training cycle for resident crews. By then, the next American resident crew-member (Lucid) was already in training for launch in spring 1996. Dunbar therefore flew to Mir as a Mission Specialist on STS-71, and assisted Payload Commander Ellen Baker with the fifteen separate biomedical investigations on Thagard and his Russian crewmates, divided into seven disciplines: cardiovascular and pulmonary functions; human metabolism; neuroscience; hygiene; sanitation and radiation; behavioural performance and biology; fundamental biology and microgravity research. Dunbar would also return to Mir as a crew-member of STS-89 – the eighth Shuttle docking mission, which delivered Andy Thomas and returned David Wolf.[18]

Two Yuris and a Shannon

Shannon Lucid was officially named as the second NASA astronaut to long-duration flight to Mir on 30 March 1995. She had already begun a Russian language course upon request from Chief Astronaut Robert Gibson in 1994, though this was not confirmation of a long visit to the station, but just a prudent precaution should she be assigned to a Shuttle visiting mission. After three months spent on an intensive language course, Dave Leestma (Director of Flight Crew operations) asked if she was ready to accept assignment to a long-duration mission. Her training in Russia began in January 1995, in the depths of winter.

When Lucid was named to the prime crew for a long-duration mission, her back-up was named as John Blaha, who had been her Commander on two earlier Shuttle missions (STS-43 and STS-58). The original planning document saw her launched with the Mir-21 resident crew in December 1995, but it was decided to launch her on STS-76 in spring 1996 for a planned five-month visit. She would then be replaced by the next American to live on Mir, originally Jerry Linenger (who was later replaced by John Blaha), and return home on the Shuttle mission that delivered him.

Although some Soyuz training was required for emergency drills, most of the training in Russia was aimed at preparing her for the long residence on the space station. 'Every morning I woke at 5 am to begin training [at TsPK]. I spent most of my day in classrooms, listening to Mir and Soyuz system lectures – all in Russian, of course. In the evenings I continued to study the language, and struggled with

A Shannon and two Yuris. Shannon Lucid shares a light-hearted moment at meal-time with her Russian colleagues during her record-breaking flight on Mir in 1996.

workbooks written in technical Russian. At midnight I finally fell exhausted into bed. I worked harder during that year [1995] than at any other time in my life ... Going to graduate school whilst raising [a family] was child's play in comparison.'[19]

Lucid was launched on STS-76 on 22 March, and two days later arrived on Mir for what was planned as a 140-day visit, surpassing Thagard's time by 56 days. During the mission, however, technical problems with the Shuttle Solid Rocket Boosters from an earlier flight delayed Lucid's return – although had an emergency occurred she could have returned with her Russian colleagues on the docked Soyuz. She had trained to work mainly with the EO-23 crew (Yuri Onufriyenko and Yuri Usachev – 'the two Yuri's', as she called them) but due to the delay she had to work with the EO-24 crew (Valeri Korzun and Alexandr Kaleri). But no problems were encountered during transition, and she continued working with her new colleagues without major interruption to her programme.

Lucid's first days were spent onboard Mir learning to move around the station, and in April, around the final module, Priroda, in which most of her experiments were stored. During the residency she placed her personal belongings in the Spektr module, where she slept each night. She found that one of the most pleasant aspects of life on Mir was in taking meals with the crew, and despite feeling that the repetitive menu would dampen her appetite, she actually found that she was hungry for each meal, consisting of a mixture of American and Russian dehydrated and tinned food. A detailed time-line scheduled the work (called Form 24), and whilst the Russians mainly worked on maintenance systems, Lucid conducted the experiments for NASA. The most unpopular aspect of life on Mir was the daily exercise routine, which she deemed very difficult and boring, although it was necessary to maintain muscle condition during the long flight for up to two hours a day, every day. She tried to alleviate the boredom of running on the treadmill by tuning in to her Walkman, but realised that the few tapes that she had chosen before the flight had almost no fast pace, and so used those from the large music collection placed on Mir after ten years of occupancy. She helped with housekeeping and in storing the trash in the attached Progress resupply vehicle, but the clutter on the station still proved to be a hindrance. During each evening, alone in her separate sleeping area, she took time to read some of the books that she had taken with her, or sent e-mails home. Sleep came early after such a long working day. The routine on the station rarely changed, but Lucid never considered it monotonous 'I was living a scientist's dream. I had my own lab, and worked independently for much of the day. Before one experiment became dull it was time to start another, with new equipment and a new scientific field ... I discussed my work at least once a day' with NASA support staff in Houston or Moscow, who coordinated her activities.

Her role in each experiment was in conducting the onboard procedures, after which the data and samples would be returned to Earth for analysis after the mission. However, she believes that her experiences on the station clearly demonstrated the value of human research on space stations. Because she was on hand to record subtle phenomena with still photographs and movies, she could also use her scientific experience to immediately examine the results of an experiment, and, if necessary, modify the procedure to obtain better results or be ready to perform in-flight repairs. Of her twenty-eight experiments, only one failed to yield results, due to a breakdown of equipment. Her research field included experiments in advanced technology, Earth sciences, fundamental biology, human life sciences, microgravity research, and space sciences.

As a child, Lucid was interested in exploration, but by the time she grew up she realised that most of the world had been explored. Then, after reading about Robert Goddard and his rocket experiments in the 1920s, and also reading science fiction, she thought that exploring the Universe would be 'neat', as it certainly would not have been explored by the time she was an adult. Indeed, she was herself a pioneer; and in flying the Mir mission she was a female pioneer in long-duration spaceflight. After handing over to John Blaha on Mir, she returned to Earth onboard STS-79 and landed on 26 September, after a flight of 188 days 4 hrs in space, setting a new endurance record for an American astronaut (and for women). Her career totalled more than 223 days on five spaceflights.

When the recovery crew entered the Shuttle after it landed, they expected Lucid to be still in the seat recliner adjusting to gravity after such a long time in space; but she defied every prediction by standing up, and, assisted by NASA technicians who held her arms, walked to the hatch and climbed out into the Crew Transporter. She was then expected to rest; but she had other ideas, calmly stating that she could easily stand up. She walked 25 feet to the transporter, and although feeling a little disorientated and light-headed she was in good form. Even President Bill Clinton told her, during a telephone call from the White House, that he could not believe she

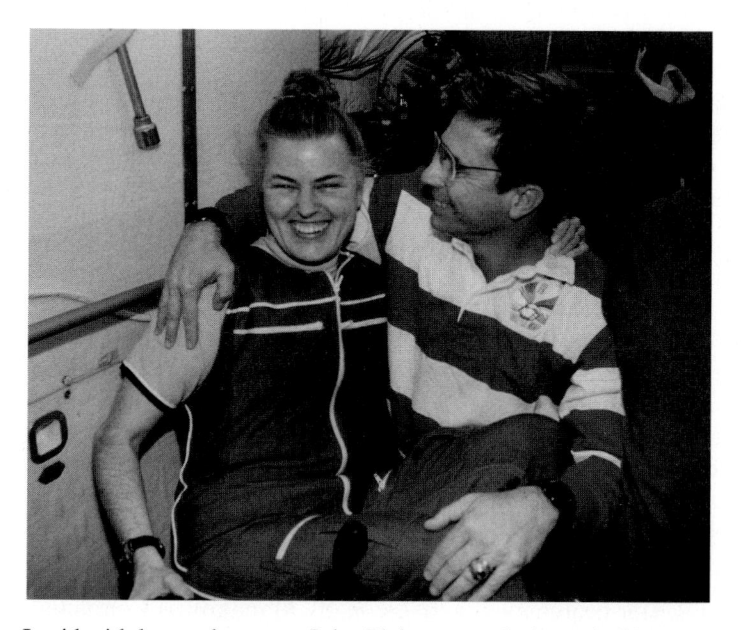

Shannon Lucid with her replacement John Blaha, towards the end of her stay on Mir in 1996. They had previously flown together on Shuttle missions in 1991 (STS-43) and 1993 (STS-58).

had walked off the spacecraft. According to colleagues she was determined to prove that she was perfectly fine, and that her mind was set. Mission completed.

Lucid was well received by the Russians. She was one of the most popular NASA astronauts to visit Mir, and received some of the highest Russian awards for her work. Following the post-flight recovery period she assumed a number of support and astronaut management roles, including that of NASA Chief Scientist from February 2002 to September 2003, after which she returned to astronaut duties at JSC in Houston.

Entente cordiale

The Soviet Union had been flying international guests on one-week visits to Salyut stations since 1978. These continued on Mir, although the emphasis changed from the flights of 'Communist guests' from Eastern Block countries – mainly for the purposes of propaganda – to fair-paying passengers from Western countries, for commercial reasons. One of the longer standing partners in these joint missions has been France, which in addition to sending a 'spationaut' to Salyut 7 in 1982 and on a Shuttle mission in 1985 has sent two astronauts on Shuttle docking missions to the Mir station and five national long-duration missions to the complex. One of these was a female French astronaut, Claudie Andre-Deshays, who later married fellow spationaut Jean-Pierre Haignere, who had logged more than 209 day on Mir during two flights.

Following the success of the French flight to Salyut 7 in 1982, the country decided to create its own cadre of spationauts (space travellers) for later assignment on the US Shuttle, the Russian space station, or, hopefully, on the Hermes mini spaceplane then under development (but later cancelled) or the European Columbus space station module. The criteria included being a French citizen aged between 25 and 35, with a scientific, engineering or medical degree and appropriate professional experience. Of the 700 applicants, seven spationauts were named on 9 September 1985 – including Claudie Deshays, selected as an experimentateur (scientist).

Claudie Haignere (née Andre-Deshays) (*b*.1957), earned a medical degree from Dijon University in 1981, specialising in biology and sports medicine. While serving in the rheumatics department at the Cochin Hospital in Paris, she conducted studies on aviation medicine. After selection as one of seven French spationauts in September 1985, she helped develop medical experiments for CNES, and continued her medical career. After completing an advanced weightlessness training programme at TsPK in 1991 she was assigned, in December 1992, as back-up to Jean-Pierre Haignere, who was assigned to the Soyuz TM-17 mission to Mir in 1993. She then helped coordinate experiments for the Euro Mir 94 mission of ESA astronaut Ulf Merbold, and in July 1994 was assigned to the 1996 French Cassiopeia mission.

Deshays was assigned to the twenty-second resident crew for the mission, planned to last for 16 days. Her training was with Gennady Manakov as Commander and Pavel Vinogradov as Flight Engineer. Just one week before launch, however, Manakov was grounded by a small medical problem, and as the Russians prefer to keep crews intact, the prime crew was replaced by the back-up crew of Korzun and Kaleri. This decision did not affect Deshays as she was only to stay for a two-week mission, independent of the resident crew. and planned to return with 'the two Yuris' currently hosting Shannon Lucid, who was awaiting her replacement and a Shuttle ride home. Launched to Mir on 17 August and docked two days later, the two crews worked on the station for the next two weeks – the first time two women had been onboard Mir, as well as two guest cosmonauts at the same time. Onufriyenko and Usachev returned to Earth on 2 September, after a flight of 15 days 18 hrs 23 min.

During the flight with two female cosmonauts and four male cosmonauts onboard for two weeks, Mir suddenly became crowded. Deshays set up her 'personal space' in the Priroda module. She was disappointed in losing the crew with whom she had worked, but she realised that the success of the mission was more important. With more than two hundred French scientists and engineers supporting her on the mission, she was able to live amicably with her new Russian crew; but she also had the opportunity to talk with her grounded Commander and Flight Engineer if she required support in completing a procedure or in repairing an item of equipment on which they had worked.

In early August more than 300 kg of equipment and experiments had been delivered to Mir onboard Progress M-32, and when Deshays arrived, her first task was to set up the experiments. These began with the science programme on 20 August, and continued with reduced activities over the weekend period, resuming the second week. The experiments – which had been developed over a period of three years – involved four fields: life sciences, physical sciences, technological experiments, and student experiments.[20] Because of Deshays' professional background in rheumatology, biomechanics and physiology, she was a perfect candidate for the experiment programme flown on the mission. Although some difficulty was experienced (some of the experiment hardware required in-flight repairs), a significant amount of data were gathered during the two weeks, although Deshays commented that she preferred a much longer flight. It had taken her a week to learn to live onboard the station and use the facilities, and a flight of several months would allow the experimenter to utilise the facilities to the full potential. She was working up to twenty hours a day and sleeping for only four or five hours, and by the end of the first week she needed a rest. The flight plan was, as during all short missions, very intensive, and there was hardly any time for rest. Longer flights have to be planned differently to pace the activities with other duties which the crew has to accomplish – such as maintenance and housekeeping – and there has to be time for relaxation.

Following her first spaceflight, Deshays returned to CNES, and in May 1998 was assigned as back-up to J.-P. Haignere (her future husband) for his second mission to Mir under the Anglo-French Perseus programme flown in 1999. Her training for this mission included EVA training and qualification as a Cosmonaut Flight Engineer for the Soyuz spacecraft and Mir space station, and during the flight she worked as a crew interface coordinator at the Russian Control Centre in Korolyov, near Moscow. In July 1999 she qualified as a Soyuz Return Commander (the first woman to do so) – a position in which she could take command of a three-person Soyuz for its return from space if required. In November 1999 she joined ESA and worked on a number of development projects, including the microgravity facilities onboard the

European Columbus (International) space station module, and supporting medical investigations and activities at the ESA Directorate of Manned Spaceflight and Microgravity, pending a future flight assignment.

'Too short' Lawrence

In March 1995, NASA's announcement of astronaut assignments to long-duration missions on Mir named Wendy Lawrence as back-up to John Blaha, scheduled to fly the fourth NASA mission on the station after Lucid and Linenger. She was expected to take up the NASA Mir 6 position, but in October 1995 she received a personal blow – as she was about to begin a year's training as back-up – when she was deemed 'too short' to fit safely inside the Soyuz Descent Module (even though she was to travel to and from Mir onboard the Shuttle), should she need to return to Earth in it. She was still involved in Shuttle–Mir operations, and was the sixth NASA Director of Operations Russia, located at Star City, in March 1996 until October that year. What surprised her was that shortly afterwards she was reinstated and cleared for Soyuz flights. She then resumed training, and was named as prime on NASA Mir 6 residency in August 1996, without first serving as a back-up. By then, Blaha had replaced Linenger to take over from Lucid, and it appeared that Lawrence would

Wendy Lawrence during water egress training – part of her training for residency on Mir.

become the second NASA female astronaut to make a long flight. Her training commenced in October 1996, but she did not complete it. The fire and collision at Mir in 1997 resulted in a new directive that all long-duration crew-members (including NASA residents) had to qualify in the Orlan EVA suit. The problem was that Lawrence was too short for that suit and had not trained for EVA, and she was replaced by David Wolf. This was a disappointment; but her preparations were not in vain, as she was assigned to fly on the Shuttle crew that should have taken her to Mir, and again on the final Mir docking mission. Lawrence was the daughter of a Mercury astronaut finalist, and would have provided a direct link between that pioneering American programme – which did not allow women to fly in space – and the new space station programme of international cooperation that allowed for such spaceflights – if you were tall enough.

Thus ended the first phase in flying women on Russian space stations. The story would continue on the new International Space Station being cooperatively developed by sixteen nations.

A NEW STATION, A NEW ERA, 1988–2004

Construction of the ISS began in autumn 1998, and required two years of activity before the first crew could take up residence in October 2000. The first four crews were named on 17 November 1997, reflecting the two-year training programme required to qualify as a resident crew-member (over that of about twelve months for a Shuttle crew). Assigned as Flight Engineer for the second radiant crew (ISS-2) was NASA astronaut Susan Helms, who would fly with fellow astronaut Jim Voss and ISS commander Yuri Usachev. Because of delays to the programme, this crew was not launched to the ISS until spring 2001. In spring 2000, however, they had flown a short visiting mission, onboard STS-101, to gain experience on the embryonic station in preparation for the first residency.

Science with the second crew

The objective of the first crew was to configure the station for permanent residency; but the American Destiny laboratory was not installed until the final six weeks of their mission, and with a lack of additional solar arrays for power, operations were limited. With the arrival of the second crew, on STS-102 in March 2001, the first real science could be conducted onboard the ISS. On this mission, the three-person crew would conduct a programme of more than forty-five experiments and technological investigations, including two on biology, one on planetary biology, thirteen on ISS environmental issues, five on radiation, one on biotechnology, four on Earth observation, five on the growth of protein crystals, seven on medical investigations, two on materials science, two technological tasks, and two under the education programme.[21]

Helms was an experienced astronaut with four Shuttle flights to her credit between 1994 and 2000 and a brief visit to the ISS in 2000; but living on the station with her two male colleagues would be a different challenge. Memories of the Mir missions were still fresh, and the ISS was still a relatively new station. When asked if she would volunteer for a long-duration mission, her overriding concern was to return to orbit, even though while training she would be almost isolated in Russia. Her two colleagues were experienced on Mir systems and procedures. Usachev had flown two missions to Mir totalling more than 375 days, and had logged more than 30 hours on six EVAs. Jim Voss was a veteran of three Shuttle mission, and though he did not fly to Mir he had performed back-up duties for two NASA astronauts on Mir (Foale and Thomas) before assignment to the ISS. It was the flight of all three of them on STS-101 that made them realise that they were a great team.

During STS-101, only four days were spent inside the station, preparing it for permanent occupation; but when the Shuttle crew of four ended their work and began packing to go home, the ISS-2 crew-members wanted to stay longer, realising that the STS-101 mission was but a taste of what was to come the next year when the space station would become their home for more than five months.[22] Onboard the station during the longer mission in 2001, Helms had her own space for privacy. At the beginning of the mission she had realised that there was a huge difference between a short flight on the Shuttle and a long flight on a space station. Living in space is not like living on Earth, and the approach to work has to be different on a long mission. Helms believes that her upbringing as one of three daughters of a USAF pilot father and a working schoolteacher mother, helped her to aim high. She was a member of the first graduate class of the USAF Academy to accept women, chose a career as an engineer in the Air Force, and brought her philosophy of military organisation to her long mission. In addition, when she arrived on the ISS in

Susan Helms participates in winter survival training in Russia in readiness for her mission as a member of the ISS-2 resident crew in 2001.

March 2001 it was more than a living area in space; it was her home. With a long training programme in Russia, the pre-flight and post-flight isolation and debriefing, and the mission itself, she would not be at home in Houston for many months, and she therefore packed her possessions and put them into storage: 'I was very disconnected from Earth . . . When we had to go and fly for six months, I effectively closed down my Earth life. I acted like it was military deployment . . . I lived in space. It was my home'.[23]

During 18 April 2001 the crew undocked the Soyuz TM-31 spacecraft from the Zarya FGB nadir rear port to the aft port of Zvezda, to make way for the visiting crew who later in the month arrived on Soyuz TM-32 to replace the aging TM-31 Soyuz in which they would return. Although guests are welcome on long flights, they disrupt the regular routine of a long-duration crew. The first Soyuz taxi crew to the station also included the first 'space tourist', American businessman Dennis Tito. In addition to the visiting crew, the second resident crew hosted STS-100 (delivering the station's RMS system) in April, STS-102 (delivering the American Quest airlock module) in July, and the undocking of one Progress resupply vehicle (M-44) and arrival of another (M1-6).

Although Helms participated with Voss in a Shuttle-based EVA at the beginning of their mission, she did not perform any spacewalk as a member of the ISS-2 crew, although Voss and Usachev performed a 19-minute internal EVA to relocate a docking cone from a storage location to an operational port inside the node of Zvezda. A major feature of the ISS-2 mission was the tasks using robotic workstations installed during STS-102 visits, and tested before the SSRMS was installed on STS-100. Helms and Voss used the station's RMS to assist in the docking of the Quest airlock module to the station during the STS-104 mission.

In August 2001 the mission ended with the arrival of the ISS-3 crew onboard STS-105. The ISS-2 crew landed with the STS-105 crew on 22 August 2001, having handed over to the next resident crew after a flight of 167 days 6 hrs 40 min from launch to landing. Helms now had a career total of more than 211 days accrued on five missions, and after a twelve-year career at NASA she realised that it was time to step aside and let others take over the pioneering exploration role on the ISS. Faced with the option of remaining at NASA in a management or administrative role, she decided to return to the USAF, though with twenty-two years of service she was eligible for retirement. 'The Air Force has always been so supportive of the things I wanted to do, and I guess I felt the time had come to come back and help the military space programme.' In July 2002 she retired from NASA and took a position as Chief of the Space Control Division at USAF Space Command.

Seats on Soyuz

Due to delays in providing an adequate Crew Rescue Vehicle at the ISS, the availability of the Russian Soyuz to serve in that role was fortuitous. By having a Soyuz docked to the station, crews could remain onboard without the need of a docked Shuttle, using the Soyuz as a crew rescue vehicle should the need arise. There were problems of certification in using the Soyuz due to crew size, but these were obviated by sending non-Russian crew-members to TsPK to qualify as Soyuz Return

Commanders or crew-members, and selecting crew-members that would fit inside the Soyuz TM or its replacement TMA. There was also the need, in the early days, to move Soyuz craft from one port to free up another, and to exchange them at the end of their six-month design life. The opportunity to fly other crew-members to and from the station just to exchange Soyuz craft therefore offered flight experience for unflown cosmonauts, additional opportunities for scientific research, much-needed cash for the Russian programme, and, as at least one crew-member had to be experienced, a final flight for a veteran cosmonaut.

In July 1997, the members of the first ISS Soyuz exchange crew were Musa Musabayev (Commander) and Nadezhda Kuzhelnaya (Flight engineer). However, over the next three years, crew assignments changed as a result of events in the Mir and ISS programmes. When US businessman Dennis Tito negotiated his deal to fly on a Soyuz, Mir was still in orbit. With Tito assigned to a crew to fly to Mir, a back-up crew was needed, and Musabayev was relieved of his role on the ISS taxi flight in September 2000, to become back-up to the planned flight to Mir in 2001. His place on the taxi crew, with Kuzhelnaya, was taken by veteran cosmonaut Viktor Afanasyev. However, it soon became clear that a flight to Mir was impossible before it re-entered, and so the deal was revised to fly Tito to the ISS. Because of these changes, Kuzhelnaya was removed from the flight, due to what she later felt were more than operational reasons. During an interview in 2001, she said that it was 'a political decision ... By the time it was made, Tito's crew had not even been approved by Rosaviacosmos' (the Russian Space Agency).[24]

Kuzhelnaya served as a FE back-up on the next taxi mission (TM-33), but never regained her flight seat. Over the next four years the seats were taken up by European crew-members, another space tourist, and, following the *Columbia* accident in February 2003, by resident crew-members. With no prospect of a flight assignment, and in what might be considered a lost opportunity to fly another woman in space, she left the cosmonaut team in May 2004 to take a student position at the Russian aviation school for civil aircraft pilots in Ulyanovsk, and to work as second pilot in TU–134s for Aeroflot, in Sheremetyevo. The loss of *Columbia* prompted a Russian decision to not assign women to spaceflights, and this probably contributed to her decision to leave the programme. She was close to becoming only the fourth Russian woman to fly in space, and because of her departure there were no female cosmonauts active in the Russian programme in 2004.

Andromede

The next woman to fly to the ISS was veteran French astronaut Claudie Haignere (formerly Deshays), who had been named to the mission in December 2000. She commenced training for her mission – Andromede – at TsPK in January 2001, and was qualified in ISS basic training (150 hours) by ESA that March. Her previous Soyuz training had qualified her as Flight Engineer 1, Soyuz TM-33 – commanded by Afanasyev (backed up by Sergei Zalyotin), with Konstantin Kozeyev as Flight Engineer 2 (backed up by Kuzhelnaya) – was flown to the ISS on a ten-day mission beginning on 21 October 2001. Andromede included an experimental programme created by the French space agency, CNES.

Claude Haignere onboard the ISS during her Andromede mission in 2001. With her are Soyuz Commander Viktor Afanasyev and Flight Engineer Konstantin Kozeyev.

Because of her experience in four training cycles for Mir assignments, Haignere was well accustomed to the operational phase of a Soyuz launch, docking, undocking and landing, and to life on a space station. In June 2001 the three crewmembers began working together as a crew, including working at the CNES facility in Toulouse, France, familiarising themselves on the experiment programme, and then at TsPK in July for fifteen four-hour sessions in the Soyuz simulator. This was followed by fifteen sessions in the docking simulator and re-entry simulator, and sessions in the Russian element mock-ups. In August they visited NASA JSC for familiarisation on the US segments. As busy as the preparation was, it was preparation for only for a few days in space, although Haignere believed that her mission was an important step towards European participation in the ISS, helping to establish the station's scientific facilities and accruing operational experience years before the launch of the Automated Transfer Vehicle and the Columbus science module. 'I think that we can show young people that Europe is a real space power and an important partner in the Space Station project . . . We hope that young people will see that a career in science and technology can be fascinating and fulfilling. Like all young people they will create their own future. But we can show them what that future could be.'[25]

During her stay onboard the station, Haignere completed her science and technology research programme in Earth observations, ionospheric studies, material sciences and technological experiments linked to the development of ATV control centre operations. She also served as guide during a TV tour of the station modules and facilities. The landing, in the older TM-32, took place successfully on 31 October after a flight of 9 days 18 hrs 58 min. Post-flights reports were full of praise for Haignere's work. Her 'relational skills and expertise are widely known, Europe could not have chosen a better ambassador.'[26] Upon landing, Roger-Gérard Schwartzenberg (the French minister of research, with responsibilities for space) expressed his esteem and admiration for Haignere whose skill, dedication and courage was 'a model for us all, and projected an image of France at its best ... [Haignere] is the face of Marianne, the figurehead of our French Republic. It is fitting that the first French astronaut to fly to the ISS should be a woman and a scientist – indeed, a doctor of medicine and science, who is helping to advance research and space exploration.'

For Claudie Haignere, however, it was time to move on. Following the completion of her mission assignments on 17 June 2002 she was appointed Minister for Research and New Technologies in the French Government.

The first ISS Science Officer

When Peggy Whitson, aged nine, watched Armstrong and Aldrin walk on the Moon, she decided that she wanted to be a pilot; and after the first female astronauts were selected by NASA in 1978 she set her heart on becoming an astronaut.

Peggy Whitson (*b.*1960) is a graduate of Iowa Wesleyan University, from where she received a BS degree in biology and chemistry in 1981. Her PhD in biochemistry was earned from Rice University in 1985. As a Research Associate in 1986 she worked at NASA JSC, and in 1988 worked for the medical sciences firm Krug International, which was also a JSC contractor. In 1989 she joined NASA in the biomedical operations and research branch, working as a research biochemist. Some of her medical research payloads flew onboard STS-47 (Spacelab J) in 1992. That same year she was named as a project scientist for the Shuttle–Mir Phase 1A programme, working on several missions (STS-60, STS-63, STS-71 and the Mir 18 and Mir 19 increment crews), as well as assignments as Deputy Chief of the Medical Sciences Division at JSC. In 1995 she served as Co-chairman of the US–Russian Mission Science discussions until her selection to astronaut training in May 1996. In 1998 she completed a two-year training and evaluation programme, and was assigned to the Astronaut Office Operations Planning Branch, serving as lead of the Crew Test Support Team in Russia (Russian Crusader) from 1998 to 1999, after which she was assigned to the ISS training group. Following her first spaceflight in 2002 she was appointed Deputy Chief of the Astronaut Office, effective from November 2003.

When selected to the prime crew of the fifth ISS residence crew on 23 March 2001, Whitson was completing her role as back-up FE for the third expedition crew, drawing on almost a decade of work with the Russians. The fifth expedition, STS-111, was launched to the station on 5 June 2002, docking two days later. The crew spent 171 days 3 hrs 33 min in space, and returned to Earth on STS-113 on 7 December 2002. During this residency, the station was expanded by the installation of the mobile Base System, S1 and P1 truss segments, with Whitson assisting in the operation of the SSRMS. In addition, she was named, in flight, the first NASA ISS Science Officer, and conducted twenty-one investigations in human life sciences and microgravity sciences, as well as a variety of commercial payloads,. Her tasks also included the activation and checkout of the Microgravity Sciences Glove Box. The Expedition 5 programme included eleven experiments in bioastronautics research, nine in the physical sciences, five in space product development, and four in educational fields. In addition, on 16 August 2002 Whitson completed (as EVA2) an EVA of 4 hrs 25 min, with her station Commander, Valeri Korzun, during which they installed micrometeoroid shielding on the Zvezda Service Module.

The crew was visited by a Soyuz taxi crew in October 2002, but though 'tourists' were at first assigned to the mission, none were flown, due to the non-completion of financial arrangements in time for the flight. One of those briefly in line for the mission during 2002 was former NASA Associate Administrator for policy and plans Lori B. Garver, who was then working for DFI International as a consultant. Garver's trip was based on sponsorship and not personal wealth, and did not progress to the signing of contracts.

In her pre-flight interview, Whitson said that she considered herself to be a scientist, builder and explorer, and that one of the most appealing aspects of being an astronaut was that since selection she had learned many new skills. Her ISS training was demanding, as she had to familiarise herself with all the US systems to become the 'crew expert', and also train as a robotics operator and qualify for EVA using Russian hardware and procedures.

During the mission she regularly sent home e-mails describing her activities on the station and life in space, and though she would have liked to write to everyone individually she frequently posted letters on the NASA human spaceflight website.[27] These included such mundane affairs as when she asked Korzun to cut her hair –

Peggy Whitson – NASA's first ISS Science Officer – onboard the ISS during 2002.

which he nervously agreed to do – after which, she cut her colleagues' hair. During most days she was busy with routine and mundane housekeeping and maintenance chores, with good measures of 'fun' such as EVA, robotics and science. As the mission drew to an end she had mixed feelings about leaving the station. She was eager to see family and friends, but it was 'hard to let go of the idea of living here'. She fully expected to spend the rest of her life trying to find the right words to explain and convey only a fraction of what she had experienced and enjoyed during her first spaceflight.

The future of ISS operations

Whitson was the most recent female to take up a long-duration mission on the ISS. The following year, the loss of *Columbia* drastically altered not only Shuttle operations but also ISS residency and Soyuz flight operations, reducing the resident crew from three to two pending the return to flight of the Shuttle and resumption of ISS construction. At the time of writing (2004) this is planned for no earlier than summer 2005. There are currently several female astronaut in training for flights onboard the Shuttle to participate in this construction phase. As for future female assignments on long-duration residencies, American astronaut Sunita Williams (selected in 1998) is in training (2004) at TsPK in Russia for assignment to an ISS three-person resident crew planned for a six-month tour. If this remains on schedule she will resume long-duration spaceflight operations by women. It can be expected that once full operations resume, then other women will be assigned to residency crews to continue the work begun by Lucid, Kondakova, Helms and Whitson.

Female space-station yuhangyuans?

In July 2004, China indicated that female astronauts (yuhangyuans) will participate in the creation of a small Chinese laboratory within the next decade. After the flight of the first Chinese man in space in October 2003, the All-China Women's Federation lobbied for the inclusion of female astronauts in the programme. The reports implied that the first Chinese female astronaut will visit a permanent space station, as a researcher or technician, around 2010; and beginning in 2005, prospective candidates with particular qualities, including good health and stamina, will be sought in the country's high schools. The candidates will initially be trained as pilots and then in spaceflight technology, before training for the mission itself. According to the director of China's manned space programme, Huang Chunping, 'It is hoped that Chinese women will realise their space dream as soon as possible.' Other reports indicate that it is important for the Chinese to demonstrate their commitment to equality, and that there is a pool of female engineers who, if sufficiently fit, might also qualify.

It is clear that with safe, regular and economic access, permanent space stations are the key to sustained access to space. They are a platform for technological and medical research and operational experience, and a stage for the commencement of journeys beyond Earth orbit into the deeper reaches of space.

REFERENCES

1 Bert Vis, 'Soviet women Cosmonaut Flight Assignments, 1963–1989, *Spaceflight*, **41**. 11, November 1999, 474–480.
2 Rex D. Hall and David J. Shayler, *Soyuz: A Universal Spacecraft*, Springer–Praxis, 2002, pp. 281–316.
3 Bettyann H. Kelvles, *Almost Heaven*, Basic Books, 2004, pp. 79–91.
4 'Russians in space', Voice of Russia archives, 2001.
5 Details of Savitskaya's EVA can be found in David J. Shayler, *Walking in Space*, Springer–Praxis, 2004, pp. 233–234.
6 Gordon R. Hooper, *The Soviet Cosmonaut Team*, vol. 2, 'Cosmonaut Biographies', GRH Publications. 1990, p. 123.
7 Colin Burgess, 'Svetlana's story', *Spaceflight*, **44**, 2002, 295–299.
8 Bert Vis, 'Mir crews', in *Mir: the Final Year*, British Interplanetary Society, 2001, pp. 3–6.
9 Helen Sharman, *Seize the Moment*, Victor Gollancz, 1993.
10 Private communication from Tony Quine to David J. Shayler, 28 October 2004.
11 http://aeroweb.lucia.it/~agretch/RAFAQ/Protasova.html.
12 Ref. 9, pp. 151–168.
13 Rex D. Hall, David J. Shayler and Bert Vis, *Russia's Cosmonauts*, Springer, (due in 2005).
14 Ref. 2, p. 144.
15 Ref. 2, pp. 145–151.
16 Neville Kidger, 'Mir Mission Report, for Euro Mir 94 and EO-17 residence on Mir', *Spaceflight*, **37**, 1–3, January–March 1995; and 6, June 1995.
17 Bert Vis, 'Russian with a foreign accent: non-Russian cosmonauts on Mir', in *The History of Mir, 1986–2000*, British Interplanetary Society, 2000. pp. 92–93.
18 David J. Shayler, 'American Flights to Mir (Space Shuttle)', in *The History of Mir, 1986–2000*, British Interplanetary Society, 2000, pp. 71–85.
19 Shannon Lucid, 'Six months on Mir' *Scientific American*, May 1998, 26–35; also, NASA JSC Oral History, Shannon Lucid, 17 June 1998.
20 Neville Kidger, 'Mir Visited (Mission Report)', *Spaceflight*, **38**, 11, 1996, 365–368; also, David M. Harland, *The Story of the Space Shuttle*, Springer–Praxis, 2004.
21 David J. Shayler, *The International Space Station: From Imagination to Reality*, British Interplanetary Society, 2002 – notably the papers 'The ISS expedition crews' by Neville Kidger (Expedition 2, pp. 66–72), and 'Research in orbit' by Andy Salmon (Expedition 2, pp, 120–123).
22 Susan Helms, 'Expedition Two Pre-flight Interview', http://spaceflight.nasa.gov/station/crew/exp2/inthelms.html.
23 Stacey Knott, 'Pioneering the last frontier', Air Force Space Command Public Affairs, 27 Match 2003, online: http://www.peteron.af.mil/hqafspc/news/News_-Aso/nws_tmp.asp?storyID = -3-074.

24 Yuri Karash (Space.com Moscow correspondent), 'Training of a female cosmonaut', 26 January 2001: http://www.space.com/news/spaceagencies/womancosmonaut010126.html.
25 Andromede press brochure, ESA, October 2001.
26 CNES news magazine, No. 14, January 2002, p. 8.
27 http://spaceflight.nasa.gov/station/crew/exp5/lettershome1.html.

Earth orbit and beyond

By 2004, interest in human spaceflight beyond Earth orbit had generated formal studies and proposals to redirect the American human spaceflight efforts to complete the ISS, retire the Shuttle, develop a new human ferry capability for Earth orbit, support a return to the Moon, and plan for human flights to Mars – all within the next fifty years. There is no certainty of support, however, and the cheques need to be signed before we embark on such a venture, and we are still probably decades away from journeying to the Red Planet.

AND SO TO MARS ...

There are, however, a number of issues and research fields that are being addressed to investigate the prospect of extended-duration spaceflight and the personal exploration of planets other than our own. Some of the pioneering research addresses the health issues in sending crews away from Earth for prolonged missions, and whether women as well as men will be crew-members.

An all-woman Mars crew?

In July 2000, Geoffrey A. Landis – a scientist at the NASA Glenn Research Centre, and a writer of science fiction – wrote a 'Viewpoint' opinion piece for the July 2000 issue of *Space Policy* in response to the question: 'Would an all-woman crew be the ideal personnel for a Mars mission?' In his response, Landis pointed out that an all-female crew would generally be smaller and lighter, and would require less water, air, food and living volume than would an all-male crew. The spacecraft might therefore be smaller and have less mass, which in turn reduces the amount of fuel required to launch towards Mars – a reflection of a 1964 study which suggested that by selecting small crew-members the overall expense of the Mars expedition could be reduced.[1]

Landis also suggested that some of the other benefits might be rather more difficult to quantify – such as the fact that all-women groups tend to choose a non-confrontational approach to solve interpersonal problems, which would help prevent conflict on long trips to and from Mars. However, the disadvantage is that humans lose up to 2% of their bone mass per month in a weightless state, and that there is a scientific argument that women are more susceptible to bone loss than men. But Landis countered this argument by pointing out that Shannon Lucid was actually subjected to less bone loss than many men (all Russians) after a similar extended-duration flight. He also suggested that it is irrelevant, as a Mars mission would probably have induced artificial gravity as a countermeasure, and suggested that both America and Russia made a serious error at the very beginning of their manned space programmes, in that they should have selected women as more logical candidates for space missions. Serious thought, he said, should be given to sending all-women crews to Mars. Unfortunately, we will have to await developments to see whether anyone – male or female – goes to Mars in our lifetime; but the suggestion is interesting.

The Extended Duration Orbiter Medical Project

This project (EDOMP) – which operated from 1989 to 1995 – allowed longer Shuttle flights of up to eighteen days by flying additional cryogenic tanks in a special pallet, and afforded the opportunity to extend scientific and engineering research beyond the 7–10-day Shuttle/Spacelab mission prior to commencing the Shuttle–Mir and ISS programmes. This was part of a continuing investigation into extended-duration spaceflight, as well as the identification of critical risks and approved countermeasures. In the findings it was suggested that some bone loss is irreversible, especially in older and female crew-members. One of the most worrying aspects of extended flight is exposure to radiation, with the orbit of the ISS exposing the crew to neutron particles, cosmic rays and other forms of radiation as it passes through the magnetosphere. However, upon leaving low Earth orbit, the crews are prone to higher risks. This could be cumulative and in some cases permanent if the doses reach sufficiently high levels. This could result in an increase in cancers, cataracts, hereditary ailments and sterility. As a result, more rational experiments are being flown to the ISS to investigate the phenomena more closely. All other factors affected by spaceflight eventually return to normal by the second or third week back on Earth, but recovery time is linked to the time spent in space – and this is without additional stress due to failure or accident, or physiological and psychological issues. Basically, an increase in time spent in space will increase the risks.

Bed rest experiments continue

In early 2005, ESA will begin a 60-day ground-based Female Bed Rest Study to gather data on how female subjects adapt to weightlessness. Most studies of this type have featured male participants, as relatively few women have flown in space, and there is relatively little data on how female bodies adjust to spaceflight. This study will therefore help to advance the knowledge of the gender differences in experiences of extended exposure to weightlessness. Twenty-four volunteers will be split into three groups of eight. One group will be the control, with no extra stimuli; another will exercise; and the third will have a nutrient supplement. The experiment will include a 21-day pre-bed-rest period prior to the collection of baseline data, and a

20-day post-bed-rest period for carrying out similar tests to compare with the re-test data. By using physical exercise and nutrition in different groups, the results of counteracting the adverse affects can be determined. This study is a cooperative venture between the space agencies of France (CNES), America (NASA) and Canada (CSA), and ESA, to be conducted at the clinical research facility at the Rangueil hospital in Toulouse, France.[2]

International cooperation – or confrontation?

One issue highlighted during the Shuttle–Mir programme was the cultural differences between the Russians and the Americans – including the treatment of women in society. Several astronauts noticed the dramatic difference between training in Russia as opposed to the United States, and equally, some of the Russians found it difficult to accept women in the programme. Bonnie Dunbar, Shannon Lucid and Wendy Lawrence were American pioneers of international training for long-duration missions, apart from their vast experience on short Shuttle flights. Dunbar found it difficult at times, but Lucid has said that she never felt that she was treated any differently because she was female. In her oral history she said that the Russians were discussing why there were relatively few women in the Russian programme, and that she had commented that this was strange, as their attitude towards her was perfectly amicable. They replied that it was acceptable for an American woman to be an astronaut – but many Russians find it difficult to accept women as cosmonauts.

Chauvinism, sexual inequality, prejudice and politics are human failures that have serious consequences on Earth and should not be carried into space; but with an international crew isolated inside a spacecraft for two or three years, it would be impossible to isolate them from developments on Earth and from disagreements between themselves. The success of these long missions will depend upon the individuals, both male and female, working together to become a crew – almost a family – and hopefully living together without too much disagreement.

Family or space?

One issue that has future application from research on the ISS and in ground studies – and one that will be addressed as the programme is extended and larger crews fly longer and more distant missions – is that of raising families in space. Shannon Lucid found that regular contact with her family on Earth was important. Though at times difficult, both technically and emotionally, the psychological advantage of talking to loved ones is evident. These are some of the happiest memories of spaceflight held by all crew-members. The challenge will be married astronauts (men or women) who are away from their young children for months at a time – perhaps on a mission to Mars. Will this mean sending only single people to Mars until a base can be established and a family environment created? There has been a case of a space explorer choosing between his family and a long training period for an extended-duration flight and the flight itself. Three years away from his family was too much to bear, and he finally decided to return to his family.

During an interview in 2004,[3] Rhea Seddon explained that radiation is a big concern – 'not the absorption, but the amount you are allowed to get. Very different for a young woman than an older man. It would be truly unfortunate if the decision was made where don't have to shield the spacecraft very well if we are only going to send older men, so we just eliminate women from the crew. I think it would be difficult for NASA to justify that.'

As for babies born off the Earth, Seddon could foresee them developing in reduced gravity; but whether that will be in zero g or the ⅓-Earth gravity of Mars, and what the consequences will be, is not yet known The general rule in the current programme is that pregnant women are not flown, because the risks are unknown, and further research is required. Keeping babies in space beyond the age of 1 is not the best option, as at that age they are beginning to walk, and need gravity to develop strong bones and muscles and a sense of balance. These are the challenges beyond the ISS and initial flights to Mars, but fundamental research being carried out on the space station is helping to address these and other important issues.

THE ISS AND WOMEN'S HEALTH

Although construction of the ISS is now well underway, there are many dissenters who claim that it is a waste of money and of no great benefit to humanity. However, this is not a view held by proponents of the ISS, who believe that medical research carried out in a space-based laboratory could lead to important developments for human health. In his testimony to the House Committee on Science, Space and Technology in June 1993, Dr Michael DeBakey, an eminent heart surgeon (who at the time was both Chancellor and Chairman of the Department of Surgery of the Baylor College of Medicine), was reported by Marsha Freeman, in her article 'Space Station to Open New Biomedical Frontiers', as saying: 'The space station is not a luxury, any more than the medical research center at Baylor College of Medicine is a luxury ... Present technology on the Shuttle allows for stays in space of only about two weeks ... [On Earth] we do not limit medical researchers to only a few hours in the laboratory and expect cures for cancer. We need much longer missions in space – months to years – to obtain research results that may lead to the development of our knowledge and breakthroughs'.[4]

As Dolores Beasley, responding on behalf of NASA's Associate Administrator for Biological and Physical Research, Mary Kicza, remarked: 'The ISS will be a superb scientific laboratory. Scientists from all disciplines will be able to use microgravity as a research tool to gain new knowledge that will be beneficial for Earth-based uses. Science-driven human exploration of the Universe is a NASA mission, and results of research to protect the health of astronauts will also be very valuable for enhancing health care on Earth.'[5]

This response was also echoed by former NASA Administrator Dan Goldin, who, in his forward to the NASA Women's Outreach Initiative document 'There's Space in My Life for my Health', wrote: 'As a husband, a father of two daughters, and a grandfather, few subjects are as important to me as women's health. That is why I am so proud to present this overview of how NASA technologies, originally developed for our space and aeronautics programs, are improving the health of women around the world. In addition to space exploration and aeronautics development, NASA is charged with transferring its technology to the public to help life on Earth. Our visionary scientists and entrepreneurs have made giant leaps in technology application. Who would have dreamed that we could learn to map human tissue by mapping the distant stars?'[6]

The women's health issues commonly cited as potential beneficiaries of space-based research are ovarian and breast cancer, and osteoporosis. One scientist carrying out this type of research is Dr Jeanne Becker, a cell biologist who, in addition to giving testimony to the House Committee on Science, Space and Technology, was the guest speaker (on 28 July 1993) at a Congressional luncheon on 'Women's Health Issues and Space-Based Research.' It is also her belief that: 'From the ISS we will gain a great deal of information – not just in the advancement of research into women's health, but in human health in general. We have a unique opportunity to study the effects of gravity on so many bodily functions, from the cellular level (that is, normal and abnormal cell growth, gene expression, production of proteins and other factors, tissue engineering) to whole body effects (such as fluid shifts, bone changes, cardiac effects, hormone alterations).'[7]

From outer space to inner space: combating cancer

Dr Becker – who until recently was Associate Professor in the Department of Obstetrics and Gynaecology, and Medical Microbiology and Immunology at the University of South Florida (USF) College of Medicine in Tampa, Florida – has been working on the development of three-dimensional tissue models of breast and ovarian cancer using NASA's Rotating Wall Vessel (RWV), or, as it is more commonly known, the NASA (rotating-wall) Bioreactor. The Bioreactor is a special tissue culture chamber designed and developed by the Microgravity Sciences and Application Division at NASA's JSC, and enables scientists to grow cells that form themselves into structures similar to tissues found in the human body. As Dr Becker described in her interview with Nicola Humphries, the Bioreactor 'is a very simple instrument – you can think of it like a Coke can. You put your media inside, which is the liquid material that the cells grow in, [and] you put the cells inside the media on little beads ... the carrier beads that give the cells something to hang on to. You then evacuate all of the air out of the container [cylinder], and then you start to rotate it ... The cylinder, which is lying horizontally, is attached at the base and is rotated by a little pulley. It has no internal parts, so when the cylinder starts to rotate, the fluid inside, that contains the cells, starts to act like a solid body and assumes a similar rotational velocity as the wall of the vessel. Everything inside the fluid is in a kind of constant free-fall [neutral buoyant state], but it never reaches the bottom because [the cylinder] keeps rotating. So the gravity vector is constantly changing and it produces a very quiescent culture environment ... very quiet, and everything is just in free-fall ... You have one cell that grows to two, that grows to four, and pretty soon you have these little masses of cells that start to form, and because there is no internal moving parts the cellular constructs are maintained. Now these little masses of cells [spheroids] get larger and they get heavier. So in order to keep them in this 'free-fall' state you have to increase the rotation ... you have to speed it up. And that's one of the limitations of this technology. It's great to grow little masses of cells of up to a size of say 0.5 cm, but after that the masses get too big to keep in suspension ... the [cylinder] is whipping around really quickly, and your 'mass' that you have grown will actually fall out of suspension, hit the side of the vessel and shear apart. When this technology goes onboard the space station there is a further reduction in gravity relative to what we can simulate in a [ground-based] vessel, and we should be able to grow larger pieces of tissue.'[8]

Dr Becker's work with the Bioreactor, which began in the early 1990s, has been funded by NASA under a combined collaborative study between the University of Southern Florida and the Biotechnology Program at NASA JSC. Originally approached to grow breast tumour cells, Dr Becker realised, having thought about the problem, that 'primary breast cancer cells, which come right out of a patient, are very hard to grow.' She therefore began with an ovarian tumour cell model before later growing some breast cancer cells. As a result of her work in culturing cancer tumours, she believes that the Bioreactor 'affords a much better way of approaching how [cancer] drugs may work in the patient.' Consequently, she is very interested in developing an understanding of how the three-dimensional model works for improved chemo-sensitivity testing as a means of assessing the ability of certain drugs to kill a particular type of cancer cell within a patient. She also intends to use the Bioreactor to study endometriosis – a painful condition that can cause infertility in some women – but this work has not yet been initiated.

On 10 August 2001, Dr Becker and several of the NASA JSC flight operations/cell scientists were at the Cape, in a large field about three miles from the launch-pad, from where they witnessed the launch of the STS-105 *Discovery* mission to the ISS. The ISS replacement crew (ISS-3) carried with them the ovarian tumour cells which formed the basis of Becker's experiment on 'Growth of Human Ovarian Tumour Cells Aboard the ISS.' Following the undocking of *Endeavour*, as it prepared to return the ISS-2 crew to Earth, the cells – which had been transferred to the ISS – were cultured in the Cellular Biotechnology Operations Support System (CBOSS), maintained by ISS-3 Commander Frank Culbertson. After two weeks the cells were fixed for analysis back on Earth, although they would have to await the next Shuttle–ISS mission (STS-108) for their ride home. The cells, together with the ISS-3 crew, returned to Earth onboard the Space Shuttle *Endeavour*, which landed at the Cape on 17 December 2001. Two weeks earlier, ground control studies were conducted at JSC. (Ground-based ovarian tumour cells were grown as a control for baseline comparison with the cells cultured on the ISS.) The results of Becker's experiment are undergoing analysis, and have not yet been published. Additional experiments onboard the ISS will be subject to flight availability for cellular biotechnology experiments.

In 1990 a commercial licence for the Bioreactor was awarded by NASA to Synthecon Inc, in Houston, Texas, who market it as the Rotary Cell Culture System (RCCS) at an average cost of $4,000. To date, Synthecon – whose European distributor is Cellon SA, in Luxembourg – has sold RCCS units to the value of more than $5 million. As Dr Becker noted: 'The patent is owned by NASA, and the

Bioreactor is manufactured through another company ... It is now more commercially available [for] scientists to use it ... The NIH [National Institutes of Health] has a lab dedicated to Bioreactor research, [and] much of this work is focused on studying replication of the AIDS virus.'

Other benefits to women's health that have been derived from NASA-inspired technology include the use of charge-coupled devices (CCDs) in the early detection of breast cancer. CCDs are compact high-technology silicon chips that convert light directly into an electronic or digital image. During the development of the Space Telescope Imaging Spectrograph (STIS), installed on the HST in February 1997 during the STS-82 *Discovery* servicing mission, scientists working at the NASA Goddard Space Flight Center discovered that the CCDs available to them did not meet the standards required for the HST. NASA therefore placed a contract with Scientific Imaging Technologies Inc (SITe), in Beverton, Oregon, to develop a more advanced, low-cost, supersensitive CCD.

In meeting its contractual obligations to NASA, SITe realised that the 'new' STIS CCD technology could be adapted for the 'digital spot mammography market.'[9] The requirements for the HST, as cited in NASA's 'HST and Women's Health: NASA Technology in Your Doctor's Office', were high resolution, to reveal fine details; wide dynamic range, to capture, in a single image, structures spanning many levels of brightness; and low-light sensitivity, to shorten exposures and reduce X-ray dosage. These are essentially the same as the requirements for mammograms. The STIS CCDs were limited to the ultraviolet, visible and infrared wavelengths, so to extend their range into the X-ray region they were sensitised with a special phosphor coating.

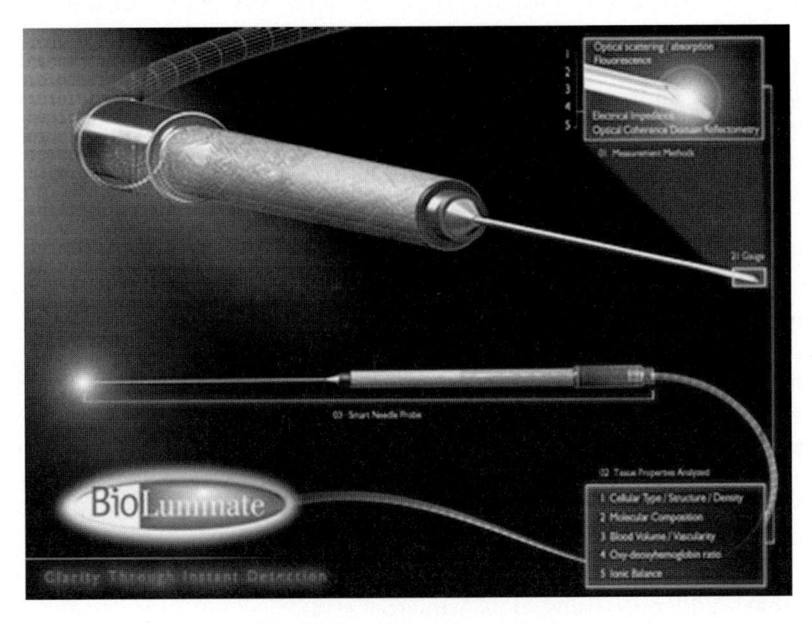

The BioLuminate Smart Probe.

One US company that has adopted the STIS CCDs (without the phosphor coating) as part of its breast imaging equipment is the Hologic Corporation, in Bedford, Massachusetts, whose commercially available Lorad breast biopsy system offers women 'a quick, easy, and less painful method for determining if a breast lump is malignant.' However, Debra Burnham, Product Support Manager, has stated that Hologic uses the device without the phosphor coating.[10] Instead of subjecting a female patient to a surgical biopsy involving a general anaesthetic and a scalpel (for cutting into the breast and removing a tissue sample), leaving a large scar, Hologic's Lorad Stereotactic Breast Biopsy System subjects the patient to only a local anaesthetic and a biopsy needle, leaving only a small puncture wound. This modified procedure is performed with the female patient lying prone on her front with one breast protruding through an opening in a specially designed table. Then, by means of the stereotactic X-ray imaging device that 'sees' the breast from two different angles, the system's computer calculates the position of the 'lump', which enables the operator (doctor) to insert the biopsy needle and extract a tiny tissue sample for analysis. Dr David Dershaw, Director of Breast Imaging at the Memorial Sloan-Kettering Cancer Center in New York, has stated that 'The woman who has gone through a needle localisation procedure and formal surgical biopsy on a prior occasion, and now comes in to have the same thing done, but has it done as a stereotactic biopsy, is about the most appreciative patient you can imagine, because you've taken a long, drawn-out, anxiety-ridden and expensive event and made it shorter, easier to schedule, and more comfortable. She has no surgical wound.'

It is also hoped that another NASA-inspired development – the Smart Probe for the detection of breast cancer, will significantly improve diagnosis time for women undergoing a mammogram. The Breast Cancer Smart Probe Project was set up as a collaborative venture between NASA Ames Research Center and Stanford University School of Medicine, utilising 'information technology being developed by NASA to assist astronaut-physicians in responding to emergencies during long spaceflights.'[11] The 'probe' – a spin-off of a 'smart' tool used for neurosurgery – uses special neural-net software that enables it to be 'taught' how to recognise a suspected breast abnormality and then predict the progress of the cancer. Dr Stefanie Jeffrey – Chief of Breast Surgery at Stanford University School of Medicine, and co-developer of the Smart Probe with Dr Robert Mah, Principal Investigator of the Smart Systems Research Laboratory at NASA Ames – has remarked that in the development of the Smart Probe she was 'using my experience of breast cancer clinical care and research, but really expect that this probe will be used much more on Earth ... well before it is needed in space. Hopefully, the concept will be applied in space for real-time measurements of many organ systems and associated disorders. Dr Mah should get the great majority of credit for this probe. It is his patent and software program. I work with him on choosing sensors [on] which he has done the majority of the research, setting up experiments and analysing data ... My main research interests are in breast cancer genomics ... I hope to relate our physiological findings that we measure with the Smart Probe to gene-expression patterns that we identify using cDNA microarrays.'[12]

In April 2000, NASA awarded an exclusive Smart Probe technology licence to the

Dublin-based Silicon Valley company, BioLuminate Inc, which planned to manufacture a commercially available 'probe' for the early detection of breast cancer. A second exclusive licence was also obtained by BioLuminate from the Los Alamos National Laboratory (LANL), for critical measurement technology used in the probe. BioLuminate then entered into partnership with the Lawrence Livermore National Laboratory (LLNL) to support and accelerate the detailed commercial design of the Smart Probe,[13] leading to a reduction in the size of the 'probe' from that originally envisaged. The complete (BioLuminate) breast cancer detection system – the BioLuminate Smart Probe – consists of a small disposable 20–21 gauge needle-like probe, a computer, optical components, conversion electronics, and graphics display.[14]

The BioLuminate Smart Probe would be used if or when an initial screening – either by mammogram or by physical examination of the breast – reveals a 'lump'. The needle-like probe (smaller than a blood-test needle) is inserted into the breast and, by means of sensors at the tip, real-time measurements of several of the known indicators of cancer are taken in both 'normal' breast tissue and at the position of the suspected 'abnormal' tissue. The BioLuminate Smart Probe, which does not remove tissue, therefore provides an instantaneous (real-time) answer as to whether the 'abnormality' is benign or malignant, and so relieves the patient's anxiety. This is in contrast to a large-core needle biopsy, in which a tissue sample is taken and then sent to a pathology laboratory for analysis. The one-minute test, which is 'expected to exceed the accuracy achieved by the core needle procedure, and approach the high levels realised by surgical biopsies', would also allow the doctor to commence treatment immediately.

Human clinical trials with the BioLuminate Smart Probe have already begun, and twenty-four patients have so far been tested. BioLuminate has also been awarded its first grant from America's National Cancer Institute (NCI). Richard Hular, Chairman of BioLuminate, expects the BioLuminate breast cancer detection system, featuring the BioLuminate Smart Probe, to be commercially available within two to three years. Since this is a new platform technology, it can be used for other solid cancers such as prostate cancer. Development of the other applications would begin after approval of the breast cancer probe by the Food and Drug Administration.

NASA is also collaborating with the NCI to enable earlier detection and better diagnoses of many other cancers. In June 1999 the two agencies held their first joint workshop on 'Sensors for Biomolecular Signatures', and on 13 April 2000 they signed a formal Memorandum of Understanding (MoU). Responding on behalf of NASA's Associate Administrator for Biological and Physical Research, Mary Kicza, Dolores Beasley explained: 'The National Aeronautics and Space Administration and the National Cancer Institute are collaborating on the development of innovative and minimally invasive sensing technologies, bioinformatics, and disease intervention strategies. The two agencies share a common interest in the development of minimally invasive approaches for detecting and interpreting biomolecular signatures that signal the emergence of disease and radiation damage in the living body. NASA's needs relate to the health maintenance challenges posed by long-duration spaceflight. NCI's priorities highlight the importance of detecting

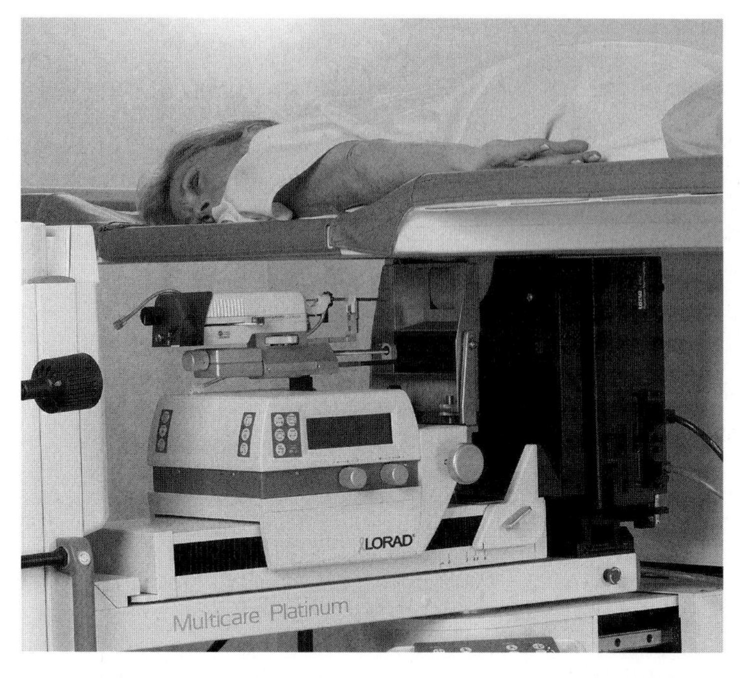

A patient being screened with Hologic's Stereotactic Breast Biopsy System.

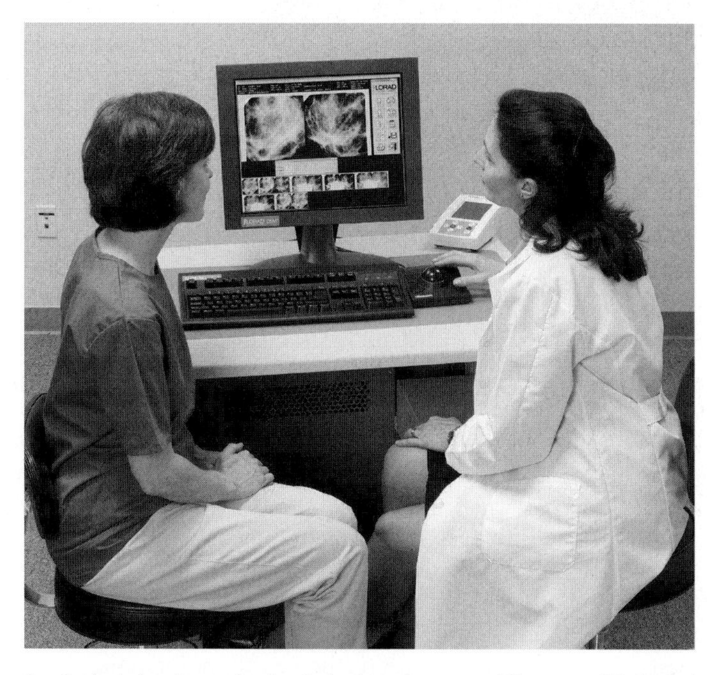

Analysing the data obtained during the scan. (Courtesy Hologic.)

the earliest signatures of cancer and immediate and effective treatment of emerging disease.'

While the medical agencies within the US appear, in general, to be aware of NASA's medical research and space-derived technology, this does not seem to be the case in either Britain or the rest of Europe. When the UK's Institute of Cancer Research (ICR) was invited to comment on this subject, its Press Officer responded with the statement: 'The Institute does not have an official line on space-based research, but you might find individual scientists have opinions. I have done some asking round for you, and have drawn a blank so far.'[15] This response came as no surprise to Dr Becker, who commented: 'I think that it is unfortunate not to raise the awareness of [this work] ... The same thing happens in America. As a scientist working on this technology, there are still people in science [and] in some of the governmental agencies that think this stuff is really 'way out there' – and that's here, where we are developing it. So I am not surprised that the UK's ICR doesn't have any line on it. It's mostly because of a lack of awareness.'

This lack of awareness was also echoed by Richard Fry, Managing Director of Cellon SA, who felt that most of the people in Europe who use the Bioreactor (RCCS from Synthecon) do so 'without thinking about, or knowing about the space link with the system.'[16] However, it is encouraging that scientists are using the

Dr Jeanne Becker, Associate Director of the National Biomedical Research Institute.

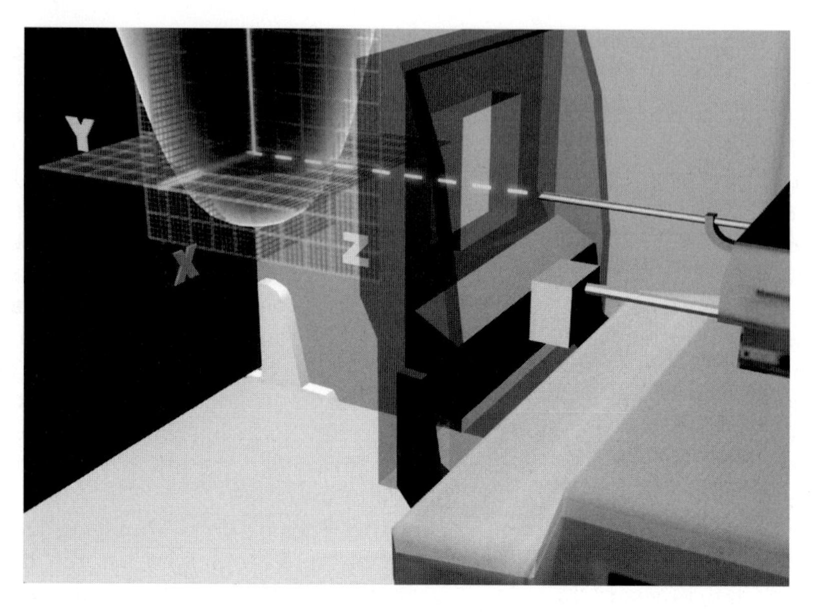

The precise needle-guidance system in Hologic's Stereotactic Breast Biopsy System. (Courtesy Hologic.)

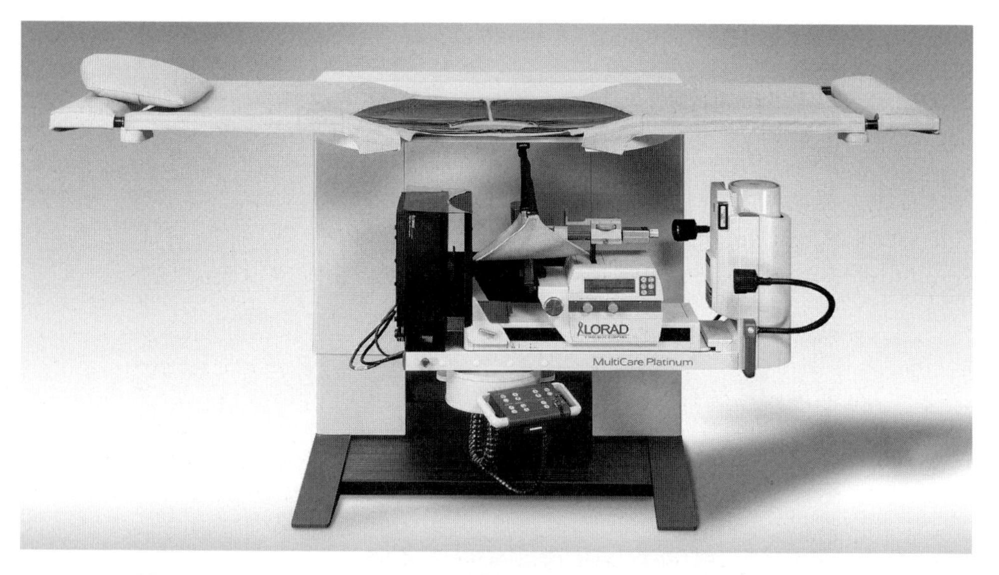

The Stereotactic Breast Biopsy System hardware. (Courtesy Hologic.)

technology, generally without the 'NASA hard-sell' – which can only be to the benefit of human health. Dr Emily Holton, Gravitational Research Branch Chief at NASA Ames, has succinctly remarked: 'Most groups are focused on curing sick people on Earth. NASA is focused on changes in healthy people. This is a paradigm shift ... Is the ISS essential for advancing cancer and osteoporosis knowledge?

Probably not. Will the ISS shed light on what happens during spaceflight that might be relevant to Earth issues. Definitely.'[17]

From outer space to inner space: osteoporosis

When astronauts are exposed to weightlessness in space, their bodies undergo a number of physical changes. One of these changes – loss of bone mass – is similar to the brittle-bone disease osteoporosis. Although men – who have a 30% greater peak bone mass than women – can suffer from the disease, osteoporosis is more common in women. Post-menopausal women are particularly vulnerable, and by the age of 75 may have lost as much as one-third of their skeletal structure.[18] Because of the many physiological changes resulting from menopause, investigation into the cause of 'thinning' bones is more difficult. But this is not the case with astronauts who, in general, tend to be young and healthy. Therefore, when they suffer from osteoporosis-like symptoms in space, it may be easier to identify, at a cellular level, the mechanism that is causing their bones to lose calcium as they adapt to being weightlessness.

Bone is generated by cells called osteoblasts, and reabsorbed by osteoclasts, in a balanced continuous cycle termed 'remodelling'. It has two primary purposes: structural support (the ultimate size of the skeleton of an adult appears to be genetically driven); and participation in calcium metabolism (calcium in the skeleton acts as 'reservoir' for other physiological systems).

In terms of structural support, the skeleton has to provide an articulate framework that can support the mass of a creature's body against the pull of gravity. It is the product of these two components – mass and gravity – that we experience as 'weight', which our bones, over the course of time, have evolved to withstand. In space, therefore, the weight-bearing parts of a skeleton are put at risk because they are no longer used under load (against the force of gravity). As a result, osteoblast activity at these specific sites is depressed, while osteoclast activity may or may not be stimulated.[19] This then causes an imbalance in the remodelling process, since both cell types (osteoblasts and osteoclasts) are required for 'bone turnover' in the adult skeleton, which in turn leads to a 'site-specific' loss of bone mass. In post-menopausal women, the remodelling process becomes unbalanced because their bodies have stopped producing the hormone oestrogen. (Without hormone replacement therapy – which itself is not totally effective – a post-menopausal woman can lose up to 3% of her bone mass annually.) However, NASA's Outreach Space Life Sciences pamphlet for Osteoporosis notes: 'Researchers do not yet know exactly what type or amount of exercise, hormones, or drugs might prevent bone loss or promote bone formation. Some combination of sex hormones, growth hormones, and exercise seems to be the key to preventing bone mass loss associated with ageing and post-menopausal hormone changes on Earth.'

One of the most promising drugs for the treatment of osteoporosis appears to be the protein osteoprotegerin (OPG), which in 1977 was 'discovered' by research scientists at the global biotechnology company Amgen Inc, in Thousand Oaks, California. While working on an internal genomic drug discovery programme, the Amgen scientists found that OPG – which is a naturally occurring protein in humans and other mammals – plays a key role in the bone remodelling process by controlling osteoclast activity. They also discovered that 'OPG binds with another protein, OPG-ligand [which is now referred to as RANK ligand], and by doing so, prevents osteoclasts from removing too much bone.'[20] As a result of this discovery, Amgen began an OPG Development Program to evaluate 'the injected protein's ability to safely and effectively treat osteoprosis and cancer metastasis to bone.'[21] The pre-clinical (ground-based) trials were aimed at determining the effectiveness of OPG to prevent bone loss, and used rats, mice and non-human primates as test subjects. The results indicated that extra OPG increases bone density when compared to normal OPG levels, and a decrease in OPG results in an increase of osteoclasts, which in turn leads to a reduction in bone density.

For the ground-based pre-clinical trials with 'unloaded' mice (gravitational loading is removed from the rodent's weight-bearing bones, which is akin to being 'weightless' in space) Amgen was assisted by BioServe Space Technologies – a non-profit, NASA-sponsored commercial space centre based at the University of Colorado in Boulder, Colorado. Encouraged by the results from the ground-based tests, Amgen and BioServe decided to conduct the 'unloaded' OPG tests in space, which would act as an experimental model for accelerated osteoporosis. On 5 December 2001, STS-108 *Endeavour* began its mission to the ISS; and accompanying the ISS replacement crew (ISS-4) were the two dozen laboratory mice that formed the basis of the Amgen–BioServe OPG experiment. Housed in the Commercial Biomedical Testing Module (CBTM) in the Shuttle's mid-deck, twelve of the mice had been given a single pre-launch injection of OPG, while the other twelve had been given a placebo. (The same experiment was also being carried out on Earth as a 'control for the effects of gravity.' However, unlike the ISS-4 crew, the mice did not transfer to the ISS, but remained in the Shuttle until it landed, with the returning ISS-3 crew, on 17 December. The twenty-four 'space mice' were then returned to the research scientists at Amgen and BioServe for examination into 'the effects of spaceflight and the ability of OPG to mitigate the effects of spaceflight on osteoclast/osteoblast signalling.' This work is still in progress.

To date, Amgen has conducted Food and Drug Administration Phase I human clinical trials with OPG, and is currently in the process of carrying out FDA Phase II human trials. However, with regard to the prescribing of drugs in general, Dr Holton offers a word of caution: 'As a pharmacologist, I understand the multitude of adverse actions that any drug can have. Thus, I am a firm believer in using exercise or normal physiology to address issues before reverting to drugs. I am very concerned about the overmedicated elderly in this country [America].'

One of the problems of space-induced osteoporosis which NASA has had to resolve, and which has potential benefits in the diagnosis and monitoring of Earth-bound osteoporosis, is that of quantifying the loss of bone mass by weightless astronauts. However, conventional bone density tests, which measure only bone mineral density as an indication of bone strength, do not distinguish fragile bones from strong bones, since some bones of low mineral density can function well without risk of fracture, while other more dense bones are more susceptible to fracture. What NASA required was a non-intrusive way of measuring the biomechanical properties of bone (that is, the bending stiffness of bone, which is a product of bone mineral content and bone shape). In 1977, therefore, NASA Ames, in collaboration with Stanford University, began work on an instrument to measure bone bending stiffness. Then, in 1989, with clinical trials of a workable device having been completed at Stanford University orthopaedic hospital, a small business company – Gait Scan Inc, based in Ridgewood, New Jersey – joined Ames and Stanford with the sole purpose of developing a practical and affordable instrument for the commercial market.[22] The resultant device – the Mechanical Response Tissue Analyser (MRTA) – is a direct, non-invasive, portable instrument that measures the bending stiffness of bone by using a brief, low-frequency vibration. Its operation is described in the NASA Press Release, 'New Technology Used to Develop Medical Instrument', which states: 'A technician places a small probe on the skin surface of the limb to be tested (either the ulna or tibia), which rests on a stable support. The patient feels a 'buzz' that lasts less than five seconds. The frequencies from the resonating bone are detected at the same site as the stimulus, and analysed by unique software in an attached computer. The result is an accurate measurement of the bending stiffness of the bone.'[23] Dr Sara Arnaud – a Principal Investigator in the Gravitational Research Branch at NASA Ames – noted: 'The MRTA doesn't replace the bone density-measuring technology we now have ... but it does provide an excellent non-radiation measure of long bone strength. It will, I am sure, find its place in the resources of physicians treating bone disease.'

In addition to its collaborative work with the DHHS and the NIH, NASA is also working with the National Institute of Arthritis and Musculoskeletal and Skin Disease (NIAMS), which is participating in research relating to spaceflight through its Osteoporosis Institute Centers. In the UK, however, the National Osteoporosis Society (like the UK's Institute for Cancer Research), 'does not have a position on space-based research.'[24] But unlike the ICR, the NOS is 'aware' of NASA's medical research, as indicated in the response from its Press Officer: 'Members of our Scientific Advisory Group are always interested to hear about research trials revolving around anything to do with bone loss and osteoporsis. There was an article published in *Osteoporosis Review* – the professional magazine produced by the NOS – about spaceflight and osteoporosis.' This article – entitled 'Skeletal Unloading: Space Flight and Osteoporosis' – appeared in the Autumn 2001 edition of *Osteoporosis Review*.[25] It was written by Steve Coates – a former RAF Engineer Officer who had entered the medical profession. In 1999, as a medical student, Coates was 'privileged to spend a summer on a research elective with NASA, looking at aspects of manned spaceflight' – his particular interest being 'the loss of bone density induced in astronauts by the effects of microgravity.' In the acknowledgements in his article, Coates offers thanks to 'the anonymous woman who, suffering for osteoporosis herself, kindly offered me sponsorship for my research at NASA.'

Dr Roger Francis – Reader in Medicine (Geriatrics) at the Freeman Hospital in Newcastle-upon-Tyne, and Editor of *Osteoporosis Review*, later reviewed Coates' article: 'Like many people of my generation, I avidly followed the early ventures into space by pioneers such as Yuri Gagarin and John Glenn. I also vividly remember the tentative first steps on the Moon taken by Neil Armstrong. Many people of all ages have enjoyed watching the fictional adventures of Captain James T. Kirk and the crew of the starship *Enterprise*. In reality, one of the major barriers to prolonged space travel is bone loss and the development of osteoporosis. In this issue, Steve Coates reviews the possible mechanisms and prevention of bone loss associated with microgravity. Although this research is important to NASA in planning further exploration in space, it may also provide valuable insights into osteoporosis associated with immobilisation.'[26]

From outer space to inner space: ageing

According to a NASA pamphlet on ageing: 'The population [of the US] is getting older. There are twice as many Americans over 65 as there were forty years ago. By 2050, the number of Americans 85 or older will increase 600 percent.'[27]

Therefore, with an ageing population, the US – together with the rest of the Western world – will have to resolve a number of socio-political issues – one of which involves the medical requirements of the elderly. However, given that there are similarities between some of the weightlessness effects on the human body and ageing, it is hoped that space-based research will provide an insight into the geriatric conditions of bone loss, degeneration of the muscles and cardiovascular system, and impaired sleep.

The first, and to date, only space-based geriatric study took place during the nine-day STS-95 *Discovery* mission in October–November 1998, and involved veteran Mercury astronaut, Senator John Glenn, who, having become the first American to orbit the Earth in February 1962, was making his second spaceflight after an interim of 36 years. At the age of 77, Glenn was also the oldest person to travel into space. According to Carla Garnett, in her article 'Next Flight, Please – Glenn', he said: I hope that by comparing what happens to me at my age in space with what happens to younger astronauts and with older people right here on Earth, maybe we can not only increase the ability of younger people to go on longer spaceflights, but also perhaps we can help eradicate many of the frailties of ageing – things like balance and muscle system changes, osteoporosis and sleep disorders.[28]

However, even though NASA and the NIH's National Institute on Ageing had jointly sponsored the 'geriatric study', many people regarded Glenn's flight on the Shuttle as merely a joyride for an influential politician and an 'all-American hero.' Glenn's flight also reopened the debate surrounding Jerrie Cobb and the other twelve women from the Lovelace Class of 1961, who still remembered Glenn's testimony at the congressional hearings and his role in preventing their access to space. As Jerri Truhill remarked: 'I was invited to the launch by NASA, and declined, as we all did, except Jerrie Cobb. [Glenn] shot us down in Congress.[29] A similar sentiment was also expressed by Wally Funk: 'Frankly, this was a political thing, and I hope he didn't waste too much of the tax-payers' dollars when he took a young astronaut's slot who was working for STS-95.'[30] Gene Nora Jessen, however, was more magnanimous: 'I'm a strong supporter of our country's space programme, and believe that NASA needs all the good press it can dredge up. If John Glenn's planned flight lends favourable publicity and prestige to the space programme, that's all to the good.'[31]

Many Americans also believed that if John Glenn could make two spaceflights (the second being under the auspices of the 'geriatric study'), then NASA should 'right the wrong' and send Jerrie Cobb – who was still carrying out her humanitarian flights in the Amazon and had a recorded medical history – into space. Taking up this cause was the National Organisation for Women, which mounted a campaign (aimed at NASA) to have Jerrie Cobb participate in the 'geriatric study'. As Patricia Ireland, NOW's President, remarked in a press release: 'If NASA wants to study the effects of space travel on ageing, then it is imperative to make these studies on women. After all, women are the majority of the elderly.'[32]

Therefore, given that the ageing population is predominately female, what benefit, if any, do women receive from Glenn's space-based results? Responding to this question in her interview with Nicola Humphries, Dr Jeanne Becker – now Associate Director of the National Biomedical Research Institute (a NASA-funded institution) – considered that elderly women 'do get a benefit from some of the results ... There is a difference physiologically, of course ... As a model for the ageing population I think that [NASA] needs to send a woman to compare the results to [those] they got with Glenn.' And what, if any, are the flaws of sending an aged man into space and then applying the results to a woman? Again, Dr Becker, in her interview with Nicola Humphries, responded by saying: 'Well, there are differences, and you have to take into account the post-menopausal status of the woman. Obviously their hormones are different ... You can't directly compare everything you get with male data to female data. It's sort of like oranges and tangerines, which is not like apples and oranges. It's not totally opposite; there are some similarities, but there are some real basic differences. You are not going to get a handle on what those differences are unless you actually send a woman up.'

Currently the oldest female astronaut that NASA has sent into space is Dr Shannon Lucid (currently NASA's Chief Scientist). At the age of 54, in 1996 she spent 188 days on Mir. During her post-flight news conference, she responded to a question relating to 'the human life science medical tests' that had been carried out upon her return: 'I have lost some calcium in my bones, but it was just what was predicted pre-flight. It was no more than the average. Actually, my bones are in really good shape. I started out at a very high level so I'm still at a very high level. A lot of the scientific investigation results haven't come back yet. I just happen to know about the calcium loss in the bones because that's a quick one to get the results back on. But everything else, that standard blood test they do, everything came back within the normal range. I am doing exercises to regain the strength in my legs and arms.'[33]

While Lucid's long-duration spaceflight has provided valuable medical data, she was not a 'geriatric' female. Jerrie Cobb, however, would fit the description; but she was yet again denied her trip into space (as part of the 'geriatric study'). Like the rest of the Lovelace Class of 1961 – who were still keen to achieve their dream of flying into space – Cobb remained an Earth-bound observer with no prospect of making a spaceflight, while Senator John Glenn was given approval to undertake his second space mission. Moreover, flying with Glenn on STS-95 was 46-year-old Dr Chiaki Mukai, who was also making her second trip into space. In July 1994, on the STS-65 *Columbia* mission, Mukai had become Japan's first female astronaut – an achievement that Glenn, at the congressional hearings in July 1966, had helped deny Jerrie Cobb when she could have been America's first female astronaut!

TO BOLDLY GO ...

Following Glenn's second mission into space, the media began circulating reports that NASA was evaluating the possibility of carrying out a dedicated all-female life sciences mission. In an article published in *The Times* 'Weekend',[34] Anjana Ahuja reported that in March 1999, 'Dan Goldin, NASA's top man, let slip that the first all-woman Shuttle mission crew was under consideration. It could be ready for lift-off in three years' time. She also included a quote from Dr Arnauld Nicogossian, NASA's Associate Administrator for Life Sciences, as to the rationale for such a mission: 'We are studying whether we should accelerate the process of getting more women in space. We want to improve the healthcare we deliver to men and women astronauts, and this might include a dedicated life sciences mission which might or might not be all-female ... We are still evaluating whether there are gender issues we need to address, and one of the questions is whether we need to do a study in space just on women. At the moment, mission 107 could potentially be dedicated to studying this topic, but no decision has been made yet.'

Although Eileen Collins, at the time of the speculation, was about to become the first female Shuttle Commander, Mark Prigg, in an article in the *Sunday Times*, reported that 'The woman most likely to command the all-female mission is Susan Still, 38, who has already co-piloted the Shuttle on two missions.'[35]

Responding to the idea of an all-female crew – which was viewed by some as another cynical publicity study by NASA, Dr Becker felt that 'they would be equal. I don't want to sound sexist because I think it should be both, but as a gesture I think that it's an amazing idea to send up an all-female crew, [although] in fairness it should be joint. I think it should be more than one [woman] out of six. I think more women should get to go. This was a sentiment echoed by Dr Holton: 'I think an all-female flight is a great idea, but I don't select the flight crews. The main focus of the Shuttle flights seems to be on the construction of the ISS rather than life sciences. Once the station is completed, then perhaps this idea could be put forward.' But Dr Bonnie Dalton, Deputy Director for the Astrobiology and Space Research Directorate at NASA Ames, considered that it would be 'silly': 'Since when have we [women] become so privileged?'[36]

In August 1999, as an 'Element of NSBRI Project 99-4 (Cooperative Agreement NCC 9-58), Enabling a Broader Segment of the Population to Explore, Live and Work in Space in the 21st Century', the National Space Biomedical Research Institute (NSBRI) held a two-day workshop in Houston on 'Gender-Related Issues in Space Flight Research and Health Care.' The task of the workshop's panel – which included astronaut Dr Ellen Baker and former astronaut Dr Rhea Seddon –

was (as cited in the report) to 'review the following, with an emphasis on gender-related issues':[37]

- The knowledge base of human spaceflight.
- The status of current investigated or proposed medical countermeasures, for maintaining crew health in flight and post-flight rehabilitation.
- Existing and proposed human–machine interface standards and requirements, addressing aspects of training and space operations relevant to crew-members' health, wellbeing and safety.

Having been briefed by NASA scientists and managers on the relevant data, issues, policies and plans, the panel produced, as instructed, a report that both addressed the issues and made recommendations in the following three areas:

Research and contermeasures

- From the data presented there are a few areas that show definite measurable gender differences.
- Although no (female) spaceflight data exists, differences could be predicted for several systems (such as post-menopausal bone loss, iron intake requirements, and muscle strength and endurance).
- Spaceflight data in other areas have not been collected, and prediction of gender differences is not possible, but studying them is important for long-term health, safety and performance.
- Due to limited in-flight research opportunities, ground-based research should be strengthened. (Women should not be excluded from studies for reasons of cost or convenience).
- Where ground-based studies indicate gender differences, the number of females should be increased to ensure findings achieve a statistical significance.
- NASA's partners on the ISS should be made aware of the gender-related issues and encouraged to collaborate in this research.

Health care

- All women are accepted into the astronaut programme during their child-bearing years.
- Many female astronauts have postponed having children to establish their careers, but due to uncertainty in the allocation of flight crews for the Shuttle and ISS missions, the planning of a pregnancy is difficult.
- Once assigned to a flight crew, a woman should not become pregnant if she wishes to fly with 'her' crew, but delays in her flight will mean that her attempts to become pregnant can be delayed by months or perhaps years.
- A woman not currently assigned to a flight crew who wishes to become pregnant will be removed from the flight crew selection list while attempting to conceive or until she returns from maternity leave.
- For the wellbeing of these women and to return them to flight-crew status as

quickly as possible, the provision of assisted reproductive technologies should be available.

- For the prevention of pregnancy, individual counselling on the best methods of birth control should always be available.
- The unknown risk to mother and foetus if conception occurs just prior to flight, or in flight, should be stressed.
- The pharmacodynamics of birth-control pills in zero g (for which there are currently no data) should be studied to ensure efficacy. (Birth-control pills are also used by flight crew-members for non-contraceptive benefits, including maintenance of bone density, reduction in the risk of ovarian cysts, ovarian cancer, anaemia, benign breast disease, and for the prevention or controlling of menstruation.)

Human–machine interface

- It should be a goal that all people selected to be astronauts be able to perform all associated tasks, regardless of size and gender. (At the time of the workshop, NASA had already begun a study to identify 'gaps' in existing equipment and task design that prevented this goal from being achieved.)
- NASA should not accept designs that restrict crew assignment to tasks, flights or vehicles, or compromise safety.
- Individuals – male and female – selected to the Astronaut Corps based on NASA requirements should be assured that all aspects of the job are open to them.
- Poor design in current EVA suits and Shuttle egress suits, or lack of appropriately sized off-the-peg suits, should in future not be used as an excuse to exclude women from certain jobs or to provide them with equipment that is less than optimal for their own performance.

The report concluded with the following statement: 'There are a number of areas of research that should focus on the study of female subjects, both on the ground and in flight. There are several health care issues unique to the female astronaut population that NASA must address. A firm commitment to equipment and task design to optimise job performance and safety is required. None of these recommendations require that all-female crews be flown. The capabilities and talents of both genders will be needed for the monumental task of exploring our Universe.'

Consequently, NASA has not flown an all-female mission, and the crew of the STS-107 *Columbia* mission, launched on 16 January 2003, contained only two female astronauts: Kalpana Chawla and Laurel Clark. Tragically, the crew perished when *Columbia* broke up during its return to Earth on 1 February 2003.

One observable effect of weightlessness on the human body that effects both male and female astronauts is the 'puffy head' syndrome, caused by the redistribution of fluids in the body and resulting in more fluid pooling in the chest and head.[38] In a question-and-answer session, Dr Holton remarked: 'This shift [of fluid to the head] seems to be perceived by many crew-members as a 'stuffy head' or sinus infection, and they take decongestants to try to eliminate the discomfort.'[39]

However, in his *Sunday Times* article Mark Prigg quotes Kathryn Clark, NASA's Chief Scientist for the ISS: 'On our mixed flights, the male astronauts have complained about feeling puffed up as their bodily fluids float in zero gravity ... Women have to put up with these feelings every month, and can easily cope with them.'

In November 2002 – around two years after the NSBRI workshop – NASA sponsored a conference at the University of Missouri (with DHHS, NIH, and the President's Deputy Associate Science Officer). Attending this conference was astronaut Dr Ellen Baker, who, as a keynote speaker, presented 'an insightful and informative talk on the role of women to the NASA astronaut programme.'[40] It was noted, during the workshop, that 'Women now represent a significant percentage of NASA's astronaut corps, and will play an important role in the International Space Station program and in other space exploration activities this century – perhaps even missions to Mars ... We currently have little, if any, data relating to long-term health, safety and performance of women in space.'

The workshop consisted of six working groups that 'focussed on issues in the musculoskeletal, cardiovascular, immunological, neurovestibular, reproductive and human behaviour risk areas', and had to answer the question, 'Does sex matter?', as applied to specifically to NASA's mission for extended human exploration of space.' The working groups' conclusions were as follows:

Musculoskeletal physiology The existing flight and bed rest database on female astronauts and female bed rest subjects is quantitatively and qualitatively deficient. Studies with adequate numbers of female and male subjects will be required to determine whether gender differences exist in muscle or bone responses to microgravity.

Cardiovascular physiology The cardiovascular system (the heart, veins and arteries) interacts with every other system of the body (the vestibular systems, which control sense of balance and direction of motion), and there will need to be an integration of cardiovascular research with other research.

Immune function Sex influences on the immune system in space is a crucial area of study to ensure crew health and safety and to further knowledge on the effects of sex on the immune system.

Neurovestibular/neuroscience Vestibular maladaption produces changes in human performance that are potentially life-threatening. The importance of solving these problems cannot be overstated. However, there is little or no information regarding sex and gender effects on neurovestibular functions (part of the nervous system responsible for balance mechanisms), even in terrestrial environments.

Reproductive biology Research priorities in reproductive biology laid down in a NASA document twenty-five years ago remain important and largely unexplored. Now that long-duration spaceflight has arrived and planning for exploration-class missions is underway, it would not be prudent to further delay these investigations.

Human behaviour and performance Given the growing awareness of the critical nature of behaviour and performance issues for long-duration space missions, understanding sex differences is crucial.

With President Bush's recent space initiative calling for a human return to the Moon by 2015, followed by later manned missions to Mars, these gender-related recommendations will hopefully be undertaken in earnest. Dr Becker has noted: 'Since the President's management [has] changed towards a large emphasis on planetary exploration, efforts on microgravity research are being focused on development of countermeasures to overcome effects of long-term space exploration in order to keep our astronauts healthy.'

Although most of the body's physiological functions adjust to 'space-normal' conditions (they stabilise) within about 1½ months of continuous exposure to the zero-g environment of space, there are two space-borne effects which do not stabilise: radiation (cumulative effects), and bone adaptation (bone and calcium loss). These are potentially the greatest obstacles in the human exploration of space.

Solar radiation varies in ouput, and is particularly prevalent following a large solar flare; while cosmic radiation emanates from outside the Solar System as a continuous stream of high-frequency, high-energy particles. Any such radiation can damage the chromosomes, resulting in cancers. On Earth, however, we are protected by the atmosphere (ultraviolet light, for example, is filtered by the ozone layer). The standard unit of measurement of radiation dosage is the rem (Röntgen Equivalent Man, in which the Röntgen is the standard unit of measurement of radiation). The currently accepted safe limits – based upon scientific research into peak and prolonged exposure of people working in hazardous radiation environments – is 5 rem per year, while the permissible maximum dose over a lifetime must be less than 500 rem. For a mission to Mars, the radiation damage is 'predicted to provide an approximately lifetime cancer risk for 30 year-old males of about 28% as compared to 20% on Earth. This is unacceptably high, and scientists are trying to reduce it to about 23%. Because the radiation cancer risk to women is projected to be substantially greater – largely as breast and ovarian cancer – mission planners lean towards all-male crews.'[41] There is a rationale for women being at greater risk, but men are also at risk of genetic damage. Dr Dalton has noted: 'We know, from studies with rats, that the testes are effected negatively ... But remember that the metabolic rate of rats is manyfold that of humans. Things happen faster in a rat'. In addressing the issue of space radiation, mission planners are examining various methods to shield crews from the harmful effects.

Bone adaptation in microgravity, for both male and female, results in approximately 1% bone loss per month (implying that astronauts on a thirty-month mission to Mars would suffer a 30% bone loss). The bone and calcium losses also give rise to space-induced osteoporosis and the risk of kidney stones. However, as previously noted: 'Changes during spaceflight are limited to the weight-bearing part of the skeleton.' Therefore, as a countermeasure to 'bone adaptation', astronauts carry out physical exercises in space. Dalton notes that bone loss is recovered after return to Earth, 'but the period of time for recovery varies from person to person, and is also the result of their dedication to exercise countermeasures', while Holton believes that 'the exercises in space do not load the skeleton to the same levels that are achievable on Earth. From my perspective, the greatest bone losses to date in space are found in individuals who exercised a lot and ate little, so that they lost significant amounts of weight. These changes are similar to those that occur in individuals on Earth who under-eat and over-exercise. The bones in space will probably reshape themselves and be quite appropriate for the space environment. I doubt that they ever really disintegrate, but I'm not sure. The body tends to conserve bone mineral, and the mineral might not change, although the amount may be less at some sites than others. Hibernating bears have been shown to lay down mineral during hibernation, and that mineral is redistributed when the bear arouses from hibernation. Perhaps the same process will be used in space, and returning to higher gravity environment will be like arousing from hibernation.[42]

Another solution to the exercise countermeasures for bone adaptation in microgravity would be to subject the crew to some form of artificial gravity. Dalton has remarked: 'We may have to provide some artificial means, although a rotating vehicle is still far too costly.' Holton stated: 'We do not yet have that data to determine if an onboard centrifuge is necessary. NASA has developed a roadmap that outlines the studies that are to be done to answer this question.' A similar sentiment has been expressed by Joseph P. Kerwin, Science Pilot on the first Skylab mission (SL-2), who considered that the question of artificial gravity is too complex for a simple answer.[43]

The effects of radiation and the importance of gravity on human physiology are of concern in the inevitable issue of human conception, foetal development, childbirth, and child development in space. Dalton has noted: 'Researchers at NASA Ames have shown that a 'force' is required to align bone cells and support their growth'; while Holton has noted: 'In reality, our bones are very much like rocks. They are the result of eons of pressure. Since gravity is ubiquitous on Earth, we had no reason to consider it a major physiological factor until we were able to access space, where the loads are far less. We're certainly learning much more about the importance of gravity to life on Earth through these experiments as well as gaining an understanding of how life might evolve with decreased gravity ... I've had a lot of fun putting together and presenting a paper on gravity's effects on living systems for cells through to humans. I was fortunate to be invited to London to give this presentation.'[44] Holton has also stated: 'Many of the reflexes in humans develop to oppose gravity, such as those required for standing erect, walking, throwing and catching a ball. These reflexes degrade in space, as you cannot 'fall on your face', and gravity creates an arc to the flight of a ball on Earth. I'm very curious to know if these reflexes would even develop in a mammal born and raised in space with minimal gravity input. I'm fascinated by the plasticity of physiological systems (bone, nerves, muscle) in their ability to adapt to differences in loading in a very efficient manner. During spaceflight, bones appear to restructure to efficiently adapt to the change in loading, and gravity receptors in the inner ear increase the number of nerve connections probably as a means of amplifying a smaller gravity signal, and so on.'

Kerwin has agreed with the concerns about conception and the effects of microgravity during the early development of the foetus: 'They are human beings too'; and Eileen Collins has spoken about the possible effects of radiation: 'I got pregnant about a month after I got back from my flight [on STS-63 in February 1995]. You do not want to fly in space while you are pregnant. The major concern is the radiation. For example, you wouldn't get an X-ray while you were pregnant. On a spaceflight you are exposed. I don't have the numbers, but ... I've heard [that it is] twenty times the amount of radiation. And it's a different type of radiation. There's solar radiation and cosmic radiation. It's not harmful to a grown adult, but any living thing that is in the process [of] cell division would be harmed. We're not sure if a developing embryo would survive. If it did survive, what type of problems would we have?'[45]

Inevitably, if humans colonise space, babies will be born in space; and when it happens, nurses will probably be in attendance. Working to ensure that there is a nursing fraternity in space is the aptly named Space Nursing Society, founded in 1991. One of its founder members was the late nursing theorist, Martha Rogers, who in 1970 first mooted the idea of the nursing profession expanding to meet the new challenges of the emerging space-age. On its website, the SNS states its principles: 'Excellence in the delivery of health care to space clients (be they space travellers, ground support staff or their families), advancement of the aerospace nursing profession, representation of all aerospace nurses, and influence in health policy and standards.' To achieve these objectives it provides: 'A forum for the discussion and exploitation of issues related to nursing in space and its impact upon our understanding of Earth-bound nursing through conference participation and its newsletter *Expanding Horizons*.' Nurses have served people wherever they have settled, so it is reasonable to expect them to follow mankind into space.[46]

THE JOURNEY CONTINUES

On 6 May 2004, NASA announced the selection of eleven new astronauts, including the first Mission Specialist Educator (MS-E) astronauts – teachers selected to help inspire the next generation of space explorers by instructing them on current and planned programmes of space exploration. The eleven formed the nineteenth astronaut selection since 1959, and were the first selected to focus on the vision of space exploration, in a programme designed to return to the Moon and journey beyond. The new astronauts included Dorothy M. Metcalf-Lindenberger (*b.*1975), a former science teacher selected as an MS-E, and Shannon Walker (*b.*1965), with a PhD in space physics, selected as an MS. They were joined on their training and evaluation programme by Japanese astronaut Naokao Yamazaki (*b.*1970), an aerospace engineer. These three women are the latest to be selected for the journey to space – a journey that began many years earlier in school, where many future space

Veteran astronaut John Young introduces the latest group of NASA astronauts, including science teacher Dorothy Metcalf-Lindenberger, space physicist Shannon Walker and, among several international candidates, Japanese astronaut Naokao Yamazaki.

explorers are inspired by their teachers and members of their families. The desire to learn continues throughout their lives and careers, and includes interests in other fields of specialisation and participation in a wide range of sporting and social activities. Although space explorers require the skills to fly the machines that take them on their journey into space, they also need to be prepared for the unexpected, and sufficiently talented to be multi-skilled in areas that are not necessarily their specialised fields. Women have many times proven that they have these skills.

REFERENCES

1 David Portree, 'Romance to Reality: Moon and Mars Plans', http://membersa.ol.com/dgportree/ex00c.htm.
2 ESA Press Release PR 45-2004, 3 August 2004.
3 AIS interview, Rhea Seddon, 21 April 2004
4 *Space Station to Open New Biomedical Frontiers*, Marsha Freemen, 21st Century, autumn 1998.
5 Questions for Dolores Beasley (Responding on behalf of Mary Kicza), questionnaire from Teresa Kingsnorth, 16 March 2003.
6 'There's Space in My Life ... for my Health', NASA Public Affairs, Women's Outreach Initiative, (undated).
7 Correspondence from Dr Jeanne Becker, 15 March 1999.
8 Dr Jeanne Becker, interview by Nicola Humphries, Orlando, Florida, July 1999.

9 'HST and Women's Health: NASA Technology in Your Doctor's Office', NASA Facts, 27 August 1997.
10 Correspondence from Debbie Burnham, 24 May 2004.
11 NASA's Smart Systems Research Lab, http://ssrl.arc.nasa.gov/breastcancer.html.
12 Correspondence from Dr Stefanie Jeffrey to Teresa Kingsnorth, 20 April 2003.
13 'Smart Probe' for breast Cancer Detects Malignant Tumours Instantly', press releas, 10 January 2001.
14 http://www.bioluminate.com.
15 Correspondence from Institute of Cancer Research, 25 February 1999.
16 Correspondence from Richard Fry, 24 March 2004.
17 Questions for Dr Holton, questionnaire from Teresa Kingsnorth, 14 March 2003.
18 Emily Holton, 'Examining the Effects of Space Flight on the Skeletal Systems, in *Human Physiology in Space: A Curriculum Supplement for Secondary Schools*, by Barbara F. Lujon and Ronald J. White, *c.*1990.
19 'Essentials of Biology: Osteoporosis, NASA Outreach Space Life Sciences pamphlet, (undated).
20 'Commercial Biomedical Testing Module (CBTM)', NASA Fact Sheet, November 2001.
21 'Amgen and NASA team up to tap Weightless Environment of Space as Lab to test Osteoporosis Treatment', Amgen Media Release, November 2001.
22 'A Boon for Bone Research', in NASA Spinoff, 1996. http://www.sti.nasa.gov/tto/spinoff1996/24.html.
23 'New Technology Used to Develop Medical Instrument', NASA Press Release, 28 February 1995.
24 Correspondence from NOS, 7 August 2002.
25 'Skeletal Unloading: Space Flight and Osteoporosis', *Osteoporosis Review* (*Journal of the National Osteoporosis Society*), **9**, 3, Autumn 2001.
26 'Space: the final frontier?', *Osteoporosis Review* (*Journal of the National Osteoporosis Society*), **9**, 3, Autumn 2001.
27 'Essentials of Biology: Ageing', NASA Outreach Space Life Sciences pamphlet, (undated).
28 Carla Garnett, 'Next Flight, Please – Glenn', 'Discovery Crew visit NIH, Recount Historic Space Journey', http://www.nih.gov/news/NIH-Record/01_26_99/story01.htm.
29 Correspondence from Jerri Truhill, 1998.
30 Correspondence from Wally Funk, 1998.
31 Correspondence from Gene Nora Jessen, 9 June 1998.
32 'NOW Launches Campaign to Send Jerrie Cobb into Space', NOW Press Release, 28 October 1998.
33 'NASA Shuttle–Mir: Lucid Post-flight News Conference', http://spaceflight.nasa.gov/history/shuttle-mir/ops/crew/lucid/shannon3.html.
34 Anjana Ahja, 'Women: the final frontier', *The Times* 'Weekend', 17 April 1999.
35 Mark Prigg, 'Space girls may beat men to Mars', *The Sunday Times*, 14 March 1999.
36 Questions for Dr Dalton, questionnaire from Teresa Kingsnorth, 8 March 2003.

37 NSBRI workshop report 'Gender-Related Issues in Space Flight Research and Health Care', http://www.reston.com/nasa/nsbri/09.30.99.gender.issues.html
38 Barbara F Lujoan and Ronald J White, *Human Physiology in Space: A Curriculum Supplement for Secondary Schools*, (undated).
39 Question and answer session with Emily Holton, http://quest.arc.nasa.gov/people/bios/women/ehqa.html.
40 Sex, Space and Environmental Adaptation: A National Workshop to Define Research Priorities Regarding Sex Differences in Human Responses to Challenging Environments, University of Missouri, Columbia, 12–14 November 2002.
41 'Dodging a bullet', in 'NASA using space incubator to understand breast cancer', http://science.nasa.gov/newhome/headlines/msad01oct98_1.htm.
42 Conversation with Emily Holton, http://quest.arc.nasa.gov/woman/archive/1-14-97eh.text).
43 Correspondence from Dr Joseph P. Kerwin to Teresa Kingsnorth, 24 February 2003.
44 Women of NASA QuestChat Archive, featuring Emily Holton, Gravitational Research Branch Chief, NASA Ames Research Center, Moffett Field California, http://quest.arc.nasa.gov/woman/archive/12-07-99eh.html.
45 Interview with Eileen Collins (via telephone), 15 May 1996 (interviewer not known), transcript given to Nicola Humphries during her research trip to the America in March 1999.
46 http://www.geocities.com/spacenursingsociety/.

Conclusion

When asked what advice she would give to a schoolgirl thinking of a career in science, veteran astronaut Rhea Seddon commented that although not every girl should have a career in science they should pursue it if they are interested, and not be held back because they are female or because they consider that there are too many other women in a particular field. Seddon has had three careers. She was trained as a clinician, work as an astronaut and flew in space, and then became a physician executive, the first two of which she used to define a whole new career as an administrator.

From the guidance of a veteran astronaut to the hopes of an unflown astronaut. When Yvonne Cagle was asked the same question (during an interview on 23 February 2004) she replied: 'The sky is no longer the limit, so if the Universe does not limit you, don't put limits on your dreams ... Keep the boundaries open, and space is a great platform where you can pursue a career as an astronaut.' The ability to realise that those dreams are possible have been a long time in the making.

On a record made during the 1960s, the American singer James Brown sang, 'It's a Man's Man's Man's World'; and for the women living in the democratic post-war countries of the West, this was certainly true. With the cessation of hostilities at the end of the Second World War there was no longer a necessity for women to carry out work that had traditionally been the preserve of men. Society – and in particular, the West's new superpower, the USA – now viewed women as 'Mom'. In the Cold War climate of the 1950s and 1960s the new (American) social order was one in which 'Dad' went to work, 'Mom' stayed at home (tending to the domestic demands of house and family), and the (male) military machine sought to defend this Utopian idyll against the menacing threat of Communism.

It was against this sociopolitical ideology of gender-assigned roles, in which women were subservient and no longer had access to the 'new technology' (jet aircraft), that the thirteen women who had passed the 'astronaut tests' were expected to conform. Unfortunately for them they were born too early for society to accommodate the changes in the social order that were needed for America to sanction sending a woman into space. This ideology became so ingrained that of all the Soviet space 'firsts', the flight by female cosmonaut Valentina Tereshkova was the only one that NASA did not counter with an American flight. It would take the feminist movement of the 1960s and 1970s for American women to be granted access to space, courtesy of NASA, and then, initially, only as scientists, not pilots.

A few women were able to push the boundaries of acceptability by means of wealth and political influence, and in aviation circles the most notable was Jackie Cochran. As America's premier female pilot, Cochran had a desire to become the first woman in space, but her age and health ruled her out of contention. Consequently she did not help promote the cause of 'women in space' – leaving us to conjecture what might have happened had she been younger and therefore eligible for such a flight.

In the Soviet Union women were assigned a variety of roles in society that also reflected the Russian belief that women are not fully equivalent to a man in health and workload. In Russia, women can drive trams or trolley buses on rails but not ordinary buses, medical rules prevent women from being metro drivers (alhough there are a few), and chauvinism and cultural dogma has prevented many aspiring female cosmonauts from reaching their goal of flying into space. Some of them, however, persevered for many years before finally accepting that they would not fly in space. In Russia, the development of the space station programme has been one of national pride; but it has also caused bitter disappointment due its dependence on the Americans. Although the Russians successfully launched seven space stations only two Russian women visited them, and currently there are no Russian female cosmonauts in training for flights to the ISS.

Since the introduction of the American Space Shuttle, women have flown into space as Mission Specialists, Payload Specialists, Pilots and Commanders. They have also perished alongside their male colleagues in the two Shuttle accidents involving *Challenger* and *Columbia*. Today we await the return to flight of the Space Shuttle and mission STS-114 (tentatively scheduled for May 2005), which will be commanded by Eileen Collins. Space Shuttle flights will end in 2010, so there are now only a limited number of places for access to space by this system. It remains to be seen how many women secure those seats, and how their role in the new post-Shuttle era of US spaceflight will evolve. The selection of a new group of astronauts in 2004 featured the first teacher-educator astronaut candidates and further Mission Specialist candidates, including women. They will be the first group to train for flights that do not feature the Space Shuttle. They will pioneer a new phase of spaceflight training that could, hopefully, see expanded female spaceflight operations and achievements.

Also awaiting the Return-to-Flight of the Space Shuttle is the ISS (which has yet to be completed). Two American women have performed long stays, and one has also acted as station Science Officer. One of the arguments produced by proponents for construction of the ISS was the potential of space-based medical research to produce breakthroughs in human health. Currently, in the fight against breast cancer, NASA-inspired technology has been incorporated into Hologic's commercially available Lorad MultiCare Stereotactic Breast Biopsy System, and it is hoped that within a couple of years the BioLuminate Smart Probe (for breast cancer detection) – which also uses NASA-inspired technology – will be commercially available. Another area of concern is the medical requirements of the West's ageing population, of which the majority are elderly women. One affliction that many of these women suffer from is the 'brittle-bone' disease, osteoporosis, which is similar to the loss of bone encountered by astronauts when they are weightless in space. As the only space-based 'geriatric study' has been conducted on an elderly man, there is a strong argument that NASA should at least send an elderly female into space. NASA has a pool of such women – some of the group of thirteen women who passed the same astronaut tests as the elderly male who undertook the 'geriatric study'!

Women have played a significant role in supporting the development of space technologies, despite great obstacles associated with traditional gender role stereotyping (particularly in the USA and Russia). Today these obstacles are far less obvious, and crews sent into space include women working alongside their male colleagues in all roles. (Women assigned to Shuttle crews view themselves as members of a team, and not, as the media insist on labelling them, 'women astronauts'.) We therefore await the return to the Moon and the first female footprints upon its surface; and, in the longer term, mixed-crew missions to Mars and beyond. Such ventures will take the best of both genders and all races, and will draw upon the spirit, ambition and achievements of many women over centuries of striving for recognition in the fields of astronomy, aeronautics, science, medicine, technology and spaceflight.

As we look to the future and venture further into space, we also look back at the efforts and sacrifices of those who went before, and to the historic flight of one woman in June 1963. As long as women fly into space, they will always follow Valentina.

Appendix 1: Spaceflight chronology, 1963–2003

1963	Jun 16	Valentina Tereshkova becomes the first woman in space as solo pilot of Vostok 6; a joint flight with Valery Bykovsky onboard Vostok 5. Tereshkova completes 48 orbits in 70 hours.
1982	Aug 19	Svetlana Savitskaya is launched as Flight Engineer (FE) onboard Soyuz T-7 for a one-week visiting mission onboard Salyut 7. Duration: 189 hours, setting new female space endurance record.
1983	Jun 18	Sally Ride becomes the first American woman to enter space as Mission Specialist (MS), on a satellite deployment mission onboard STS-7 (*Challenger*). Duration: 146 hours.
1984	Jul 7	Svetlana Savitskaya becomes the first woman to make two space flights, as FE onboard Soyuz T-12 on a week-long visiting mission to Salyut 7. On 25 July she becomes the first woman to perform an EVA (3 hrs 35 min). Duration: 283 hours, surpassing her own female space endurance record.
	Aug 30	Judy Resnik flies as MS on the satellite deployment mission STS 41-D – the maiden launch of *Discovery*. Duration: 144 hours.
	Oct 5	Sally Ride becomes the first American woman to make two spaceflights, as MS onboard the science mission STS 41-G (*Challenger*). On the same crew is MS Kathy Sullivan, who becomes the first American woman to perform an EVA (3 hrs 27 min on 11 October). Duration: 197 hours.
	Nov 8	Anna Fisher flies a week-long satellite deployment and retrieval mission as MS on STS 51-A (*Discovery*). Duration: 191 hours.
1985	Apr 12	Rhea Seddon flies as MS on STS 51-D (*Discovery*), a satellite deployment mission. Duration: 167 hours.
	Jun 17	Shannon Lucid flies as MS STS 51-G (*Challenger*), a satellite deployment mission. Duration: 169 hours.
	Oct 17	Bonnie Dunbar flies as MS onboard STS 61-A (*Challenger*), which carries the Spacelab D1 laboratory payload. Duration: 168 hours.
	Nov 26	Mary Cleave flies as MS onboard STS 61-B (*Atlantis*), a satellite

		deployment and EVA construction demonstration mission. Duration: 165 hours.
1986	Jan 28	NASA astronaut Judy Resnik (MS) and civilian teacher Payload Specialist (PS) Christa McAuliffe are among seven astronauts killed 73 seconds after launch onboard *Challenger* during STS 51-L. They are the first female space explorer fatalities either in training accidents or during a mission. Duration: 73 seconds until loss of vehicle.
1989	May 4	Mary Cleave flies as MS STS-30 (*Atlantis*), which deploys the Venus probe *Magellan*. Duration: 96 hours.
	Oct 18	Ellen Baker and Shannon Lucid serve as MS onboard STS-34 (*Atlantis*), which deploys the Jupiter probe *Galileo*. Duration: 119 hours.
	Nov 22	Kathy Thornton flies as MS onboard STS-33 (*Discovery*), a classified DoD mission. Duration: 120 hours.
1990	Jan 9	Bonnie Dunbar and Marsha Ivins fly as MS onboard STS-32 (*Columbia*), a satellite deployment and retrieval mission. Duration: 261 hours.
	Apr 24	Kathy Sullivan flies as MS onboard STS-31 (*Discovery*), the Hubble Space Telescope deployment mission. Duration: 121 hours.
1991	Apr 5	Linda Godwin flies as MS onboard STS-37 (*Atlantis*), the Compton Gamma-Ray Observatory deployment mission. Duration: 143 hours.
	May 18	Helen Sharman becomes the first citizen of the UK to fly in space, as Cosmonaut Researcher onboard Soyuz TM-12, on a week-long visiting mission to the Mir space station. Duration: 189 hours.
	Jun 5	Millie Hughes-Fulford flies as a PS onboard STS-40 (*Columbia*), the first Spacelab Life Sciences research mission. On the same mission are Rhea Seddon and Tammy Jernigan, flying as MS. Duration: 218 hours.
	Aug 2	Shannon Lucid is launched as MS onboard STS-43 (*Atlantis*), a satellite deployment mission, becoming the first woman to fly in space three times. Duration: 213 hours.
1992	Jan 22	Roberta Bondar becomes the first female Canadian citizen in space as PS onboard STS-42 (*Discovery*), flying the first Spacelab International Microgravity mission. Duration: 193 hours.
	Mar 24	Kathy Sullivan is launched as MS onboard STS-45 (*Atlantis*), flying the first Atmospheric Laboratory for Application and Science (ATLAS-1) mission. Duration: 214 hours.
	May 7	Kathy Thornton flies as MS onboard STS-49, the maiden launch of *Endeavour* and a satellite repair and EVA construction demonstration mission. Thornton also completes a 7 hr 45 min EVA (setting a female record) on 14 May, testing space construction techniques. Duration: 213 hours.
	Jun 25	Bonnie Dunbar and Ellen Baker fly as MS onboard STS-50

(*Columbia*), which completes the first Extended Duration Orbiter mission which carries the Spacelab US Microgravity Laboratory 1 payload. Duration: 313 hours – a new Shuttle endurance record, and the longest US manned spaceflight since the 84-day Skylab 4 mission in 1973–74. Both astronauts create a new female space endurance record, surpassing the record set by Savitskaya in 1984.

Jul 31 Marsha Ivins flies as MS onboard STS-46 (*Atlantis*), a satellite deployment mission and the first Tethered Satellite experiment. Duration: 191 hours.

Sep 12 Jan Davis and Mae Jemison fly as MS onboard STS-47 (*Endeavour*), the Spacelab J (Japanese) research mission. Davis flies with her then husband Mark Lee, making them the first married couple to fly in space together. Jemison is also the first African-American woman to fly into space. Duration: 190 hours.

Oct 22 Tammy Jernigan serves as MS onboard STS-52 (*Columbia*), which deploys a geophysical satellite and operates the first US Microgravity Payload (USMP-1). Duration: 236 hours.

1993 Jan 13 Susan Helms serves as MS onboard STS-54 (*Endeavour*), a Tracking Data and Relay Satellite (TDRS) deployment mission. Duration: 143 hours.

Apr 7 Ellen Ochoa flies as MS onboard STS-56 (*Discovery*), which carries the ATLAS-2 payload. Duration: 222 hours.

Jun 21 Nancy Sherlock (Currie) and Janice Voss fly as MS onboard STS-57 (*Endeavour*), which carries the EURECA retrievable satellite and the first SpaceHab payload module. Duration: 239 hours.

Oct 18 Rhea Seddon and Shannon lucid fly as MS onboard STS-58 (*Columbia*), flying the Spacelab module configured for the SLS-2 payload. Seddon is also the first female Payload Commander (PC) crew-member, and Lucid becomes the first woman to fly in space four times. Duration: 336 hours, setting new endurance record for a both a female spaceflight and for a Shuttle mission.

Dec 2 Kathy Thornton flies as MS onboard STS-61 (*Endeavour*) on the first Hubble Space Telescope service mission. She completes two EVAs (6 hrs 36 min on 5 December, and 6 hrs 50 min on 7 December) totalling 12 hrs 26 min servicing the telescope. Duration: 259 hours.

1994 Feb 3 Jan Davis flies as MS STS-60 (*Discovery*), which carries the second SpaceHab module and the Wake Shield Facility, as well as the first Russian cosmonaut (Sergei Krikalev) to be launched onboard an American spacecraft. Duration: 199 hours.

Mar 4 Marsha Ivins flies as MS onboard STS-62 (*Columbia*), which carries USMP-2. Duration: 335 hours.

Apr 9 Linda Godwin flies as MS/PC for STS-59 (*Endeavour*), which carries the first Space Radar Laboratory payload SRL-1. Duration: 269 hours.

	Jul 8	Chiaki Mukai flies as PS onboard STS-65 (*Columbia*), which carries a Spacelab research laboratory configured as the USML-2 payload. Mukai also becomes the first Japanese woman to fly into space. Duration: 353 hours, which sets both a new Shuttle mission endurance record and a new female space endurance record.
	Sep 9	Susan Helms flies as MS on STS-64 (*Columbia*), which carries the Lidar-In-Space Technology Experiment (LITE). Duration: 262 hours.
	Oct 4	Yelena Kondakova flies as FE on Soyuz TM-20 to the Mir space station. As a member of the Mir-17 resident crew, she is the first long-duration Russian female crew-member, landing on 22 March 1995. She is also only the third Russian female cosmonaut in space after Tereshkova (1963) and Savitskaya (1982, 1984). Duration: 4,064 hours (169 days 8 hours), setting a new female space endurance record.
	Nov 3	Ellen Ochoa flies as MS/PC onboard STS-66 (*Atlantis*), which carries the ATLAS-3 payload. Duration: 262 hours.
1995	Feb 5	Eileen Collins becomes the first female Shuttle Pilot onboard STS-63 (*Discovery*), the third SpaceHab mission and first Shuttle–Mir rendezvous (Near Mir Mission). Janice Voss flies as MS on the same mission. Duration: 198 hours.
	Mar 2	Tammy Jernigan and Wendy Lawrence fly as MS onboard STS-67 (*Endeavour*), which carries the astronomical observation payload ASTRO-2. Duration: 399 hours, setting a new Shuttle endurance record.
	Jun 27	Bonnie Dunbar and Ellen Baker fly as MS onboard STS-71 (*Atlantis*) the first Shuttle–Mir (SM-1) docking mission. Duration: 235 hours.
	Jul 13	Nancy Currie and Mary Weber fly as MS onboard STS-70 (*Discovery*), a TDRS deployment mission. Duration: 214 hours.
	Oct 20	Kathy Thornton and Catherine Coleman fly as MS onboard STS-73 (*Columbia*), the second USML mission. Thornton also serves as Payload Commander for the mission. Duration: 381 hours.
1996	Mar 22	Linda Godwin flies as MS onboard STS-76 (*Atlantis*), the third Shuttle–Mir docking mission. Also onboard is Shannon Lucid, making a record fifth flight into space, who flies as MS and transfers to the Mir resident crew for a six-month mission (see 16 September 1996) onboard the Russian space station. STS-76 mission duration: 221 hours. Godwin also performed a 6 hr 02 min EVA on 27 March.
	Jun 20	Susan Helms flies as MS/PC onboard STS-78 (*Columbia*), which carries the Life and Microgravity Spacelab (LMS) laboratory. Duration: 405 hours, setting a new Shuttle duration record.
	Sep 26	Shannon Lucid (NASA Board Engineer lands onboard STS-79 (*Atlantis*) after a mission of 4,516 hours (188 days 4 hrs), during

which she served as a member of the Mir-21 and Mir-22 crews onboard the Russian space station. She is replaced onboard the station by John Blaha. The only NASA female astronaut to complete a long-duration mission to Mir, she also establishes a new female world endurance record for a single spaceflight.

Nov 19 Tammy Jernigan flies as MS onboard STS-80 (*Columbia*), which carries the Orfeus–SPAS free-flying pallet satellite and the Wake Shield Facility. A planned EVA by Jernigan (and T. Jones) on 28 November was cancelled due to problems with the hatch. Duration: 423 hours – the longest Shuttle flight to date.

1997 Jan 12 Marsha Ivins flies as MS onboard STS-81 (*Atlantis*), the fifth Shuttle-Mir docking mission (SM-5). The mission delivered Jerry Linenger and returned John Blaha. Duration: 244 hours.

Apr 4 Susan Still (Kilrain) becomes the second female Shuttle pilot and Janice Voss flies as MS/PC onboard STS-83 (*Columbia*), the planned sixteen-day Spacelab Material Science Laboratory. The mission is terminated early due to a fuel-cell problem and is re-flown as STS-94. Duration: 95 hours.

May 15 Eileen Collins serves as Pilot on the STS-84 mission (*Atlantis*), the sixth Shuttle–Mir docking mission (SM-6). Russian cosmonaut Yelena Kondakova is also a member of the Shuttle core crew flying as an MS – the first female cosmonaut to fly onboard the American Shuttle. The mission delivered resident astronaut Mike Foale and returned Jerry Linenger. Duration: 221 hours.

Jul 1 Susan Still (Pilot) and Janice Voss (FE/PC) are crew-members on the STS-94 (*Columbia*) re-flight mission which carries the MSL science payload. Duration: 370 hours.

Aug 7 Jan Davis flies as MS/PC onboard STS-85 (*Discovery*), which deployed and retrieved the Crista–SPAS free flying satellite. Duration: 284 hours.

Sep 25 Wendy Lawrence flies as MS onboard STS-86 (*Atlantis*), the seventh Shuttle–Mir docking mission (SM-7) which returned resident crew-member Mike Foale and delivered Dave Wolf to replace him. Duration: 259 hours.

Nov 19 Kalpana Chawla flies as MS onboard STS-87 (*Columbia*), flying the fourth USMP payload. Chawla, a naturalized US citizen, becomes the first female from India to fly into space. Duration: 376 hours.

1998 Jan 22 Bonnie Dunbar flies as MS/PC onboard STS-89 (*Endeavour*), the eighth Shuttle–Mir docking mission (SM-8). The mission delivered Andy Thomas and returned Dave Wolf. Duration: 211 hours.

Apr 17 Kathryn Hire flies as MS onboard STS-90 (*Columbia*), the Neurolab mission which was also the last scheduled Spacelab mission. Duration: 381 hours.

Jun 2 Wendy Lawrence and Janet Kavandi fly as MS onboard STS-91 (*Discovery*), the ninth and final Shuttle–Mir docking mission (SM-

		9). The mission returns the seventh and final NASA Mir-resident crew-member, Andy Thomas. Duration: 235 hours.
	Oct 29	Chiaki Mukai flies as PS onboard STS-95 (*Discovery*), a SpaceHab research mission that also features the return to space, after 36 years, by 77-year-old former Mercury astronaut John Glenn. Duration: 213 hours.
	Dec 4	Nancy Currie flies as MS onboard STS-88 (*Endeavour*), the first Shuttle–ISS assembly mission. Duration: 283 hours.
1999	May 27	Ellen Ochoa and Tammy Jernigan fly to ISS as MS onboard STS-96 (*Discovery*), the second Shuttle–ISS mission, which also delivers logistics to the developing station. Also onboard is Canadian astronaut Julie Payette. On 28 May Jernigan completes an EVA of 7 hr 55 min, setting a female record for a single EVA. Duration: 235 hours.
	Jul 23	Eileen Collins flies as Commander of STS-93 (*Columbia*), a mission to deploy the Chandra X-Ray Observatory. She becomes the first woman to command an American space mission. Cady Coleman also flies on the mission as MS. Duration: 118 hours.
2000	Feb 11	Janice Voss and Janet Kavandi fly as MS onboard STS-99 (*Endeavour*), the Shuttle Radar Topographical Mission (SRTM). Duration: 269 hours.
	May 19	Susan Helms and Mary Weber fly as MS to ISS onboard STS-101 (*Atlantis*), the third Shuttle–ISS mission, which also delivers logistics in preparation for the first resident crew that October. Duration: 237 hours.
	Oct 11	Pamela Melroy, as pilot of STS-92 (*Discovery*), becomes the third female Shuttle pilot. The mission is the fifth ISS mission, continuing station construction in preparation for the first resident crew. Duration: 309 hours.
2001	Mar 8	Susan Helms flies as FE on the ISS-2 crew launched on STS-102 (*Discovery*) to return on STS-105 (*Discovery*) in July after a 163-day residency. On 11 March she completes an EVA of 8 hrs 56 min, setting a new female single EVA duration record. Duration for the second resident ISS crew from launch to landing: 4,014 hours (167 days 6 hrs).
	Jul 12	Janet Kavandi flies as MS on STS-104 (*Atlantis*), the tenth Shuttle–ISS mission. Duration: 306 hours.
	Dec 5	Linda Godwin flies as MS STS-108 (*Endeavour*), the twelfth Shuttle–ISS mission, and on 11 December completes an EVA of 4 hrs 12 min. Duration: 283 hours.
2002	Mar 1	Nancy Currie flies as MS onboard STS-109 (*Columbia*), the fourth Hubble Service Mission. Duration: 262 hours.
	Apr 8	Ellen Ochoa flies as MS onboard STS-110 (*Discovery*), the thirteenth Shuttle–ISS mission. Duration: 259 hours.
	Jun 5	Peggy Whitson (ISS-5 NASA Science Officer) is launched on STS-

		111 (*Endeavour*) for a six-month residency on the ISS, returning on STS-113 (*Endeavour*) in December. On 16 August she completes an EVA of 4 hrs 25 min. ISS-5 duration from launch to landing: 4,435 hours (184 days 19 hrs).
	Oct 7	Pamela Melroy serves as Pilot and Sandra Magnus as MS onboard STS-112 (*Atlantis*), the fifteenth Shuttle–ISS mission. Duration: 259 hours.
2003	Jan 16	Kalpana Chawla and Lauren Clark fly as MS onboard STS-107 (*Columbia*), a SpaceHab research mission. On 1 February both are killed in the mid-air break-up of the orbiter over continental USA during the final descent, sixteen minutes from landing. Duration: 382 hours up to the moment of loss of contact.

Assigned female crew-members

As at 30 November 2004

Space Shuttle missions (to launch no earlier than spring 2005)

STS-114	Eileen Collins (Commander); Wendy Lawrence (MS)
STS-121	Lisa Nowak and Stephanie Wilson
STS-115	Heidemarie Stefanyshyn-Piper (MS)
STS-116	(no female crew-members)
STS-117	Joan Higginbotham (MS)
STS-118	Barbara Morgan (E-MS)
STS-119	No crew announced
STS-120	No crew announced

Appendix 2: Careers and experience

	Born	Country	Selected	Group	Flights	First	Second	Third	Fourth	Fifth	Hrs	EVAs	Status at 31 August 2004
Almelkina	1954	Russia	1980	IMBP	0							0	Former
Baker, E.	1953	USA	1984	NASA 10	3	1989	1992	1995			567	0	Management Astronaut
Bondar	1945	Canada	1983	CSA	1	1992					193	0	Former
Brummer	1955	Germany	1987	DFLR	0							0	Former
Cagle	1959	USA	1996	NASA 16	0	Pending						0	Active Mission Specialist
Caldwell	1969	USA	1998	NASA 17	0	Pending						0	Active Mission Specialist
Chawla	1961	USA	1994	NASA 15	2	1997	2003				641	0	Died 3 February (STS-107)
Clark	1961	USA	1996	NASA 16	1	2003					382	0	Died 3 February (STS-107)
Cleave	1947	USA	1980	NASA 09	2	1985	1989				261	0	Former
Coleman	1960	USA	1992	NASA 14	2	1995	1999				499	0	Active Mission Specialist
Collins, E.	1956	USA	1990	NASA 13	3	1995	1997	1999			537	0	Flight Cdr, STS-114
Currie (Sherlock)	1958	USA	1990	NASA 13	4	1993	1995	1998	2002		998	0	Management Astronaut
Davis	1953	USA	1987	NASA 12	3	1992	1994	1997			769	0	Former
Dobrokvashina	1947	Russia	1980	IMBP	0							0	Former
Dunbar	1949	USA	1980	NASA 09	5	1985	1990	1992	1995	1998	1,206	0	Management Astronaut
Fisher, A.	1949	USA	1978	NASA 08	1	1984					171	0	Management Astronaut
Godwin	1952	USA	1985	NASA 11	4	1991	1994	1976	2001		916	2	Management Astronaut
Haignere, C.	1957	France	1985	CNES	1	1996					378	0	Former
Helms	1958	USA	1990	NASA 13	5	1993	1994	1996	2000	2001	5,061	1	Former
Higginbotham	1964	USA	1996	NASA 16	0	Pending						0	Flight MS, STS-117
Hilliard-Robertson	1963	USA	1998	NASA 17	0							0	Died May 2001
Hire	1959	USA	1994	NASA 15	1	1998					381	0	Active Mission Specialist
Hughes-Fulford	1945	USA	1984	SLS-PS	1	1991					218	0	Former
Ivanova	1949	Russia	1980	LMI	0							0	Former
Ivins	1951	USA	1984	NASA 10	5	1990	1992	1994	1997	2001	1,340	0	Management Astronaut
Jemison	1956	USA	1987	NASA 12	1	1992					190	0	Former
Jernigan	1959	USA	1985	NASA 11	5	1991	1992	1995	1996	1999	1,463	1	Former
Johnston	1943	USA	1983	SL3 PS	0							0	Former
Kavandi	1959	USA	1994	NASA 15	3	1998	2000	2001			810	0	Active Mission Specialist
Kikuchi	1964	Japan	1989	TBS CR	0							0	Former
Kilrain (Still)	1961	USA	1994	NASA 15	2	1997	1997				470	0	Former
Klyushnikova	1953	Russia	1980	IMBP	0							0	Former
Kondakova	1957	Russia	1989	Energiya	2	1994	1997				4,282	0	Former
Kuleshova	1956	Russia	1980	Energiya	0							0	Former
Kuzhelnaya	1962	Russia	1994	Energiya	0							0	Former
Kuznetsova	1941	Russia	1962	Air Force	0							0	Former
LaComb	1956	USA	1982	USAF MSE	0							0	Former

Latysheva	1953	Russia	1980	IKI	0							0	Former
Lawrence	1959	USA	1992	NASA 14	3	1995	1997	1998			893	0	Flight MS, STS-114
Lucid	1943	USA	1978	NASA 08	5	1985	1989	1991	1993	1996	5,353	0	Management Astronaut
Magnus	1964	USA	1996	NASA 16	1	2002					259	0	Active Mission Specialist
McArthur, K.	1971	USA	2000	NASA 18	0	Pending						0	Active Mission Specialist
McAuliffe	1948	USA	1985	Teacher PS	1	1986					N/A	0	Died January 1986 (STS 51-L)
Melroy	1961	USA	1994	NASA 15	2	2000	2002				568	0	Active Pilot
Merchez	1960	Belgium	1992	ESA	0							0	Former
Metcalf-Lindenberger	1975	USA	2004	NASA 19	0	Pending						0	Ascan Educator MS
Morgan*	1951	USA	1998	NASA 17	0	Pending						0	Flight Educator MS, STS-118
Mukai	1952	Japan	1985	NASDA PS	2	1994	1998				566	0	Former
Nowak	1963	USA	1996	NASA 16	0	Pending						0	Flight MS, STS-118
Nyberg	1969	USA	2000	NASA 18	0	Pending						0	Active Mission Specialist
Ochoa	1958	USA	1990	NASA 13	4	1993	1994	1999	2002		1,068	0	Active Mission Specialist
Omelchenko	1951	Russia	1990	Journalist	0							0	Former
Payette	1963	Canada	1992	CSA	1	1999					235	0	Active Mission Specialist
Ponomoryova	1933	Russia	1962	Air Force	0							0	Former
Pozharskaya	1947	Russia	1980	IMBP	0							0	Former
Prinz	1938	USA	1978	SL2 PS	0							0	Deceased
Pronina	1953	Russia	1980	Energiya	0							0	Former
Resnik	1949	USA	1978	NASA 08	2	1984	1986				144	0	Died January 1986 (STS 51-L)
Ride	1951	USA	1978	NASA 08	2	1983	1984				343	0	Former
Roberts	1954	USA	1983	USAF MSE	0							0	Former
Savitskaya	1948	Russia	1980	Energiya	2	1982	1984				472	1	Former
Sharman	1963	UK	1989	Juno	1	1991					189	0	Former
Seddon	1947	USA	1978	NASA 08	3	1985	1991	1993			721	0	Former
Solovyova	1937	Russia	1962	Air Force	0							0	Former
Stefanyshyn-Piper	1963	USA	1996	NASA 16	0	Pending						0	Flight MS, STS-115
Stevens	1960	USA	1986	USAF MSE	0							0	Former
Stott	1962	USA	2000	NASA 18	0	Pending						0	Active Mission Specialist
Sudarmono	1952	Indonesia	1985	Shuttle PS	0							0	Former
Sullivan	1951	USA	1978	NASA 08	3	1984	1990	1992			532	1	Former
Tereshkova	1937	Russia	1962	Air Force	1	1963					70	0	Former
Thornton, K.	1952	USA	1984	NASA 10	4	1989	1992	1993	1995		973	3	Former
Voss, J.E.	1957	USA	1990	NASA 13	5	1993	1995	1997	1997	2000	1,171	0	Management Astronaut
Walker, S.	1965	USA	2004	NASA 19	0	Pending						0	Ascan Mission Specialist
Walpot	1960	Germany	1987	DFLR	0							0	Former
Weaver	1953	USA	1988	USAF WOSE	0							0	Former
Weber	1962	USA	1992	NASA 14	2	1995	2000				451	0	Management Astronaut

Whitson	1960	USA	1996	NASA 16	1	2002					4,438	1	Active Mission Specialist

Table (contd)

	Born	Country	Selected	Group	Flights	First	Second	Third	Fourth	Fifth	Hrs	EVAs	Status at 31 August 2004
Williams, S.	1965	USA	1998	NASA 17	0	Pending						0	Active Mission Specialist
Wilson	1966	USA	1996	NASA 16	0	Pending						0	Active Mission Specialist
Yorkina	1939	Russia	1962	Air Force	0							0	Former
Zakharova	1952	Russia	1980	IMBP	0							0	Former

Active: available for flight assignment/category
Flight: named to a pending mission
Management Astronaut: assigned at JSC out of crew training, but available for future crew assignment
Pending: preparing for first flight assignment

Metcalf-Lindenburger
* In 1985 Morgan was initially selected as back-up Teacher-in-Space PS to McAuliffe, before selection to the NASA astronaut programme.

Appendix 3: Spaceflight records and EVAs

SPACEFLIGHT RECORDS

Name	Country	Flights	Missions	dd:hh:mm:ss	Career total
STS: duration launch to wheel-stop					
Mir/ISS resident crews: duration launch to landing					
Lucid	USA	5	STS 51-G	07:01:39:42	
			STS-34	04:23:40:14	
			STS-43	08:21:22:23	
			STS-58	14:00:13:33	
			NASA Mir 2	188:04:00:11	223:02:56:03
Helms	USA	5	STS-54	05:23:39:08	
			STS-64	10:22:50:58	
			STS-78	16:21:48:33	
			STS-101	09:21:10:10	
			ISS-2	167:06:40:49	211:00:09:38
Whitson	USA	1	ISS-5	184:22:14:23	184:22:14:23
Kondakova	Russia	2	Mir EO-17	169:05:21:35	
			STS-84	09:05:20:48	178:10:42:23
Jernigan	USA	5	STS-40	09:02:15:14	
			STS-52	09:20:57:16	
			STS-67	16:15:09:49	
			STS-80	17:15:54:26	
			STS-96	09:19:13:57	063:01:30:42
Ivins	USA	5	STS-32	10:21:01:38	
			STS-46	07:23:16:07	
			STS-62	13:23:17:36	
			STS-81	10:04:56:28	
			STS-98	12:21:21:00	055:21:52:49

Table (contd)

Name	Country	Flights	Missions	dd:hh:mm:ss	Career total
Dunbar	USA	5	STS 61-A	07:00:45:48	
			STS-32	10:21:01:38	
			STS-50	13:19:31:02	
			STS-71	09:19:23:11	
			STS-89	08:19:48:06	050:08:29:45
Voss, J.E.	USA	5	STS-57	09:23:45:59	
			STS-63	08:06:29:36	
			STS-83	03:23:13:39	
			STS-94	15:10:45:32	
			STS-99	11:05:29:41	048:21:44:27
Currie (Sherlock)	USA	4	STS-57	09:23:45:59	
			STS-70	08:22:21:05	
			STS-88	11:19:18:47	
			STS-109	10:22:11:09	045:15:37:00
Ochoa	USA	4	STS-56	09:06:09:22	
			STS-66	10:22:34:54	
			STS-96	09:19:13:57	
			STS-110	10:19:42:44	040:19:40:57
Thornton, K.	USA	4	STS-33	05:00:07:50	
			STS-49	08:21:18:36	
			STS-61	10:19:59:30	
			STS-73	15:21:53:18	040:15:19:14
Godwin	USA	4	STS-37	05:23:33:39	
			STS-59	11:05:50:24	
			STS-76	09:05:16:48	
			STS-108	11:19:36:45	038:06:17:36
Lawrence	USA	3	STS-67	16:15:09:49	
			STS-86	10:19:22:15	
			STS-91	09:19:55:04	037:06:27:08
Kavandi	USA	3	STS-91	09:19:55:04	
			STS-99	11:05:29:41	
			STS-104	12:18:36:39	33:20:01:24
Chawla	USA	2	STS-87	15:16:34:59	
			STS-107	15:22:20:00*	31:14:54:59
Seddon	USA	3	STS 51-D	06:23:56:31	
			STS-40	09:02:15:14	
			STS-58	14:00:13:33	30:02:25:18
Baker, E.	USA	3	STS-34	04:23:40:14	
			STS-50	13:19:31:02	
			STS-71	09:19:23:11	28:14:34:27

Davis	USA	3	STS-47	07:22:31:13	
			STS-60	08:07:10:13	
			STS-85	11:20:28:09	28:02:09:35
Haignere, C.	France	2	Mir (Cassiopeia)	15:18:23:37	
			ISS (Andromeda)	09:20:00:22	25:14:23:59
Melroy	USA	2	STS-92	12:21:43:47	
			STS-112	10:19:58:44	23:17:42:31
Mukai	Japan	2	STS-65	14:17:56:08	
			STS-95	08:21:44:56	23:15:41:04
Collins E.	USA	3	STS-63	08:06:29:36	
			STS-84	09:05:20:48	
			STS-93	04:22:50:18	22:10:40:42
Sullivan	USA	3	STS 41-G	08:05:24:33	
			STS-31	05:01:17:07	
			STS-45	08:22:10:24	22:04:52:04
Coleman	USA	2	STS-73	15:21:53:18	
			STS-93	04:22:50:18	20:20:43:36
Savitskaya	Russia	2	Salyut 7 (Soyuz T-7)	07:21:52:24	
			Salyut 7 (Soyuz T-12)	11:19:14:36	19:17:07:00
Kilrain (Still)	USA	2	STS-83	03:23:13:39	
			STS-94	15:10:45:32	19:09:58:11
Weber	USA	2	STS-70	08:22:21:05	
			STS-101	09:21:10:10	18:19:31:15
Clark	USA	1	STS-107	15:22:20:00*	15:22:20:00
Hire	USA	1	STS-90	15:21:50:56	15:21:50:56
Ride	USA	2	STS-7	06:02:25:33	
			STS 41-G	08:05:24:33	14:07:50:06
Cleave	USA	2	STS 61-B	06:21:06:21	
			STS-30	04:00:57:30	10:22:03:51
Magnus	USA	1	STS-112	10:19:58:44	10:19:58:44
Payette	Canada	1	STS-96	09:19:13:57	09:19:13:57
Hughes-Fulford	USA	1	STS-40	09:02:15:14	09:02:15:14
Bondar	Canada	1	STS-42	08:01:15:42	08:01:15:42
Fisher A.	USA	1	STS 51-A	07:23:45:59	07:23:45:59
Jemison	USA	1	STS-47	07:22:31:13	07:22:31:13
Sharman	UK	1	Mir (Juno)	07:21:14:20	07:21:14:20

Resnik	USA	2	STS 41-D	06:00:57:05	
			STS 51-L	00:00:01:13	06:00:58:18
Tereshkova	Russia	1	Vostok 6	02:22:40:48	02:22:40:48
McAuliffe**	USA	1	STS 51-L	00:00:01:13	00:00:01:13

* STS-107 duration: from launch to loss of signal
** STS 51-L duration: from launch to loss of signal
NASA Mir: US long-duration crew-member
EO-17: Russian long-duration crew (Mir)
ISS-5: International Space Station resident crew
Mir (Juno): international visit to Mir or ISS (project name)
Soyuz T-7: (Soyuz T) ferry/visiting mission to Salyut 7

EVA EXPERIENCE

Name	Country	EVAs	Missions	hh:mm	Career total
Thornton	USA	3	STS-49	07:45	
			STS-61	06:36	
			STS-61	06:50	21:11
Godwin	USA	2	STS-76	06:02	
			STS-108	04:12	10:14
Helms	USA	1	STS-102/ISS-2	08:56	08:56
Whitson	USA	1	ISS-5	04:25	04:25
Savitskaya	Russia	1	Soyuz T-12/Salyut 7	03:35	03:35
Sullivan	USA	1	STS 41-G	03:27	03:27

Appendix 4: NASA Shuttle missions with female crew-members, 1983–2004

Mission STS designation	Orb. Seq.	Vehicle	OV Seq.	Date of Launch	Date of landing	Duration dd:hh:mm:ss	Crew position	Astronaut name (Flt)	Nationality/ agency/institution
STS-7	07	*Challenger*	02	1983 Jun 18	1983 Jun 24	06:02:25:33	MS2/FE	Ride, S.	American/NASA
STS 41-D	12	*Discovery*	01	1984 Aug 30	1984 Sep 5	06:00:57:05	MS3	Resnik, J.	American/NASA
STS 41-G	13	*Challenger*	06	1984 Oct 5	1984 Oct 13	08:05:24:33	MS2/FE	Ride, S. (second)	American/NASA
							MS1	Sullivan, K.	American/NASA
STS 51-A	14	*Discovery*	02	1984 Nov 8	1984 Nov 16	07:23:45:59	MS2/FE	Fisher, A.	American/NASA
STS 51-D	16	*Discovery*	04	1985 Apr 12	1984 Apr 19	06:23:56:31	MS1	Seddon, R.	American/NASA
STS 51-G	18	*Discovery*	05	1985 Jun 17	1985 Jun 24	07:01:39:42	MS3	Lucid, S.	American/NASA
STS 61-A	22	*Challenger*	09	1985 Oct 30	1985 Nov 6	07:00:45:48	MS1	Dunbar, B.	American/NASA
STS 61-B	23	*Atlantis*	02	1985 Nov 26	1985 Dec 3	06:21:06:12	MS2/FE	Cleave, M.	American NASA
STS 51-L	25	*Challenger*	10	1986 Jan 1	n/a	00:00:01:13	MS2/FE	Resnik, J (second)	American/NASA
							PS2	McAuliffe, S.	American/SFP
STS-30	29	*Atlantis*	04	1989 May 4	1989 May 8	04:00:57:30	MS3	Cleave, M (second)	American/NASA
STS-34	31	*Atlantis*	05	1989 Oct 18	1989 Oct 23	04:23:40:14	MS1	Lucid, S. (second)	American/NASA
							MS3	Baker, E.	American/NASA
STS-33	32	*Discovery*	09	1989 Nov 22	1989 Nov 28	05:00:07:50	MS3	Thornton, K.	American/NASA
STS-32	33	*Columbia*	09	1990 Sep 1	1990 Jan 20	10:21:01:38	MS1	Dunbar, B. (second)	American/NASA
							MS2/FE	Ivins, M.	American/NASA
STS-31	35	*Discovery*	10	1990 Apr 24	1990 Apr 29	05:01:17:07	MS3	Sullivan, K. (second)	American/NASA
STS-37	39	*Atlantis*	08	1991 Apr 5	1991 Apr 11	05:23:33:39	MS-1	Godwin, L.	American/NASA
STS-40	41	*Columbia*	11	1991 Jun 5	1991 Jun 14	09:02:15:14	MS2/FE	Jernigan, T.	American/NASA
							MS3	Seddon, R. (third)	American/NASA
							PS2	Hughes-Fulford, M.	American/VAMC
STS-43	42	*Atlantis*	09	1991 Aug 2	1991 Aug 11	08:21:22:23	MS1	Lucid, S. (third)	American/NASA
STS-42	45	*Discovery*	14	1992 Jan 22	1992 Jan 30	08:01:15:42	PS1	Bondar, R.	Canadian/CSA
STS-45	46	*Atlantis*	11	1992 Mar 24	1992 Apr 2	08:22:10:24	MS1/PC	Sullivan, K. (third)	American/NASA
STS-49	47	*Endeavour*	01	1992 May 7	1992.16.05	08:21:18:36	MS4	Thornton, K. (second)	American/NASA
STS-50	48	*Columbia*	12	1992 Jun 25	1992 Jul 9	13:19:31:02	MS1/PC	Dunbar, B. (third)	American/NASA
							MS2/FE	Baker, E. (second)	American/NASA
STS-46	49	*Atlantis*	12	1992 Jul 31	1992 Aug 8	07:23:16:07	MS2/FE	Ivins, M (second)	American/NASA
STS-47	50	*Endeavour*	02	1992 Sep 12	1992 Sep 20	07:22:31:13	MS3	Davis, J.	American/NASA
							MS4/SMS	Jemison, M.	American/NASA
STS-52	51	*Columbia*	13	1992 Oct 22	1992 Nov 1	09:20:57:16	MS3	Jernigan, T. (second)	American/NASA

STS-54	53	*Endeavour*	03	1993 Jan 13	1993 Jan 19	05:23:39:08	MS3	Helms, S.	American/NASA
STS-56	54	*Discovery*	16	1993 Apr 8	1993 Apr 17	09:06:09:22	MS3	Ochoa, E.	American/NASA
STS-57	56	*Endeavour*	04	1993 Jun 21	1993 Jul 1	09:23:45:59	MS2/FE	Voss, J.E.	American/NASA
							MS3	Sherlock, N.	American/NASA
STS-58	58	*Columbia*	15	1993 Oct 18	1993 Nov 1	14:00:13:33	MS1/PC	Seddon, R. (third)	American/NASA
							MS4	Lucid, S. (fourth)	American/NASA
STS-61	59	*Endeavour*	05	1993 Dec 2	1993 Dec 12	10:19:59:30	MS4	Thornton, K. (third)	American/NASA
STS-60	60	*Discovery*	18	1994 Feb 3	1994 Feb 11	08:07:10:13	MS1	Davis, J. (second)	American/NASA
STS-62	61	*Columbia*	16	1994 Mar 4	1994 Mar 18	13:23:17:36	MS3	Ivins, M. (third)	American/NASA
STS-59	62	*Endeavour*	06	1994 Apr 9	1994 Apr 20	11:05:50:24	MS3/PC	Godwin, L. (second)	American/NASA
STS-65	63	*Columbia*	17	1994 Jul 8	1994 Jul 23	14:17:56:08	PS1	Mukai, C.	Japanese/NAS-
DA									
STS-64	64	*Discovery*	19	1994 Sep 9	1994 Sep 20	10:22:50:58	MS3	Helms, S. (second)	American/NASA
STS-66	66	*Atlantis*	13	1994 Nov 3	1994 Nov 14	10:22:34:54	MS1/PC	Ochoa, E. (second)	American/NASA
STS-63 (SM-RV)	67	*Discovery*	20	1995 Feb 3	1995 Feb 11	08:06:29:36	Pilot	Collins, E	American/NASA
							MS1	Voss, J.E. (second)	American/NASA
STS-67	68	*Endeavour*	08	1995 Mar 2	1995 Mar 18	16:15:09:49	MS2/FE	Lawrence, W.	American/NASA
							MS3/PC	Jernigan, T. (third)	American/NASA
STS-71 (SM-1)	69	*Atlantis*	14	1995 Jun 27	1995 Jul 7	09:19:23:11	MS1	Baker, E. (third)	American/NASA
							MS3	Dunbar, B. (fourth)	American/NASA
STS-70	70	*Discovery*	21	1995 Jul 13	1995 Jul 22	08:22:21:05	MS2/FE	Sherlock, N (second)	American/NASA
							MS3	Weber, M.	American/NASA
STS-73	72	*Columbia*	18	1995 Oct 20	1995 Nov 5	15:21:53:18	MS1	Coleman, C.	American/NASA
							MS3/PC	Thornton, K. (fourth)	American/NASA
STS-76 (SM-3)	76	*Atlantis*	16	1996 Mar 22	1996 Mar 31	09:05:16:48	MS3	Godwin, L. (third)	American/NASA
						n/a	MS4	Lucid, S. (fifth)*	American/NASA
							(up only)		
STS-78	78	*Columbia*	20	1996 Jun 20	1996 Jul 7	16:21:48:33	MS2/FE	Helms, S. (third)	American/NASA
STS-79 (SM-4)	79	*Atlantis*	17	1996 Sep 16	1996 Sep 26	n/a	MS4	Lucid, S (fifth)*	American/NASA
							(down only)		
STS-80	80	*Columbia*	21	1996 Nov 19	1996 Jul 12	17:15:54:26	MS1	Jernigan, T. (fourth)	American/NASA
STS-81 (SM-5)	81	*Atlantis*	18	1997 Jan 12	1997 Jan 22	10:04:56:28	MS3	Ivins, M. (fourth)	American/NASA
STS-83	83	*Columbia*	22	1997 Apr 4	1997 Apr 8	03:23:13:39	Pilot	Still, S.	American/NASA
							MS1/PC	Voss, J.E. (third)	American/NASA
STS-84 (SM-6)	84	*Atlantis*	19	1997 May 15	1997 May 24	09:05:20:48	Pilot	Collins, E. (second)	American/NASA
							MS4	Kondakova	Russian/RSA

Table (contd)

Mission STS designation	Orb. Seq.	Vehicle	OV Seq.	Date of Launch	Date of landing	Duration dd:hh:mm:ss	Crew position	Astronaut name (Flt)	Nationality/ agency/institution
STS-94	85	*Columbia*	23	1997 Apr 4	1997 Apr 8	03:23:13:39	Pilot	Still, S. (second)	American/NASA
							MS1/PC	Voss, J.E (fourth).	American/NASA
STS-85	86	*Discovery*	23	1997 Aug 7	1997 Aug 19	11:20:28:09	MS1/PC	Davis, J.(third)	American/NASA
STS-86 (SM-7)	87	*Atlantis*	20	1997 Sep 25	1997.Oct 6	10:19:22:15	MS4	Lawrence, W. (second)	American/NASA
STS-87	88	*Columbia*	24	1997 Nov 19	1997 Dec 5	15:16:34:59	MS1	Chawla, K.	American/NASA
STS-89 (SM-8)	89	*Endeavour*	12	1998 Jan 22	1998 Jan 31	08:19:48:06	MS3/PC	Dunbar, B. (fifth)	American/NASA
STS-90	90	*Columbia*	25	1998 Apr 17	1998 May 3	15:21:50:56	MS2/FE	Hire, K.	American/NASA
STS-91 (SM-9)	91	*Discovery*	24	1998 Jun 2	1998 Jun 12	09:19:55:04	MS2/FE	Lawrence, W. (third)	American/NASA
							MS3	Kavandi, J.	American/NASA
STS-95 DA	92	*Discovery*	25	1998 Oct 29	1998 Nov 7	08:21:44:56	PS1	Mukai, C. (second)	Japanese/NAS-
STS-88 (2A)	93	*Endeavour*	13	1998 Dec 4	1998 Dec 15	11:19:18:47	MS2/FE	Currie, N. (third)**	American/NASA
STS-96 (2A.1)	94	*Discovery*	26	1999 May 27	1999 Jun 6	09:19:13:57	MS1	Jernigan (fifth)	American/NASA
							MS2/FE	Ochoa, E. (third)	American/NASA
							MS4	Payette, J.	Canadian/CSA
STS-93	95	*Columbia*	26	1999 Jul 23	1999 Jul 27	04:22:50:18	CDR	Collins, E. (third)	American/NASA
							MS1	Coleman, C. (second)	American/NASA
STS-99	97	*Endeavour*	14	2000 Feb 11	2000 Feb 22	11:05:29:41	MS2/FE	Kavandi (second)	American/NASA
							MS3/PC	Voss, J.E. (fifth)	American/NASA
STS-101 (2A.2a)	98	*Atlantis*	21	2000. May 19	2000 May 29	09:21:10:10	MS1	Weber, M. (second)	American/NASA
							MS4	Helms, S. (fourth)	American/NASA
STS-92 (3A)	100	*Discovery*	28	2000 Oct 11	2000 Oct 24	12:21:43:47	Pilot	Melroy, P.	American/NASA
STS-98 (5A)	101	*Atlantis*	23	2001 Feb 7	2001 Feb 20	12:21:21:00	MS2/FE	Ivins, M. (fifth)	American/NASA
STS-102 (5A.1)	103	*Discovery*	29	2001 Mar 8	2001 Mar 21	n/a	MS4 (up only)	Helms, S. (fifth)*	American/NASA
STS-104 (7A)	105	*Atlantis*	24	2001 Jul 12	2001 Jul 31	12:18:36:39	MS2/FE	Kavandi, J. (third)	American/NASA
STS-105 (7A.1)	106	*Discovery*	30	2001 Aug 10	2001 Aug 22	n/a	MS4 (down only)	Helms, S. (fifth)*	American/NASA
STS-108 (UF-1)	107	*Endeavour*	17	2001 Nov 29	2001 Dec 12	11:19:36:45	MS1	Godwin, L. (fourth)	American/NASA
STS-109	108	*Columbia*	27	2002 Mar 1	2002 Mar 12	10:22:11:09	MS2/FE	Currie, N. (fourth)**	American/NASA
STS-110 (8A)	109	*Discovery*	31	2002 Apr 8	2002 Apr 19	10:19:42:44	MS2/FE	Ochoa, E. (fourth)	American/NASA

STS-111 (UF-2)	110	*Endeavour*	18	2002 Jun 5	2002 Jun 19	n/a	MS5 (up only)	Whitson, P.*	American/NASA
STS-112 (9A)	111	*Atlantis*	25	2002 Oct 7	2002 Oct 18	10:19:58:44	Pilot	Melroy, P. (second)	American/NASA
							MS2/FE	Magnus, S.	American/NASA
STS-113 (11A)	112	*Endeavour*	19	2002 Nov 23	2002 Dec 7	n/a	MS5 (down only)	Whitson, P.*	American/NASA
STS-107	113	*Columbia*	28	2003 Jan 16	2003 Feb 1	15:22:20:00	MS2/FE	Chawla, K. (second)	American/NASA
							MS4	Clark, L.	American/NASA

Durations are from lift-off to wheel-stop

* Space station resident crew-member
** Currie, N. formerly flew as Sherlock, N.

STS 51-L duration: from launch to explosion of vehicle 73 seconds after lift-off
STS-107 data: to loss of telemetry 16 minutes from planned landing, due to mid-air break-up of vehicle

SM-1 Shuttle/Mir mission
2A ISS/Shuttle assembly mission

MS	Mission Specialist	SFP	Spaceflight Participant (Teacher-in-Space)
PS	Payload Specialist	VAMC	Veterans Administration Medical Center
SMS	Science Mission Specialist	CSA	Canadian Space Agency
FE	Flight Engineer	RSA	Russian Space Agency
PC	Payload Commander	NASDA	Japanese Space Agency

Glossary

On each Shuttle mission, crew-members are responsible for a number of systems and payloads. This selected glossary of terms used in this book indicates the diversity of skills and training required in addition to learning to fly the vehicle and living and working in space.

Systems and procedures

APU	Auxiliary Power Unit
CCTV	Closed Circuit Television
Comm	Communications
CDMS	Command and Data Management System
ECLSS	Environmental Control and Life Support System
ECS	Environmental Control System
FDF	Flight Data File
FES	Flash Evaporator System
GNS	Guidance and Navigation System
HYD	Hydraulics
IFM	In-Flight Maintenance
Inst	Instrumentation
MPS	Main Propulsion System
OMS	Orbital Manoeuvring System
PLBD	Payload Bay Doors
PDRS	Payload Deployment and Retrieval Systems
Rads	Radiators
RCS	Reaction Control System
WSC	Waste Collection System

Experiments and payloads

AASF	Advanced Automated Solidification Furnace
AGHF	Advanced Gradient Heating Facility
ADSEP	Advanced Organic Separation
APCF	Advanced Protein Crystallisation Facility

BDS	Bioreactor Demonstrator System
BIMBA	Bioserve-Instrumentation Technology Associates Materials Dispersion Apparatus
BRIC	Biological Research In Canisters
CCM	Cell Culture Module
CGBA	Commercial Generic Biomedical Experiment
CVDA	Crystal Vapour Diffusion Apparatus
CCK	Cell Culture Kit
CIBX	Commercial ITA Biomedical Experiment
CPCG	Commercial Protein Crystal Growth
CPDS	Charged Particle Directorial Spectrometer
CREAM	Cosmic Radiation Effects and Activation Monitor
CRISTA	Cryogenic Infrared Spectrometers and Telescopes for the Atmosphere
CONCAP	Consortium for Materials Development in Space Complex Autonomous Payload
DXS	Diffuse X-ray Spectrometer
ECLIPSE	Equipment for Controlled Liquid Phase Sintering Experiments
EDOMP	Extended Duration Orbiter Medical Project
ELF	Enclosed Laminar Flames
ESCAPE	Experiment of the Sun for Complementing the Atlas Payload and for Education
FAST	Facility for Absorption and Surface Tension
FFEU	Free Flow Electrophoresis Unit
GAS	Getaway Specials
GOSAMR	Gelation of Sols Applied Microgravity Research
GPPM	Gas Permeable Polymetric Materials
HFA	Human Factors Assessment
HPP	Heat Pipe Performance
ICBC	Imax Cargo Bay Camera
IDGE	Isothermal Dentritic Growth Experiment
LBNP	Lower Body Negative Pressure
LEMZ	Liquid Encapsulated Melt Zone
MEMS	Micro-Electrical Mechanical System
MEPS	Microencapsulating Electrostatic Processing System
MFD	Manipulator Flight Demonstrator
MGBX	Microgravity Science Glove Box
MRI	Magnetic Resonance Imaging
MSX	Midcourse Space Experiment
NIH-R	National Institute of Health – Rodents
PARE	Physiological and Anatomical Rodent Experiment
PCG	Protein Crystal Growth
REM	Risk Mitigation Experiment
ROMPS	Robotic Operated Materials Processing System
RRMD	Real Time Radiation Monitoring Device
SAMS	Space Acceleration Measurement System

SHOOT	Superfluid Helium On-orbit Transfer Flight Demonstration
SIMPLEX	Shuttle Ionospheric Modification with Pulsed Local Exhaust
SAREX	Shuttle Amateur Radio Experiment
SPAS	Shuttle Pallet Satellite
SPIFEX	Shuttle Plume Infringement Flight Experiment
SSBUV	Space Shuttle Backscatter Ultra Violet
SSCE	Solid Surface Combustion Experiment
STL	Space Tissue Loss
SMIDEX	Spacelab Mid-deck Experiments
SWUIS	South West Ultraviolet Imaging System
OAST	Office of Aeronautics and Space Technology
OCG	Organic Crystal Growth
OSTEO	Osteoporosis Experiment in Orbit
OSVS	Orbiter Space Vision System
TDRS	Tracking and Data Relay Satellite
TEI	Thermal Electric Incubator
TES	Thermal Enclosure System
USMP	United States Materials Processing
VFEU	Vestibular Function Experiment Unit
WCI	Wetting Characteristics of Immiscibles
WINDEX	Window Experiment
WSF	Wake Shield Facility
ZCG	Zeolite Crystal Growth

Bibliography

In addition to the cited references pertaining to specific material in this book, this bibliography includes several general reference books focusing on aspects of the involvement of women in the fields of astronomy, engineering and technology, as well as the aerospace and space programmes. Both authors have also utilised their own personal archives and world-wide network of contacts. The books listed here include titles recommended for further reading, and also selected juvenile literature (J) to introduce the subject to a younger generation.

Specific titles

1960 *Heroines of the Sky*, Hervé Lauwick (English translation), Frederick Muller

1965 *Dee O'Hara: Astronauts' Nurse*, Virginia B. McDonnell, RN, Rutledge Books

1966 *American Women of the Space Age*, Mary Finch Hoyt, Athenuem Books

1975 *'It is I, Sea Gull': Valentina Tereshkova, First Woman in Space*, Mitchell R. Sharpe, Crowell

1986 *'I Touch the Future': The Story of Christa McAuliffe*, Robert T. Hohler, Random House

To Space and Back, Sally Ride with Susan Okie, Lothrop Books

1988 *Women in Space: Reaching the Last Frontier*, Carole S. Briggs, Lerner Publications

Into the Unknown: Women History Makers, Carole Stott, Macdonald (J)

1990 *Judith Resnik, Challenger Astronaut*, Joanne E. Bernstein and Rose Blue, Lodestar Books (J)

1991 *Christa McAuliffe: Teacher in Space*, Corinne J. Naden and Rose Blue, Millbrook Press (J)

1993 *Valentina: First woman in Space*, A. Lothian, Pentland Press

Seize the Moment, Helen Sharman and Christopher Priest, Victor Gollancz

A Journal for Christa, Grace George Corrigan, University of Nebraska Press

	On the Shuttle: Eight days in Space, Barbara Bondar with Roberta Bondar, Greey de Pencier Books, (J)
1995	*Mae Jemison, a Space Scientist*, Gail Sakurai, Children's Press (J)
1996	*Une Française Dans L'Espace*, Claudie André-Deshays and Yolaine De La Bigne, Plon
1998	*Shannon Lucid: Space Ambassador*, Carmen Bredeson, Millbrook Press (J)
1999	*Female Firsts in their Fields: Air and Space*, Doug Buchanan, Chelsea House Publishers
2000	*Teacher in Space: Christa McAuliffe and the Challenger Legacy*, Colin Burgess, Bison Books
2001	*Tethered Mercury*, Bernice Trimble Steadman, Aviation Press
2002	*Women Astronauts*, Laura S. Woodmansee, Apogee Books
	Space for Women: A History of Women with the Right Stuff, Pamela Freni, Seven Locks Press
	The Woman Face of Cosmos (in Russian), Valentina L. Ponomoryova, Moscow
2003	*Almost Heaven: The Story of Women in Space*, Bettyann Holtzmann Kevles, Basic Books
	Women of Space: Cool Careers on the Final Frontier, Laura S. Woodmansee, Apogee Books

General references and periodicals

	Spaceflight, British Interplanetary Society
1961–95	*Aeronautics and Astronautics* (various volumes), NASA
1972	*Russians in Space*, Evegenny Riabchikov (English translation), Weidenfeld and Nicolson
1981	*Test Pilots*, Richard P. Hallion, Doubleday
1985	*The Real Stuff*, Joseph D. Atkinson Jr and Jay M. Shafritz, Praeger
	Space Farers of the '80s and '90s, Alcestis R. Oberg, Columbia University Press
1987	*Before Lift-off: The Making of a Shuttle Crew*, Henry S.F. Cooper Jr, Johns Hopkins University Press
1989	*Race to the Stratosphere*, David H. DeVorkin, Springer
1990	*The Soviet Cosmonaut Team* (two vols.), Gordon Hooper, GRH Publications
1992	*Men and Women of Space*, Douglas B. Hawthorne, Univelt
	The Shuttlenauts, 1981–1992: The First 50 Missions; Vol. 2, Shuttle Flight Assignments STS-1 through STS-47, David J. Shayler, AIS Publications (with unpublished updates through 2004)
1994	*US Space Gear: Outfitting the Astronauts*, Lillian D. Kozloski, Smithsonian Institute Press
	They Had a Dream: The Story of African American Astronauts, J. Alfred Phelps, Presido
1995	*Mir Hardware Heritage*, David S.F. Portree, NASA JSC 26770

1997 *Walking to Olympus: An EVA Chronology*, David S.F. Portree and Robert C. Treviño, NASA Monographs in Aerospace history, No. 7

1999 *Who's Who in Space: International Space Station Edition*, Michael Cassutt, Macmillan

2000 *Challenge to Apollo: The Soviet Union and the Space Race, 1945–1974*, Asif A. Siddiqi, NASA SP-2000-4408

The History of Mir, 1986–2000, ed. Rex Hall, British Interplanetary society

2001 *Mir: The Final Year*, ed. Rex Hall, British Interplanetary Society

2002 *The International Space Station: From Imagination to Reality (1988–2002)*, ed. Rex Hall, British Interplanetary Society

Springer–Praxis Space Science Series

2000 *Challenges of Human Space Exploration*, Marsha Freeman

Disasters and Accidents in Manned Spaceflight, David J. Shayler

2001 *The Rocket Men*, Rex Hall and David J. Shayler

Russia in Space: A Failed Frontier, Brian Harvey

2002 *Creating the International Space Station*, David M. Harland and John E. Catchpole

The Continuing Story of the International Space Station, Peter Bond

2003 *Soyuz: A Universal Spacecraft*, Rex Hall and David J. Shayler

Russian Spacesuits, Isaak P. Abrahamov and A. Ingemar Skoog

2004 *Walking in Space*, David J. Shayler

The Story of the Space Shuttle, David M. Harland

Space Shuttle *Challenger* (1986) and *Columbia* (2003) accidents

Challenger

1986 *Challengers*, Staff of the Washington Post, Pocket Books

Heroes of the Challenger, Daniel and Susan Cohen, Archway Paperback

Report of the Presidential Commission on the Space Shuttle Challenger Accident (five vols.), Washington

1987 *Challenger: A Major Malfunction.* Malcolm McConnell, Simon Schuster

Prescription for Disaster, Joseph J. Trento, Crown.

1988 *Challenger: The Final Voyage*, Richard S. Lewis, Columbia University Press

Challenger, Aviation Fact File, David J. Shayler, Salamander Books

1993 *Contest for the Heavens: The Road to the Challenger Disaster* (English edition, 1996), Claus Jensen, Harvill

1996 *The Challenger Launch Decision*, Diane Vaughan, University of Chicago Press

Columbia

2003 *Sixteen Minutes from Home*, Mark Cantrell and Donald Vaughan, American Media Books

Columbia Accident Investigation Report, Vol. 1 plus supplementary material, Apogee Books

2004 *Comm Check: The Final Flight of Columbia*, Michael Cabbage and William Harwood, Free Press

Index

Printing: Mercedes-Druck, Berlin
Binding: Stein + Lehmann, Berlin